Sustainable Agriculture Reviews

Volume 44

Series Editor
Eric Lichtfouse
Aix-Marseille University, CNRS, IRD, INRAE, Coll France, CEREGE
Aix-en-Provence, France

Other Publications by Dr. Eric Lichtfouse

Books
Scientific Writing for Impact Factor Journals
https://www.novapublishers.com/catalog/product_info.php?products_id=42242

Environmental Chemistry
http://www.springer.com/978-3-540-22860-8

Sustainable Agriculture
Volume 1: http://www.springer.com/978-90-481-2665-1
Volume 2: http://www.springer.com/978-94-007-0393-3

Book series
Environmental Chemistry for a Sustainable World
http://www.springer.com/series/11480

Sustainable Agriculture Reviews
http://www.springer.com/series/8380

Journal
Environmental Chemistry Letters
http://www.springer.com/10311

Sustainable agriculture is a rapidly growing field aiming at producing food and energy in a sustainable way for humans and their children. Sustainable agriculture is a discipline that addresses current issues such as climate change, increasing food and fuel prices, poor-nation starvation, rich-nation obesity, water pollution, soil erosion, fertility loss, pest control, and biodiversity depletion.

Novel, environmentally-friendly solutions are proposed based on integrated knowledge from sciences as diverse as agronomy, soil science, molecular biology, chemistry, toxicology, ecology, economy, and social sciences. Indeed, sustainable agriculture decipher mechanisms of processes that occur from the molecular level to the farming system to the global level at time scales ranging from seconds to centuries. For that, scientists use the system approach that involves studying components and interactions of a whole system to address scientific, economic and social issues. In that respect, sustainable agriculture is not a classical, narrow science. Instead of solving problems using the classical painkiller approach that treats only negative impacts, sustainable agriculture treats problem sources.

Because most actual society issues are now intertwined, global, and fast-developing, sustainable agriculture will bring solutions to build a safer world. This book series gathers review articles that analyze current agricultural issues and knowledge, then propose alternative solutions. It will therefore help all scientists, decision-makers, professors, farmers and politicians who wish to build a safe agriculture, energy and food system for future generations.

More information about this series at http://www.springer.com/series/8380

Ankit Saneja • Amulya K. Panda
Eric Lichtfouse
Editors

Sustainable Agriculture Reviews 44

Pharmaceutical Technology for Natural Products Delivery Vol. 2 Impact of Nanotechnology

Editors
Ankit Saneja
Product Development Cell
National Institute of Immunology
New Delhi, Delhi, India

Amulya K. Panda
Product Development Cell
National Institute of Immunology
New Delhi, Delhi, India

Eric Lichtfouse
Aix-Marseille University CNRS, IRD,
INRAE, Coll France, CEREGE
Aix-en-Provence, France

ISSN 2210-4410 ISSN 2210-4429 (electronic)
Sustainable Agriculture Reviews
ISBN 978-3-030-41844-1 ISBN 978-3-030-41842-7 (eBook)
https://doi.org/10.1007/978-3-030-41842-7

This Springer imprint is published by the registered company Springer Nature Switzerland AG.
The registered company address is: Gewerbestrasse 11, 6330 Cham, Switzerland

Preface

The science of today is the technology for tomorrow (Edward Teller).

Natural products are a golden source for drug discovery due to their high chemical diversity that has been designed by living organisms. However, classical drug administration is often poorly efficient. Here, the recent development of nanotechnologies has open new routes because nanomaterials have unique properties that improve pharmacodynamic and pharmacokinetic properties of active chemicals. This book presents recent advances in nanoscience for improving the therapeutic efficacy of natural products and discusses lipid nanoarchitectonics, inorganic particles, and nanoemulsions for delivering natural products.

In Chap. 1, Bilia et al. reviews nanotechnology applications for natural products delivery. They also present nanocarriers made of polymers or lipid constituents for the delivery of artemisinin, curcumin, andrographolide, resveratrol, honokiol, salvianolic acid B, green tea catechins, and silymarin. In Chap. 2, Gonçalves et al. highlight the use of biosynthesized nanomaterials as a viable alternative to conventional techniques. They also discuss plant extracts as a source of new nanomedicines, acting as an ally or alternative to existing therapies (Fig. 1). In Chap. 3, Saka and Chella explore the various delivery options of bioactive molecules and extracts using nanocarriers made of polymers, lipids, and inorganic materials. The role of nanocarriers for improving therapeutic efficacy is highlighted by case studies.

In Chap. 4, Barradas et al. describe the main aspects of nanoemulsion formulations, type of surfactants and oil phases, and techniques for characterizing nanoemulsions. They also discuss applications of nanoemulsions for improvement of bioactive oil bioavailability and solubilization, masking unpleasant aspects of oils, and enhancement of essential oils pharmacological activity. In Chap. 5, Chaudhari et al. discuss lipid nanoarchitectonics in natural product delivery for the treatment of cancer therapy. In Chap. 6, Meena et al. describe the properties, synthesis, advantages, and toxicities of inorganic particles made of silver, gold, iron oxide, and silica and their application for the delivery of natural products. Finally, in Chap. 7, Padhi and Behera discuss the delivery of camptothecin for increasing stability and solubility using novel drug delivery platforms.

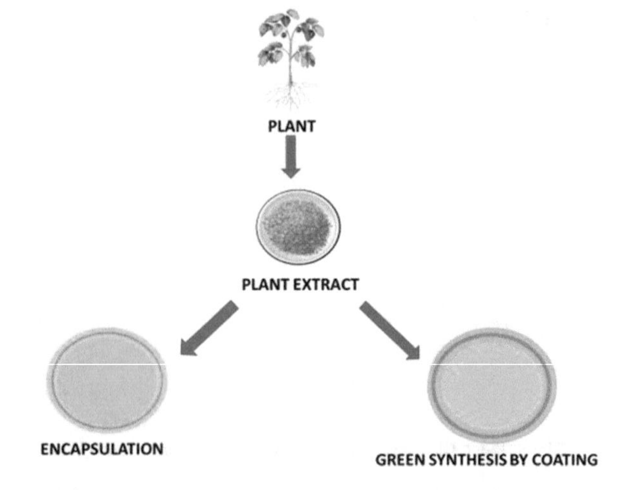

Fig. 1 Plant extracts in nanotechnology. Left: encapsulation of plant extracts to enhance therapeutic efficiency. Right: green synthesis of nanoparticles, during which plant compounds nucleate and stabilize nanoparticles. From Chap. 2 by Gonçalves et al.

We extend our sincere gratitude to all contributors who had made significant contribution to prepare high-quality chapters. We also extend our thanks to Melanie van Overbeek, Assistant Editor at Springer Nature, for her support during this whole process.

New Delhi, India — Ankit Saneja
New Delhi, India — Amulya K. Panda
Aix-en-Provence, France — Eric Lichtfouse

Contents

About the Editors

Ankit Saneja currently works as Research Associate in the Product Development Cell, National Institute of Immunology, New Delhi, India. He has received his PhD from the Academy of Scientific and Innovative Research (AcSIR) at the CSIR-Indian Institute of Integrative Medicine, Jammu. He has been recipient of several international awards such as the CEFIPRA-ESONN Fellowship, Bioencapsulation Research Grant, and PSE Travel Bursary. He has to his credit several international publications. His research interest includes exploring different types of formulations for various biomedical applications.

Amulya K. Panda has been a Scientist at the National Institute of Immunology (NII), New Delhi, for the last 29 years. Currently, he is the Director of NII. He completed his master's degree in Chemical Engineering from IIT Madras and his PhD in Biochemical Engineering and Biotechnology from IIT Delhi. He has been a Visiting Scientist with Prof. Harvey Blanch at the Chemical Engineering Department, University of California, Berkeley, USA, and in the Department of Pharmaceutical Science at the University of Nebraska Medical Center, Omaha, USA. His research interest includes bioprocess engineering, recombinant fermentation process development, high-throughput protein refolding from inclusion bodies, and vaccine delivery using biodegradable polymer particles. He has published more than 130 research papers and 10 book chapters and is an inventor of more than 30 issued or pending patents. His honors include Samanta Chandra

Sekhara Award from Odisha Bigyan Academy for the year 2000; Biotech Product and Process Development and Commercialization Award 2001 from the Department of Biotechnology, Government of India, on May 11, 2001; Young Asian Biotechnologist Prize from The Society for Biotechnology, Japan, for the year 2004; Tata Innovation Fellowship 2010 from the Department of Biotechnology, Government of India; and GB Manjrekar Award of AMI at AMI 2010 Conference at BIT Mesra in December 2010. His research group was also awarded Second Prize at Intel-UC Berkeley Technology Entrepreneurship Challenge 2008 and Third Prize at World Innovation Summit, Barcelona, on June 16, 2009. He was the President of the Association of Microbiologist of India (AMI) for the year 2017.

Eric Lichtfouse, PhD is Geochemist and Professor of Scientific Writing at Aix-Marseille University, France, and Visiting Professor at Xi'an Jiaotong University, China. He has discovered temporal pools of molecular substances in soils, invented carbon-13 dating, and published the book *Scientific Writing for Impact Factor Journals*. He is Chief Editor and Founder of the journal *Environmental Chemistry Letters* and the book series Sustainable Agriculture Reviews and Environmental Chemistry for a Sustainable World. He has awards in analytical chemistry and scientific editing. He is World XTerra Vice-Champion.

Chapter 1
Nanotechnology Applications for Natural Products Delivery

Anna Rita Bilia, Vieri Piazzini, and Maria Camilla Bergonzi

Abstract Natural products are fascinating molecules in drug discovery for their exciting structure variability and for their interaction with various biological targets, which represent the best approach to develop successful medications for many diseases. The scarce water solubility, low lipophilicity and inappropriate molecular size of many natural compounds, which undergo structural instability in biological milieu, rapid clearance and high metabolic rate, have severely limited their use in clinic. Nanomedicine represents an excellent tool to increase bioavailability and activities of natural products. Generally, nanosized delivery systems provide large surface area increasing dissolution properties and can overcome anatomic barriers. In addition, passive and active targeting can optimize the performance of the nanocarriers. Passive targeting takes advantage of the unique pathophysiological characteristics of inflamed and tumor vessels, enabling nanodrugs to accumulate in the tissues. The effect is called enhanced permeation and retention, generally obtained by the decoration with polyethylene glycol the vector surface. An intriguing strategy is to decorate the nanocarriers with special ligands in order to recognize and bind to target cells through ligand–receptor interactions. Although the active targeting strategy looks intriguing, nanodrugs currently approved for clinical use are relatively simple and generally lack active targeting or triggered drug release components.

In this review different nanocarriers made of polymers or lipid constituents mainly based on artemisinin, curcumin, andrographolide, resveratrol, honokiol, salvianolic acid B, green tea catechins, silymarin and other extracts are reported. Each nanosystem has its own advantages, disadvantages, and characteristics. Polymeric nanoparticles are solid in nature and include nanosphere and nanocapsule. They are ideal candidates to enhance the bioavailability of the natural products after various routes of administration, principally oral and parenteral, but also nasal and intraocular, as well as to cross physiological barriers including blood brain barrier. Polymeric micelles have high safety, worthy stability and low cost. Polymeric micelles are very stable in physiological media with a consequent controlled drug

A. R. Bilia (✉) · V. Piazzini · M. C. Bergonzi
Department of Chemistry "Ugo Schiff", Florence, Italy
e-mail: ar.bilia@unifi.it

A. Saneja et al. (eds.), *Sustainable Agriculture Reviews 44*, Sustainable Agriculture Reviews 44, https://doi.org/10.1007/978-3-030-41842-7_1

release of drugs, while the hydrophilic shell protects the encapsulated drug from the external medium and prevents the interaction with plasma components, resulting in in vivo long circulating properties. Dendrimers are characterized by low polydispersity, good biocompatibility, able to cross cell barriers via both paracellular and transcellular pathways, very versatile able to carry both lipophilic and hydrophilic drugs. Lipid nanocarriers include vesicles, nanocochleates, micelles, solid lipid nanoparticles and nanostructured lipid particles, emulsions with nano scale, including nanoemulsions and microemulsions. Vesicles include liposomes and niosomes are the first nano drug delivery systems that have been successfully translated into real-time clinical applications. They are extremely versatile in terms of route of administration and characteristics of the loaded drug. Due to their similarity to biological membranes provides unique opportunities to deliver drug molecules into cells or subcellular compartments. Liposomes can be converted to nanocochleates, which are unique nanovectors, after treatment with divalent ions, which are very useful for both oral and parenteral administration. Solid lipid nanoparticles and nanostructured lipid particles are easy to scale-up, low cost of production, relative nontoxic nature, biodegradable composition, and stability against aggregation or coalescence. Mostly lipid drugs can be loaded in these nanoparticles to avoid extrusion. Nanoemulsions and microemulsions are the most interesting nanostructures to essentially increase drug loading and enhance bioavailability. They both give reproducible plasma drug profile and can also be used for sustained and targeted drug delivery.

Keywords Nanosized delivery systems · Natural products · Polymeric and lipid nanosystems · Bioavailability · Efficacy

1.1 Introduction

Natural products from plants, animals, and minerals have represented for millennia the only resource to maintain health, for prophylaxis or to cure human and animal diseases. Still currently, between 65 and 80% of populations in developing countries use medicinal plants as therapeutic remedies for their primary healthcare (Cameron et al. 2011).

The tangible importance of natural products to the drug discovery process and their possible role in therapy is unquestionable and numerous strong scientific evidences have confirmed the reputation of traditional knowledge suggesting that the worldwide natural products' sources represent a varied pool of key drugs (Bilia et al. 2014a, b, 2017; 2018a, b). Interestingly, the impact of natural products in the clinic is quite marked, about 69% of anti-infective (ca. 195 molecules) are naturally derived or inspired. Among the antitumor drugs (ca. 172 molecules), 75% are represented by natural products or inspired to them (Newman and Cragg 2012). Natural products still represent a main source of drugs thanks to their enormous structural and chemical diversity, which is incomparable by any synthetic libraries (Bilia et al. 2017). In the Dictionary of Natural Products, over 300,000 natural products describing the available chemical, physical and biological properties are reported (Dictionary of Natural Products. http://dnp.chemnetbase.com). Natural products are also very attractive molecules because they can modulate multiple targets activating various signalling or functional pathways increasing their therapeutic value against multifactorial and complex diseases, especially cancer and diabetes. Conversely, the *in vivo* performance could results limited because of low hydrophilicity and poor intrinsic dissolution rate, physical or chemical instability. In addition, they may present low absorption, scarce biodistribution, first-pass metabolism, poor penetration and accumulation in the organs of the body, or trivial targeting efficacy (Bilia et al. 2017, 2018a, b).

1.1.1 Nanotechnologies Strategies to Optimize the Clinical Use of Natural Products

Diverse strategies are used to optimise the bioavailability of natural products, including the development of semisynthetic compounds or synthetic analogues, the production of prodrugs, and the technological approach for the production of appropriate formulations (Bilia et al. 2017; 2018a). The latter strategy to optimize the biopharmaceutical performance of natural products is the development of suitable drug delivery systems, in particular those nano-sized, generally between 50 and 300 nm up to 1 μ, which correspond from the size of one-half of DNA diameter and one eighth of the red blood cell diameter (Fig. 1.1). Some nanosized delivery

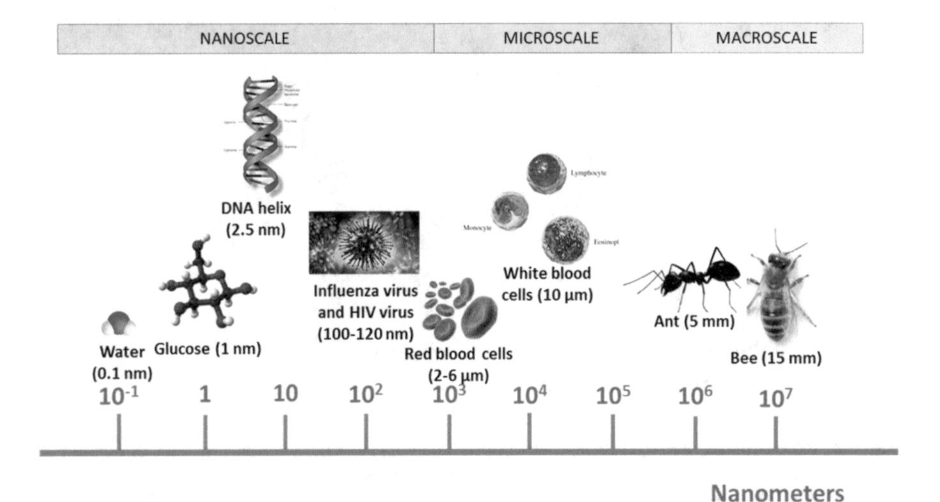

Fig. 1.1 Scale of nature. (Reproduced with permission from Bilia et al. 2017 from Thieme)

systems have already entered in the clinic because they can offer an advanced approach to optimized the therapeutic efficacy targeting definite tissues or organs or capable to cross biological barriers with the aim to increase efficacy, safety profile and compliance of drugs. An efficacious drug delivery system should possess optimal drug loading, ideal release characteristics, extended shelf-life, with a consequent considerable greater clinic effectiveness and lesser side effects (Bilia et al. 2017, 2018a).

Different types of nanomaterials can be used to prepare nanocarriers which are capable of being loaded with hydrophobic and hydrophilic drugs and generally are classified as polymer-based systems and lipid-based systems. Roles of size, shape, charge, hydrophobic/hydrophilic character, and surface chemistry of nanocarriers to optimize *in vivo* behaviour including active intracellular delivery and improved pharmacokinetics and pharmacodynamics of drugs, modifying some physiological parameters, including hepatic filtration, tissue extravasation, tissue diffusion, and kidney excretion (Fig. 1.2).

Passive and active targeting is generally obtained modifying the surface of nanosystems (Fig. 1.3). Nanocarriers are generally recognized as extraneous structures, and consequently are opsonized by the reticuloendothelial system, thus decreasing the availability of the drug.

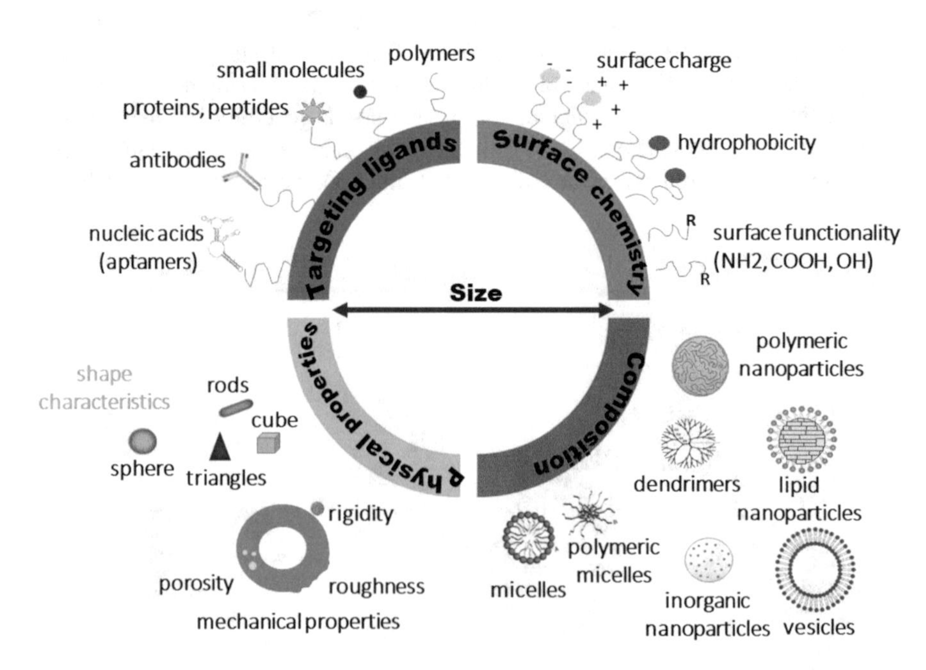

Fig. 1.2 Roles of size, shape, charge, hydrophobic/hydrophilic character, and surface chemistry of nanocarriers to optimize *in vivo* behavior including active intracellular delivery and improved pharmacokinetics and pharmacodynamics of drugs, modifying some physiological parameters, including hepatic filtration, tissue extravasation, tissue diffusion, and kidney excretion. (Reproduced with permission from Bilia et al. 2017 from Thieme)

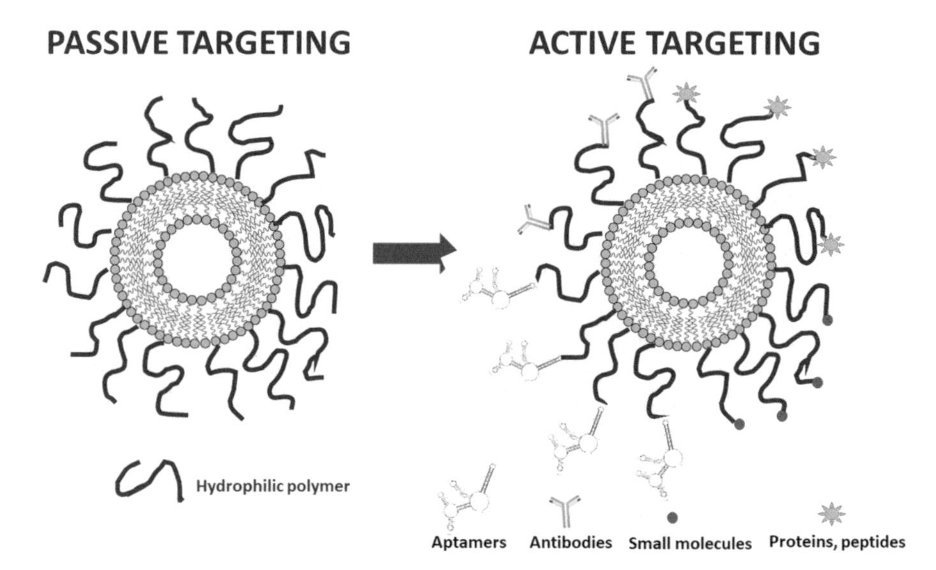

Fig. 1.3 Passive and active targeting. (Reproduced with permission from Bilia et al. 2018a from Wiley)

It has been observed that nanocarriers decorated in the superficial area with hydrophilic polymers such as polyethylene glycol are not recognised by the reticuloendothelial system obtaining a "stealth effect", resulting in an enhancement of the pharmacokinetic properties of the drug, increasing drug solubility and drug stability. In addition, these long circulating nanocarriers can better penetrate that inflamed tissues, tumours other pathological conditions in the body, because of the more permeable blood vessel walls with respect to the normal tissues and a consequent enhanced permeation and retention effect, reaching up to 50-fold accumulation in tumours compared to physiological tissues (Fig. 1.4, Bilia et al. 2017, 2018a, b). A further approach is to decorate the surface of the nanovectors with diverse elements to obtain the active drug targeting, also called ligand-mediated targeting (Fig. 1.3). Ligands are selected to target surface molecules or receptors overexpressed in diseased organs, tissues, cells or subcellular domains. Active targeting increases internalization of drugs without altering the overall biodistribution, because of the recognition of the ligand by its target substrate. Various ligands can be used, namely small molecules, antibodies, nucleic acids, proteins and peptides. Peptide-targeting fragments are ranging from 2 to 50 amino acids. Numerous peptide receptors are overexpressed in tumour cells and include luteinizing hormone-releasing hormone, bombesin and somatostatin receptor. Peptides can also target integrins, which are transmembrane receptors, whose role is vital in the adhesion between cells and surrounding tissues and are generally overexpressed in tumour neovasculature. Small molecules such as folic acid and sugars are often used as active targeting, because of the small production costs, lack of degradation, low immunogenic properties and ease of conjugation to nanocarriers (Bilia et al. 2018a).

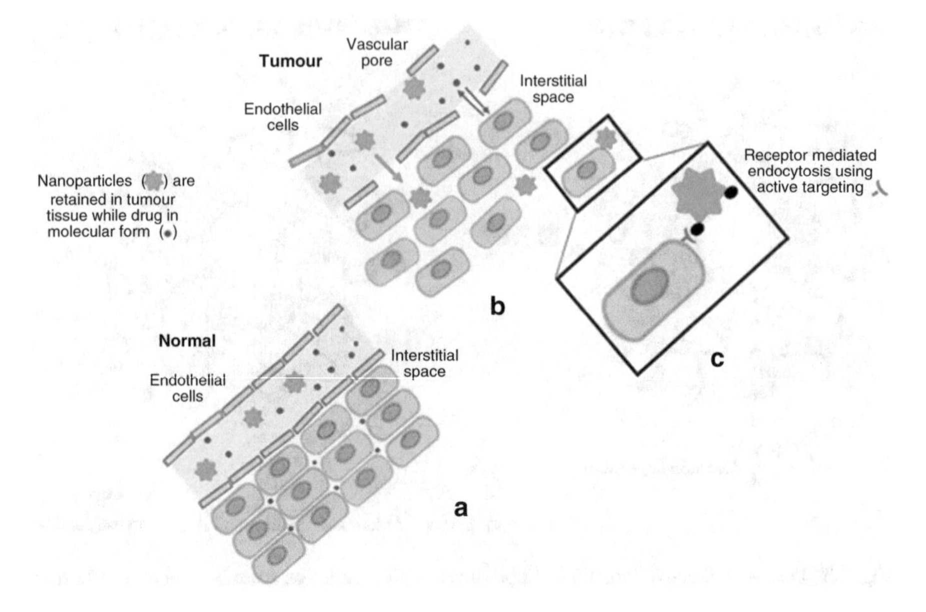

Fig. 1.4 The role of enhanced permeation and retention effects for nanocarrier behavior in healthy tissues (**a**), their accumulation in inflamed or tumor tissues (**b**), and active drug targeting of nanocarriers (**c**). (Reproduced with permission from Bilia et al. 2017 from Thieme)

1.2 Polymeric Nanocarriers

Nicolas et al. (2013) Polymeric nanocarriers are composed of natural, semisynthetic or synthetic polymers, and their selection is mainly based on their biodegradation properties, biocompatibility, surface characteristics, in order to develop of new and safe carriers with controlled and targeted drug delivery concept.

Natural polymers, also called biopolymers, are obtained from bacteria, fungi, animals, and plants, and are mainly represented by polysaccharides and proteins.

Polysaccharides derived from plant kingdom are cellulose, pectin, starch, alginic acid, carrageenan, Arabic gum. Those from animal or bacterial origin are chitosan, and gums including xanthan, gellan. Nanocarriers based on polysaccharides are very suitable for water-soluble or hydrophilic drugs. They are characterised by high stability, extraordinary safety, low cost, great biodegradability and no toxicity. Several natural proteins are also used to produce nanocarriers and include gelatin, casein, albumin and soy proteins hydrolysate. Among the synthetic polymers, polyvinyl alcohol, poly-cyanoacrylate alkyl esters, polyglycolic acid, polylactic acid, polylactic glycolic acid, and poly(N-vinyl pyrrolidone), are among the most used (Bilia et al. 2017, 2018a, b).

Polymeric nanocarriers include nanospheres and nanocapsules, polymeric micelles and dendrimers (Fig. 1.5).

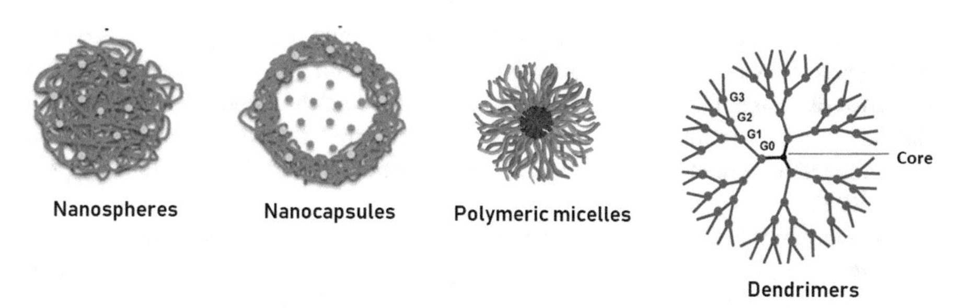

Fig. 1.5 Polymeric nanoparticles. Reproduced with permission from Bilia et al. 2018a from Wiley

1.2.1 Polymeric Nanoparticles

Nanoparticle production is based on chemical reactions of polymerization, or can be produced directly from a macromolecule or a preformed polymer. Nanoparticles include nanospheres and nanocapsules (Fig. 1.5). Nanospheres are matrix systems in which the drug is physically and uniformly dispersed, while in nanocapsules the drug is confined to a cavity surrounded by a polymeric membrane. Diverse nanocarriers have been formulated in the form of polymeric nanoparticles to increase membrane permeability, bioavailability, to obtain sustained release, or to selectively cross barriers, in particular the blood brain barrier.

Injectable nanoparticles formulated monomethoxy polyethylene glycol and poly lactide-co-glycolide and co-encapsulated with ginkgolides A, B, C and bilobalide, the active components of *Ginkgo biloba* extract were developed. A mean diameter of 123.3 ± 44.0 nm and zeta potential of −30.86 ± 2.49 mV characterised the developed nanoparticles. The encapsulation efficiency of the total ginkgo terpenes in the optimal formulation was 78.84 ± 2.06%, with a loading dose of 11.90 ± 0.31 mg of terpenes in 150 mg of polymer. The particles exhibited a sustained and synchronized release of the four components, both *in vitro* and *in vivo*. In addition, the half-life time of the four terpenoid compounds was also significantly improved after their loading into the nanoparticles (Han et al. 2012).

A further example is represented by quercetin encapsulated in polylactic acid nanoparticles whose release patterns was studied over a period of 96 h. Within the first half hour, 40–45% of the quercetin was released. This quick burst was attributed to the quercetin at the surface of the particle diffusing into the surroundings. Over the next 96 h, the release was slower and reached a maximum of 87.6%. This slower release was attributed to the diffusion of the quercetin from deeper within the nanoparticle (Kumari et al. 2010).

In a pharmacokinetic study, curcumin loaded in polylactic glycolic acid nanoparticles improved the relative oral bioavailability by 563%, compared to free curcumin. The increased bioavailability of curcumin was due to the inhibition of

P-gp-mediated efflux. The authors tested this hypothesis by adding verapamil, a P-glycoprotein inhibitor, to curcumin or curcumin loaded in polymeric nanoparticles. After 120 min of treatment, the remaining curcumin in the jejunum was evaluated. There was no significant difference in the residual curcumin content in the presence of verapamil and the values obtained with curcumin nanoparticles without the addition of verapamil. Accordingly, it was found that nanoparticles inhibit P-glycoprotein, which allows increased drug permeability and bioavailability (Xie et al. 2011).

Bioavailability of curcumin loaded in PEGylated polyester nanoparticles was in investigated in a Caco-2 cell model. PEG polymers (from 2000 to 5000 Da) and polylactic or polylactic polyglycolic acid scarcely modify the transcellular transport of nanoparticles. This was optimised using PEG with a molecular weight of 5000 and polylactic acid with a molecular weight of 40,000, which demonstrated a greater drug loading and slower release properties (Song et al. 2013).

Diverse studies suggest that polymeric nanoparticles represent very useful drug delivery systems to improve solubility and transport across physiological barriers, in particular polymeric nanoparticles are useful drug-delivery systems to cross the blood-brain barrier. The blood-brain barrier hinders the passage of systemically delivered therapeutics and the brain extracellular matrix limits the distribution and durability of locally delivered agents.

Poly alchylcyanoacrylate represent an interesting class of polymers to successfully cross blood-brain barrier.

In a recent study, nanoparticles made of poly ethylcyanoacrylate coated with polysorbate 80 were developed, and characterised in terms of dimensional analysis, polydispersity and zeta potential, morphology, encapsulation efficacy, and loading capacity (Grossi et al. 2017). Nanoparticles' distribution and fate were evaluated after intracerebral injection in healthy rats. Furthermore, their ability to reach the brain tissues and the lack of their toxicity was demonstrated after systemic administration in rats. After intracerebral injection in healthy rats, nanoparticles were distributed within the injected hemisphere and mainly interacted with microglial cells, presumably involved in their clearance by phagocytosis. Furthermore, nanoparticles were able to pass the blood-brain barrier after systemic administration in rats, and the lack of toxicity in C57/B6 mice chronically administered was highlighted (Fig. 1.6). Nanoparticles were loaded with salvianolic acid B, the bioactive constituent from *Salvia miltiorrhiza* which represents a molecule to prevent degeneration in several animal models by various biological mechanisms (Bonaccini et al. 2015), with a poor chemical stability and low bioavailability which limits its clinical application for central nervous system neuronal injury and degeneration. Encapsulation efficacy and loading capacities were 98.70% ± 0.45 and 53.3% ± 0.24, respectively. In addition, the nanoparticles were suitable for parental administration because their mean diameter smaller than 300 nm, with a polydispersity of 0.04 ± 0.03, and a ζ-potential of −8.38 mV ± 3.87. The *in vitro* release of salvianolic acid B from the nanoparticles was sustained and prolonged during 8 h, suitable for a promising clinical application (Grossi et al. 2017).

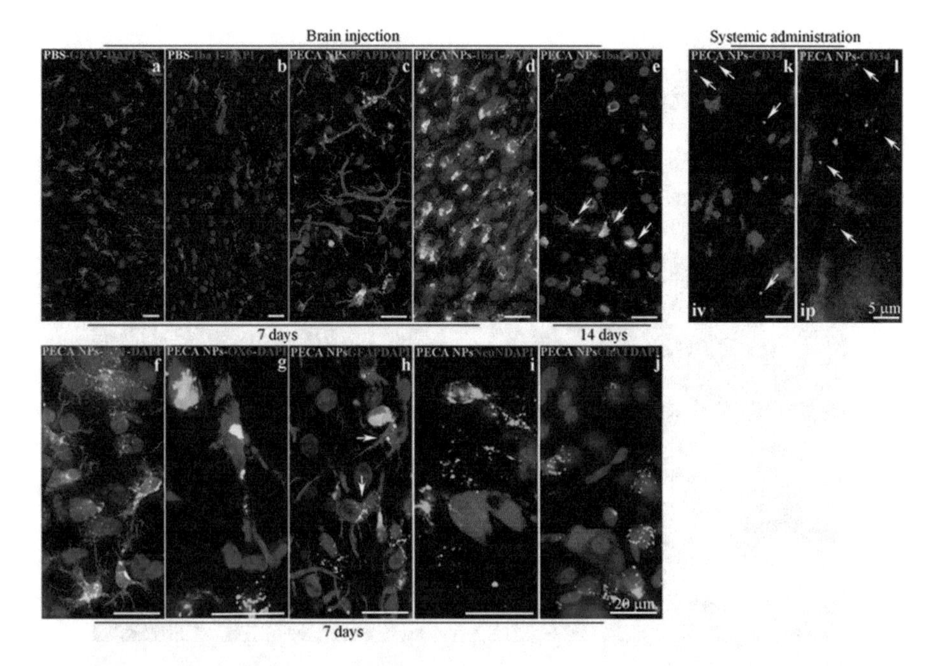

Fig. 1.6 Intracerebral injection of poly ethylcyanoacrylate nanoparticles loaded with loaded with the florescent probe and coated with polysorbate 80 and their blood brain barrier crossing abilities after systemic administration. (**c–j**) Double labelling experiments between poly ethylcyanoacrylate nanoparticles loaded with loaded with the florescent probe and coated with polysorbate 80 (green) and glial markers GFAP/Iba-1/OX6 (red, arrows) or neuronal NeuN/ChAT antibodies (red) plus 4′,6-Diamidine-2′-phenylindole dihydrochloride (DAPI, blue) in injected rats, 7 (**c**, **d**, **f–j**) and 14 (**e**) days after surgery. GFAP- and Iba 1-immunopositive cells (red) with resting morphology and no poly ethylcyanoacrylate nanoparticles loaded with loaded with the florescent probe and coated with polysorbate 80 (green) signal were detected in PBS-injected rats (**a**, **b**, **k**, **l**). The lack of colocalization between poly ethylcyanoacrylate nanoparticles loaded with loaded with the florescent probe and coated with polysorbate 80 (green, indicated by the arrows) and the brain vascular endothelium marker CD34 (red) demonstrated their ability to cross the blood brain barrier after an acute i.v. (**k**) and i.p. (**l**) administration. (Reproduced with permission from Grossi et al. 2017 from Thieme)

A further approach to penetrate blood-brain barrier is the use of albumin nanoparticles. In particular albumin when polymerised results a versatile nontoxic, non-immunogenic, biocompatible and biodegradable protein having high binding capacity of various natural products and being well tolerated without any serious side effects (Bilia et al. 2018a). A recent publication reported the production of albumin nanoparticles without the use of organic solvents to obtain useful drug-delivery systems to cross the blood-brain barrier. Human serum albumin nanoparticles have gained considerable attention owing to their high loading capacity for various drugs and the fact that they are well tolerated. Two different cross-linking methods, the chemical and the thermal ones, were investigated to produce the nanoparticles (Bergonzi et al. 2016). Nanoparticles were chemically and physically

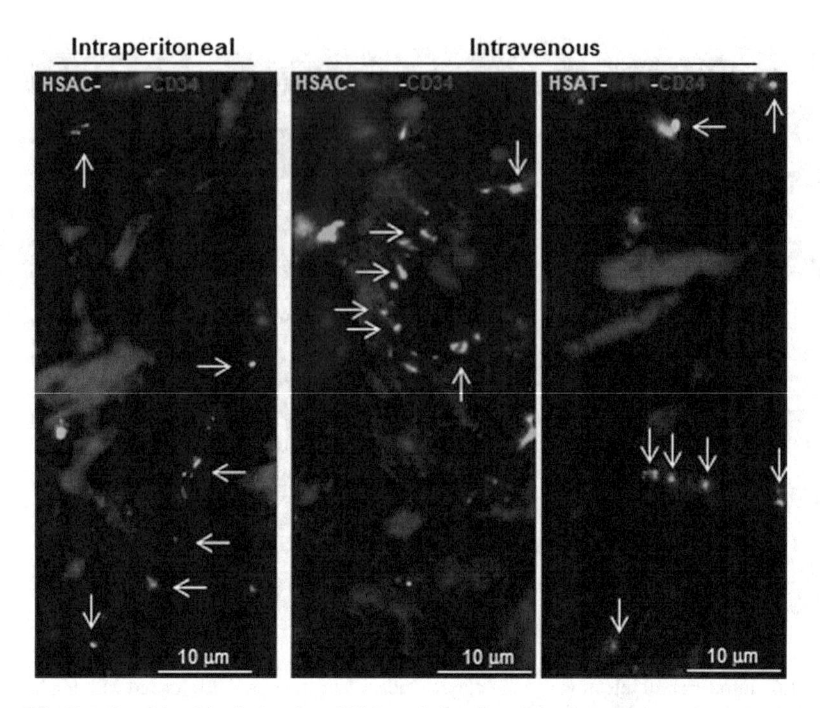

Fig. 1.7 Crossing blood brain barrier abilities of the albumin nanoparticles prepared with heat (HSAT) or chemically synthetized (HSAC) after injection in rats. The lack of co-localization of the nanoparticle loaded with a florescent probe (green, arrows) with the brain vascular endothelium marker CD34 (red), the brain vascular endothelium marker, indicates the ability of the two nano-formulations to cross the blood brain barrier after an acute systemic administration. (Reproduced with permission from Bergonzi et al. 2016 from Wiley)

characterized by dynamic light scattering, transmission electron microscopy, and high-performance liquid chromatography coupled with diode array and fluorimetric detectors. A first set of studies evaluated the *in vivo* distribution and the fate of albumin nanoparticles in healthy rats, after intracerebral injection. Then the ability of formulations to cross the blood-brain barrier and reach the brain tissues was demonstrated by intravenous and intraperitoneal administration in healthy rats (Fig. 1.7). In addition, the toxicity of nanoparticles was excluded by behavioural tests.

The toxicity of the developed carriers was estimated by behavioral tests. Nanoparticles were observed to be located in different brain tissues depending on the mode of injection, and did not induce an inflammatory response. Behavioral tests demonstrated no locomotor, explorative, or cognitive function impairment induced by the nanoparticles. Despite the similar results obtained by *in vivo* tests, thermal cross-linking method was consider superior to chemical cross-linking in terms of formulation parameters such as a decrease in costs and the time necessary for the preparation (Bergonzi et al. 2016).

In a further study, andrographolide the major diterpenoid of the Asian medicinal plant *Andrographis paniculata* having exciting pharmacological potential for treatment of neurodegenerative disorders (Casamonti et al. 2019a) was loaded into human serum albumin based nanoparticles and poly ethylcyanoacrylate nanoparticles and characterized in terms of size, zeta potential, polydispersity, and release studies (Guccione et al. 2017). The ability of free andrographolide and andrographolide loaded in both types of polymeric nanoparticles for their crossing BBB properties were investigated using an *in vitro* BBB model based on human cerebral microvascular endothelial cell line (hCMEC/D3). Free andrographolide did not permeate the blood-brain barrier model, as also predicted by in silico studies. By contrast, albumin nanoparticles improved by two-fold the permeation of andrographolide while maintaining the integrity of the cell layer, while poly ethylcyanoacrylate nanoparticles temporarily disrupted blood-brain barrier integrity (Guccione et al. 2017).

In a successive study andrographolide encapsulated in human albumin nanoparticles was investigated for the distribution in brain and activity effects in TgCRND8 mice, an Alzheimer's disease mouse model. Developed nanoparticles had proper size (mean size: 159.2 ± 4.5 nm), size distribution (nearby 0.12 ± 0.01) and ζ-potential (-24.8 ± 1.2 mV) with a remarkable encapsulation efficiency (more than 99%). In the step down inhibitory avoidance test, nanoparticles administered to TgCRND8 mice significantly improved their performance ($p < 0.0001$) reaching levels comparable to those displayed by wild type mice. In the object recognition test, treated and untreated animals showed no deficiencies in exploratory activity, directional movement towards the objects and locomotor activity. No cognitive impairments (discrimination score) were detected in TgCRND8 mice ($p < 0.0001$) treated with nanoparticles. After acutely i.v. administration or chronical i.p. administration, nanoparticles were detected in the brain parenchyma of TgCRND8 mice. The immunofluorescent analyses evidenced the presence of nanoparticles both in the pE3-Aβ plaque surroundings, and inside the pE3-Aβ plaque, indicative of the ability of these nanoparticles to cross the BBB and to penetrate in both undamaged and damaged brain tissues (Bilia et al. 2019).

Finally, human serum albumin nanoparticles coated with chitosan were developed as drug delivery systems to be used as nose-to-brain carrier. The mean particle sizes was 241 ± 18 nm and 261 ± 8 nm and the zeta potential was -47 ± 3 mV and $+ 45 \pm 1$ mV for albumin nanoparticles and albumin nanoparticles coated with chitosan, respectively. The optimized formulations showed excellent stability upon storage both as suspension and as freeze-dried product after 3 months. The mucoadhesion properties were assessed by turbidimetric and indirect method. Nanoparticles were loaded with sulforhodamine B sodium salt as model drug and the effect of chitosan coating properties was investigated by *in vitro* release studies, permeation and uptake experiments using Caco-2 and hCMEC/D3 cells as model of the nasal epithelium and blood-brain barrier, respectively. Furthermore, *ex vivo* diffusion experiments were tested using rabbit nasal mucosa and the ability of the formulations to reversibly open tight and gap junctions was explored by western blotting and RT-PCR analysing in both Caco-2 and hCMEC/D3 cell (Piazzini et al. 2019a).

1.2.2 Polymeric Micelles

Polymeric micelles are nanostructures (20–200 nm) made up of amphiphilic polymers consisting of block copolymers of hydrophobic and hydrophilic monomer units, that spontaneously self-associate in aqueous solution, forming a micelle at concentrations higher than the critical micellar concentration (Fig. 1.5). Generally, the sequence of the hydrophilic polymer (A) represents the shell, while the core is formed by an hydrophobic polymer (B) arranged to obtain a di-block or a multiblock polymer. The most common used hydrophilic portion is represented by polyethylene glycol, while poly(D, L)-lactic acid, polypropyleneoxide, poly-caprolactone, and poly(L)-aspartic acid. A very common polymers used to prepare micelles are Pluronics (polyoxyethylene polyoxypropylene block copolymers) (Bilia et al. 2017, 2018a).

Principal advantages of polymeric micelles are their high safety, worthy stability and low cost. Polymeric micelles are very stable in physiological media with a consequent controlled drug release of drugs, while the hydrophilic shell protects the encapsulated drug from the external medium and prevents the interaction with plasma components, resulting in *in vivo* long circulating properties. Moreover, the small particle size prolongs the residence time in blood circulation, bypassing the liver and spleen filtration and the glomerular elimination, and enhances cellular uptake and the ability to cross epithelial barriers. All these aspects result in increased drug bioavailability (Bilia et al. 2017, 2018a).

A study reported the formulation of a polymeric micelle of about 142 nm obtained by esterification of oleoyl chloride and polyethylene glycol 400 and loaded with curcumin was developed and tested for the inhibition of human brain(glioblastoma astrocytoma) cells proliferation, a model of glioblastoma. In addition, the sensitivity of adult human bone marrow stromal cells and regular human fibroblastic (HFSF-PI3) cell lines to formulations were also investigated. The micelle significantly suppressed the proliferation of human brain(glioblastoma astrocytoma) cells in a time- and dose-dependent manner, with the half maximal inhibitory concentration of 20 μM after 24 h and 48 h, which declined to 10 μM at 72 h ($p < 0.001$). Moreover, the viability of human brain(glioblastoma astrocytoma) was not affected by free curcumin. Cell viability of adult human bone marrow stromal cells cell line after 24 h of micelle exposure declined to 67% after treatment with 25 μM micelle ($p < 0.01$) and to 35 and 31% after treatment with 30 μM and 35 μM micelle, respectively ($p < 0.001$). The viability of regular human fibroblastic (HFSF-PI3) cell line was not significantly affected by the micelle treatment and no inhibitory effect was detected on these cells ($p > 0.05$)at the dose of 20 μM; but, at 25 μM concentration, the viability of the cells decreased to 50% ($p < 0.01$). Therefore, in concentrations suppressive for cancer cells, no harmful effects connected to micelles were observed in stem cells and normal fibroblast cells, showing the safety of this formulation as an anticancer treatment agent on normal cells (Tahmasebi Mirgani et al. 2014).

In a further study, Soluplus a polymer constituted of polyvinyl caprolactam (57%), polyvinyl acetate (30%) and polyethylene glycol (13%), and soluplus plus containing also d-α-tocopherol polyethylene glycol 1000 succinate, also known as vitamin E TPGS (ratio Soluplus/TPGS was 20:1) were employed as amphiphilic polymers for the development of micelles able to improve the oral bioavailability of poorly water-soluble drugs. Micelles were loaded with increasing amounts of silymarin, from 0.5 mg/mL to 4 mg/mL. Micelles loaded up to 3 mg/mL of silymarin (encapsulation efficiency >92%) had similar average diameter (ca. 60 nm) and homogeneity (polydispersity index ≤0.1) to the empty micelles. Solubility studies demonstrated that the solubility of silymarin increased by ca. six-fold when loaded into micelles. Micelles avoided silymarin degradation in the gastrointestinal tract and showed a slow release of silymarin observed for both types of micelles. The potential increase of the intestinal absorption of silymarin was investigated by the parallel artificial membrane permeability assay, which confirmed a significant or borderline improvement the passive diffusion of silymarin when formulated into micelles. Transport studies employing Caco-2 cell line demonstrated that mixed micelles statistically enhanced the permeability of silymarin compared to polymeric micelles and unformulated silymarin. The study eveidenced that micelles entered into Caco-2 cells via energy-dependent mechanisms (Piazzini et al. 2019b).

1.2.3 Dendrimers

Dendrimers are well-defined hyperbranched polymers which form semiglobular to globular structures (Fig. 1.5), mostly with a high density of functionalities on the surface together with a limited "volume." Dendrimer originate from the Greek word δένδρον (dendron), translated as tree. Diameters of these nanoparticles are generally less than 10 nm, but can be modulated by varying dendrimer generations (G0, G1, G2, G3 etc.).

The dendrimer is characterised by a central core surrounded by branches of repeating units. The surface can be characterised by the presence of functional groups, which have a crucial role in dendrimer characteristics. Drug molecules can be linked to superficial moieties or loaded in the interior cavities. Diverse polymers including polyamidoamine (PAMAM), polyethylenglycol, poly(L-glutamic acid), polyethyleneimine, polypropyleneimine, are used to synthetize dendrimers. They are nanocarriers with vast benefits including the possibility to select the nanometric size range, simplicity of variation by modifying their ends, easy to prepare (Bilia et al. 2017; 2018a). In order to avoid bioavailability limits of curcumin, nanoparticles made by a new dendrimer G0.5 based on polyamidoamine was developed. The developed dendrimer G0.5 had no cytotoxic effects in breast cancer MCF-7 cells and produced spherical NPs of ca. 150 nm. After the loading of curcumin [molar ratio G0.5/curcumin 1:1 (formulation 1) and 1:0.5 (formulation 2)], both dendrimers were very homogenous. Formulation 1 was further tested for the drug release properties because the highest encapsulation efficiency and loading capacity. Curcumin solubility for both formulations was strongly enhanced (ca. 415 and 150 times more soluble than aqueous solubility for formulation 1 and 2, respectively (Falconieri et al. 2017).

1.3 Lipid Nanocarriers

Lipid-based nanocarriers are based on lipids, natural glycerides, waxes or long chain fat acids or alcohols or diverse synthetic molecules represented by glycerides, sterols, aliphatic molecules, long chain acids or alcohols or their esters. Surfactants from natural (mainly lecitins) or synthetic origins are also included in the formulations. Generally, excipients are carefully chosen from food excipients, which are marketed under the denomination "Generally Recognized As Safe" (GRAS).

Lipid nanocarriers include vesicles, nanocochleates, micelles, solid lipid nanoparticles (SLN) and nanostructured lipid particles (NLC), emulsions with nano scale, including nanoemulsions and microemulsions (Fig. 1.8). Vesicles include liposomes and niosomes and are obtained using amphiphilic lipids, phosphatidylcholine derivatives and non-ionic surfactants. Liposomes can be converted to nanocochleates, which are unique nanovectors, after treatment with divalent ions (Bilia et al. 2018a).

1.3.1 Micro and Nanoemulsions

Nanoemulsions and microemulsions are formulated with an oil phase, an aqueous phase, a surface active agent and probably a co-surfactant (McClements 2012). Microemulsions are defined as homogeneous thermodynamically stable transparent dispersions of two immiscible liquids stabilized by an interfacial film of surfactants. They have droplet size above 100–500 nm and require very low energy to formulate

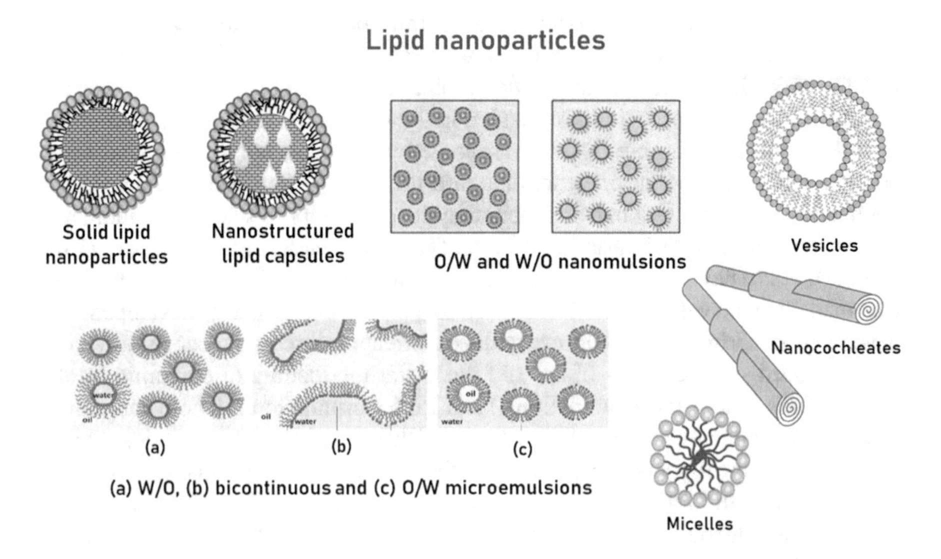

Fig. 1.8 Lipid nanoparticles. (Reproduced with permission from Bilia et al. 2018a from Wiley)

emulsion, since they form spontaneously when aqueous, oil, and amphiphilic components are brought into contact, besides having a lower production cost compared to nanoemulsions. They are characterised by three systems, thus in addition to oil in water and water in oil emulsions, a unique biphasic system called "bicontinuos systems" can be also formulated (Fig. 1.8). Nanoemulsions are non-equilibrium systems with a spontaneous tendency to separate into the constituent phases, prepared using lower surfactant concentrations than microemulsions. They have droplet covering the size range of 20–500 nm and referred to as miniemulsions, ultrafine emulsions, and submicrometer emulsions (Bilia et al. 2017; 2018a). Although both nanoemulsions and microemulsions show long-term stability, microemulsions are much more sensitive to environmental changes, such as temperature, ionic strength, composition (adding/removing molecules to/from the aqueous continuous phase). A drawback when comparing to nanoemulsions, is that microemulsion formation requires the use of relatively large amounts of surfactant, *i.e.*, their loading capacity is significantly lower than this of comparable nanoemulsion delivery systems, especially when using triglycerides as the dispersed oil phase (McClements 2012). A further and very successful approach to improve the solubility, chemical stability, and oral bioavailability of poorly water soluble molecules is self-microemulsifying drug delivery systems. They are defined as isotropic mixtures of oil, surfactant, cosurfactant and drug, which form oil-in water microemulsion with droplet size less than 100 nm when exposed to aqueous media with gentle agitation or motility of gastrointestinal tract (Kang et al. 2004). Different studies are present in the literature about the application of micro/nanoemulsion to deliver the natural compounds and recently some researches are concerning the formulation also of the plant extracts due to the high versatility of the formulations.

Astaxanthin, as other carotenoids, has a very low absorption by the human body due to its low water solubility, and it is quite unstable to high temperature, pH changes, oxygen and light, limiting its use. The formulation of astaxanthin in oil-in-water nanoemulsions using gypenosides (a natural mix of triterpenoid saponins from *Gynostemma pentaphyllum*) as natural emulsifiers was investigated and compared with a synthetic emulsifier, Tween 20 (Chen et al. 2018). Gypenosides produced nanoemulsions with a volume mean diameter (125 ± 2 nm), which was similar to those prepared using Tween 20 (145 ± 6 nm). In addition, gypenosides was able to protect the oil droplets from coalescence during thermal processing at elevated temperatures for 30 min, even if the rate of astaxanthin degradation increased with increasing temperature. All nanoemulsions showed excellent stability when stored at 5 °C and 25 °C for at least 30 days, regardless of emulsifier type. Gypenosides provided better protection against astaxanthin degradation than did Tween 20, probably due to the free radical scavenging ability of gypenosides (Chen et al. 2018).

Oil-in-water astaxanthin-loaded nanoemulsions were also prepared using ginseng saponins as natural emulsifiers (Shu et al. 2018). An oil phase (5% w/w) was obtained by dispersing 2% w/w of astaxanthin in soybean oil. The aqueous phase (95%) was prepared by dissolving saponins (0.08–1.2%, (w/w) in water. Saponins were capable of producing nanoscaled droplets (mean values were ca. 125 nm). The droplet size of the nanoemulsions decreased with increasing emulsifier concentration and homogenization pressure. The nanoemulsions were stable without droplet coalescence against thermal treatment, and over a narrow range of pH values (ranging from 7 to 9) but not in acidic conditions (pH range from 3 to 6) and in the presence of salt (Shu et al. 2018).

β-Carotene is as another important carotenoid with high instability and low oral-bioavailability. A formulation based on tea polyphenols and β-carotene was developed. It was an oil-in-water nanoemulsion with the core oil phase containing the carotenoid and the water phase containing tea polyphenols (Meng et al. 2019). The polyphenol extracts from green tea contained 20.04% of (−)-epigallocatechin, 11.17% of (−)-epicatechin gallate, 3.37% of (−)-epicatechin, 2.13% of (−)-gallocatechin gallate, 1.98% of (−)-gallocatechin and 0.51% of (+)-catechin. Corn oil was used as oily phase. During storage at three different temperatures (4, 25 and 35 °C), the nanoemulsion had a better stability and higher retention rate of β-carotene than the nanoemulsion containing only β-carotene. An *in vitro* simulated digestion assay indicated that the β-carotene recovery rates of the nanoemulsion containing also the tea polyphenols at digestion phases I and II were significantly increased compared to the nanoemulsion containing only the carotenoid. An *in vivo* absorption study showed that the nanoemulsion contain the tea polyphenols showed a higher conversion from β-carotene to vitamin A compared to the nanoemulsion containing only the carotenoid (Meng et al. 2019).

An oil-in-water nanoemulsion containing capsaicin used for management of pain and inflammatory disorders, was prepared using Tween 80 and Span 80 as non-ionic surfactants, ethanol as co-surfactant, olive oil as oil phase and water as external phase (Ghiasia et al. 2019).

The formulation had a droplet diameter of 13–14 nm and was stable for more than 8 months at both 4 °C and 45 °C. The nanoemulsion was then formulated into topical cream and gel to compare its efficacy and safety profiles with conventional cream of capsaicin. Carbopol® was selected as gel matrix base, while the cream base consisted of cetyl alcohol, cholesterol, liquid paraffin, soft paraffin and beeswax. Both formulations contained 0.075 w/w of capsaicin.

Skin irritation study showed that topical application of capsaicin nanoemulsion was safe and no sign of oedema or erythema was observed. The preparation significantly decreased inflammation of rats paw oedema compared to the conventional formulations and control group. Capsaicin nanoemulsion loaded in gel with showed very good resistance to the pain caused by heat stimulus in rats (Ghiasia et al. 2019).

Curcumin as other lipophilic small molecules, has a low hydrophilicity and intrinsic dissolution rate(s), low physical and chemical stability, rapid metabolization with low nanomolar levels of the parent compound and its glucuronide and sulphate conjugates found in the peripheral or portal circulation, low absorption, poor pharmacokinetics and bioavailability, and low penetration and targeting efficacy (Bilia et al. 2017). Three microemulsions were developed and characterized, stabilized by non-ionic surfactants macrogolglycerol ricinoleate (Cremophor EL), polysorbate 20 (Tween 20), polysorbate 80 (Tween 80), or lecithin and contained a variety of oils, namely, olive oil, wheat germoil, and vitamin E. The oral absorption of curcumin microemulsions was investigated in vitro using parallel artificial membrane permeability assay. The optimal formulation consisted of vitamin E (3.3 g/100 g), Tween 20 (53.8 g/100 g), ethanol (6.6 g/100 g), and water (36.3 g/100 g), obtaining a percentage of permeation through the artificial membrane of about 70% (Bergonzi et al. 2014).

In a further study, nanoemulsions containing 0.5% w/w corn oil containing 0.4% w/w curcumin, sodium-alginate (1.0% w/w) and 0.5, 1.0 or 2.0% w/w of surfactant (Tween 20, lecithin or sucrose monopalmitate), were developed (Artiga-Artigas et al. 2018). Nanoemulsions showed particle sizes ≤400 ± 3 nm and effectively reduced droplets interfacial tension with negative ζ-potential values (≤ − 37 mV), regardless the concentration of surfactant. Nanoemulsions with 2.0% w/w lecithin were stable during 86 days of experiment, whereas those containing Tween 20 or sucrose monopalmitate at the same concentrations were not stable after 5 days or along 24 h, respectively (Artiga-Artigas et al. 2018).

A study by Machado et al. (2019) developed a curcumin-nanoemulsion to be used as a photosensitizing agent in photodynamic therapy in an *in vitro* breast cancer model, breast cancer MCF-7 cells. The nanoemulsion containing 0.1 mg/mL showed an efficient internalization in the fibroblast and adenocarcinoma mammary line, as well as absence of cytotoxicity. The oil phase was prepared containing medium-chain-triglycerides and natural soy phospholipids. Poloxamer 188 was used as an anionic surfactant. The physical parameters of NE were: size 199 ± 0.2 nm; Zeta potential −46.3 (mV) polydispersity index 0.179. Curcumin had phototoxic effects, significantly decreased the proliferation of breast cancer MCF-7 cells and stimulating the reactive oxygen species production in combination with photodynamic therapy (Machado et al. 2019).

Green tea catechins (Polyphenon 60) and caffeine were encapsulated in a microemulsion to obtain synergistic antibacterial activity against selected pathogens (Gupta et al. 2014). Combination of two natural compounds would advocate two different mechanisms on the bacterial growth thereby providing for better antibacterial activity. Thermodynamically stable microemulsion was developed by using Labrasol as an oil phase, Cremophor EL as surfactant, and glycerol as cosurfactant. The combination of the above two natural compounds was proficient in lowering the MICs of individual agents. Higher antimicrobial effect of microemulsions can be attributed to the formation of nanodrops that increase the surface tension and thereby force themselves to merge with the lipids present in the bacterial cell membrane. Results of 2,2-diphenyl-1-picrylhydrazyl assay indicated that microemulsion system preserved the long term antioxidative potential of P60 and caffeine (Fig. 1.9). The cytotoxicity of the optimized formulation on Vero cell line by MTT assay was found to be nontoxic to mammalian cells (Fig. 1.10).

I a further study, mangiferin nanoemulsions were developed using hyaluronic acid of different molecular weights, in absence or presence of Transcutol-P (Pleguezuelos-Villa et al. 2019). Mangiferin (1%) and glycerine (3%) were added to distilled water in a glass tube in absence or presence of hyaluronic acid (1%) to obtain aqueous phases. Lipoid® S75 (5%), polysorbate 80 (1%), tocopherol (0.1%) and almond oil (10%) were also mixed with or without Transcutol-P (4%) to obtain

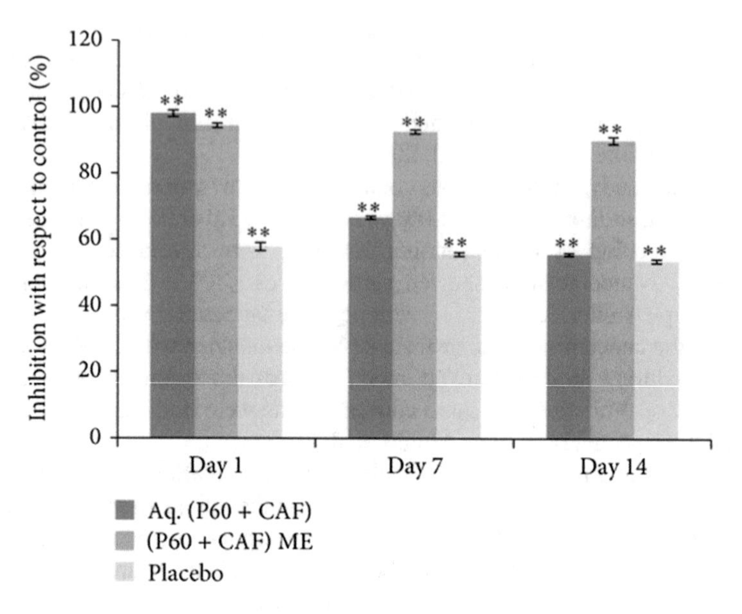

Fig. 1.9 Antioxidative effect of the aqueous green tea extract plus caffein, loaded in the microemulsion, and placebo via 2,2-diphenyl-1-picrylhydrazyl (DPPH) assay. Data are represented as percentage of inhibition with respect to control. Mean values of three independent experiments and S.E. are shown. **Significant at P < 0.005. (Reproduced with permission from Gupta et al. 2014 from Hindawi)

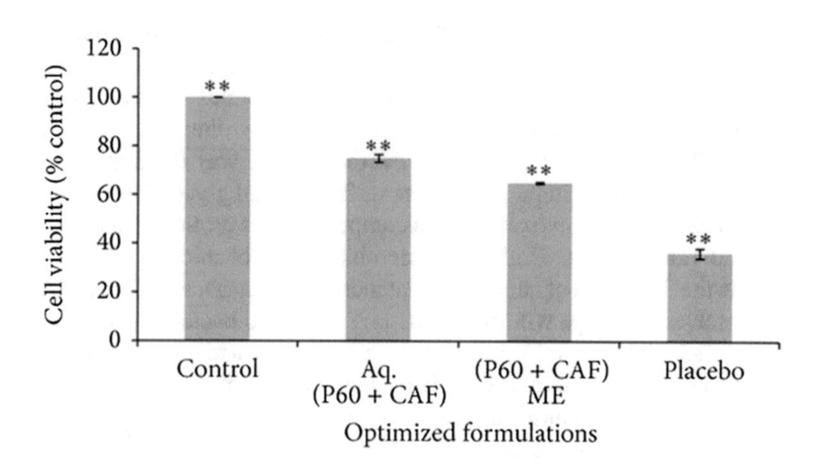

Fig. 1.10 Cytotoxicity analysis of the aqueous green tea extract plus caffein, loaded in the microemulsion, and placebo on Vero cell lines after 24 hrs via MTT assay. Data are represented as percentage of Vero cell viability. Mean values of three independent experiments and S.E. are shown. (Reproduced with permission from Gupta et al. 2014 from Hindawi)

the oil phases. Mangiferin release depended on the molecular weight of the hyaluronic acid. Permeability assays on pig epidermis showed that nanoemulsions containing low molecular weight hyaluronic acid had an improved permeation, being this effect more pronounced in nanoemulsions with Transcutol-P. Administration of mangiferin nanoemulsions on inflamed skin mice model provided an attenuation of oedema and leucocyte infiltration (Pleguezuelos-Villa et al. 2019).

Recently, Piazzini and coworkers (Piazzini et al. 2017a, Piazzini et al. 2017b) focused the studies on the development of innovative oil in water nanoemulsions to improve the solubility and the oral absorption of *Vitex agnus-castus* and *Silybum marianum* extracts. Physical characterization showed droplets' dimensions from 10 to 20 nm. Solubility of the extracts was improved considerably respect to water (about 10 times in the case of *Vitex agnus-castus* and about 100 times in the case of *Silybum marianum*). The optimized formulation of *Silybum marianum* contained 40 mg/mL of commercial extract (4% w/w) and it was composed of 2.5 g labrasol (20%) as the oil phase, 1.5 g cremophor EL as the surfactant, and 1 g labrafil as the cosurfactant (mixture surfactant/cosurfactant, 20%). Physical and chemical stabilities were assessed during 40 days at 4°C and 3 months at 25°C. Stability in simulated gastric fluid followed by simulated intestinal conditions was also assessed. *In vitro* permeation studies performed using parallel artificial membrane permeability assay and Caco-2 cell lines showed a pronounced permeability, in respect to a saturated aqueous solution (Piazzini et al. 2017b). The composition of the nanoemulsion loaded with *Vitex agnus-castus* was triacetin as oil phase, labrasol as surfactant, cremophor EL as co-surfactant and water, containing 60 mg/mL of extract. Droplets' average diameter was 11.82 ± 0.125 nm and a polydispersity index of 0.117 ± 0.019. A prediction of oral absorption of the extracts loaded in the microemulsions was tested *in vitro* using different models such as parallel artificial membrane permeability assay and Caco-2 cell line as reported in Fig. 1.11. These studies clearly demonstrated that developed drug delivery systems increased the permeation of the main constituents of the extracts compared to the aqueous solutions (Piazzini et al. 2017a).

A microemulsion system was developed and investigated to increase the solubility and the absorption of *Salicis cortex* extract (Piazzini et al. 2018a). The optimized microemulsion consisted of 2.5 g of triacetin, as oil phase, 2.5 g of tween 20 as surfactant, 2.5 g of labrasol as co-surfactant and 5 g of water. The developed formulation appeared transparent, the droplet size was around 40 nm and zeta potential values ranged from −11 to −14 mV, indicating that developed microemulsion were negatively charged. Microemulsion increased considerably the solubility of salicylic derivatives and flavanones respect to water (about 3.6 times for salicylic derivatives and about 2 times for flavanones). The maximum loading content of *Salicis cortex* extract resulted 40 mg/mL. The developed formulations were stable during the storage period and they were able to prevent the degradation of loaded molecules. I*n vitro* parallel artificial membrane permeability assay and Caco-2 tests proved that the microemulsion was successful in enhancing the permeation of extract compounds (Piazzini et al. 2018a).

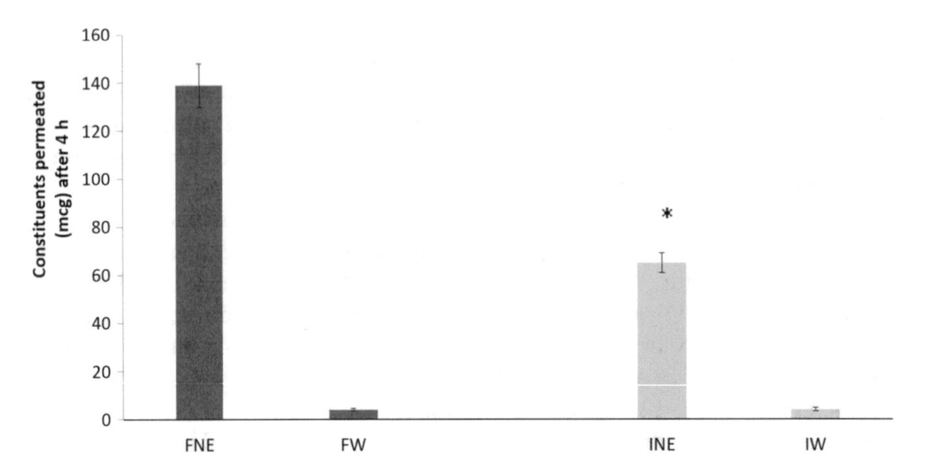

Fig. 1.11 Comparative PAMPA permeability: micrograms of total flavonoids (F) and iridoids (I) of extract of VAC permeated from nanoemulsion (NE) and from saturated aqueous solution (W). Average values ± S.D. of experiments carried out in triplicate are presented. $* p < 0.05$. (Reproduced with permission from Piazzini et al. 2017a from Taylor & Francis)

Zhang et al. (2012a) reported the preparation of self-microemulsifying drug delivery systems in sustained release pellets to enhance the oral bioavailability of puerarin. Castor oil was used as the oil phase, Cremophor EL as the emulsifier, and 1,2-propanediol as the co-emulsifier. The pharmacokinetic profile and bioavailability of the puerarin loaded in the innovative formulation and conventional tablets of puerarin were compared in beagle dogs. The absolute bioavailability of the puerarin in the nanoformulation was enhanced by 2.6-fold compared with that of the tablet. The relative bioavailability of the nanoformulation was 259.7% compared with the conventional tablet group (Zhang et al. 2012a).

Quercetin has been formulated into self-microemulsifying drug delivery systems to improve its oral bioavailability and antioxidant potential. The formulation was optimized using 40:40:20 w/w of Capmul MCM:QT (19:1)/Tween20/ethanol and showed a rapid internalization within 1 h of incubation with Caco-2 cells and a significant increase in cellular uptake by 23.75-fold in comparison with free quercetin cultured with Caco-2 cells. Five-fold enhancement in oral bioavailability compared to free quercetin suspension was found (Jain et al. 2013).

Baicalein, is another poor bioavailable flavonoid which was formulated in self-microemulsifying drug delivery systems using Cremophor RH40 (53.57%) as surfactant, Transcutol P (21.43%) as cosurfactant, and caprylic capric triglyceride (25%) as oil. The drug release rate of self-microemulsifying drug delivery systems was significantly higher than that of the baicalein suspension. The *in vivo* studies showed that the absorption of baicalein from SMEDDS resulted in about 200.7% increase in relative bioavailability compared with that of the baicalein suspension (Liu et al. 2012).

Recently, Guccione et al. (2018) reported the formulation two microemulsions and two self-microemulsifying drug delivery systems based on a complex commercial mixture containing a CO2 extract of *Serenoa repens* and two hydrophilic extracts: nettle root and pineapple stem. For both formulations stability in simulated gastric fluid and intestinal media was substantiated. *In vitro* parallel artificial membrane permeability assay evidenced an increased mucosal permeation when the formulations where compared with the raw mixture of the different extracts. In particular, for a microemulsion a 30–70% passive absorption after oral administration was predicted (Guccione et al. 2018).

1.3.2 Vesicles

Bingham first reported vesicles of natural origin, based on phospholipids in 1965. They are characterised by the presence of amphiphilic molecules represented by natural or synthetic phospholipids or non-ionic surfactants, and accordingly they are denominated liposomes or niosomes (Fig. 1.8). Many advantages are reported for vesicles over the other types of nanocarriers because both hydrophilic and hydrophobic drugs can be easily encapsulated due to an aqueous compartment and lipophilic bilayer. Advantages are numerous, enhanced bioavailability of drugs, delayed elimination of rapidly metabolizable drug, prolonged circulation life time of drugs, targeted delivery of drugs, increased stability and decreased toxicity issues of certain drugs. Disadvantages and limits are represented by their leaky in nature leading to premature drug release, poor encapsulation efficiency for hydrophilic drug, quite expensive and short half-life (Bilia et al. 2014a, b, 2017, 2018a).

Structure of vesicles is spherical, made of natural or synthetic phospholipids (mainly phosphatidylcholine) and cholesterol that acts as a fluidity buffer. Although cholesterol do not participate in bilayer formation, it can be added to phosphatidylcholine up to 1:1 or even 2:1 molar ratio of cholesterol to phosphatidylcholine. Liposomes represent the utmost investigated nanocarriers (size ranging from 20 nm to several μm), and exhibit numerous benefits, principally high biocompatibility, and easy access to surface decoration in order to obtain passive and active targeting (Bozzuto and Molinari 2015).

An *in vivo* study reported the liposomal formulation of verbascoside to enhance the chemical stability by preventing its hydrolysis. The mean particle size of the liposomes prepared was found to be around 120 nm with a polydispersity index <0.2. About 83% of encapsulated verbascoside was released from the liposomes within 24 h in an *in vitro* release test. The optimized drug delivery formulation was tested in the paw pressure test in two animal models of neuropathic pain: a peripheral mononeuropathy was produced either by a chronic constriction injury of the sciatic nerve or by an intra-articular injection of sodium monoiodoacetate. The performance of the liposomal formulation was compared with that of the free drug. For evaluating the paw pressure test in chronic constriction injury rats, a liposomal formulation administered i.p. at the dosage of 100 mg/kg showed a longer lasting anti-hyperalgesic effect in comparison with a 100-mg/kg verbascoside saline solution, as well as in the sodium monoiodoacetate models. The effect appeared 15 min after administration and persisted for up to 60 min (Isacchi et al. 2016).

Naringenin was loaded liposome for pharmacokinetic and tissue distribution studies in animal models. The liposomal carrier consisted of phospholipid, cholesterol, sodium cholate, and isopropyl myristate. It had a suitable particle size, zeta potential, and encapsulation efficiency (70.53 ± 1.71 nm, −37.4 ± 7.3 mV, and 72.2 ± 0.8%, respectively). The *in vitro* release profile of naringenin from the formulation in three different media (HCl solution, pH 1.2; acetate buffer solution,pH 4.5; phosphate buffer solution, pH 6.8) was significantly greater than the free drug. The studies with mice also revealed an increase in AUC of the naringenin-loaded liposome and approximately 13.44-fold increase in relative bioavailability was observed in animals after oral adiministartion. Studies of tissue distribution assessed that the drug was predominantly distributed in the liver (Wang et al. 2017).

An extensive study was carried out using diverse natural products with anti-inflammatory activity, including thymol, carvacrol, pterostilbene, N-(3-oxododecanoyl)-l-homoserine lactone, resveratrol, caffeic acid, and caffeic acid phenethyl ester, loaded in PEGylated liposomes to enhance their stability, solubility, and bioavailability. The study displayed that 3-oxo-C(12)-homoserine lactone and stilbene derivatives can be loaded into liposome with extraordinary percentages (50–70%), allowing the parental use of these molecules. In particular, resveratrol chemical stability was enhanced preventing the inactivation of molecule activity because a cis-trans isomerization. The loading of 3-oxo-C(12)-homoserine lactone and resveratrol in liposomes also increased their antitumor properties, inhibiting tumour growth for approximately 70% in mice (Coimbra et al. 2011).

In a further study, PEGylated liposomes were developed as drug carriers for parental administration of Salvianolic acid B in order to increase the poor chemical stability and bioavailability. Both conventional and long circulating liposomes loaded with salvianolic acid B were developed. These carriers were submitted to pharmacological studies using the paw pressure test in an animal model of neuropathic pain where a peripheral mononeuropathy was produced by a chronic constriction injury of the sciatic nerve. Salvianolic acid B was effective against mechanical hyperalgesia when administered intraperitoneally at a dose of 100 mg/kg, 15 min after administration. According to the in vivo studies, encapsulation, especially into PEGylated liposomes, increased and prolonged the antihyperalgesic activity 30 min after i.p. administration and the effect was still significant after 45 min (Isacchi et al. 2011a).

A series of studies to compare the performance of conventional and long circulating liposomes have been carried out using artemisinin, the unique sesquiterpene lactone with an endoperoxide moiety isolated from *Artemisia annua* L.

In a first research, artemisinin was loaded in conventional and PEGylated (long circulating) liposomes to investigate the biopharmaceutical properties of the two formulations. Vesicles were purified by dialysis, obtaining an encapsulation efficacy higher than 70%, and mean diameters in the range from 130 to 140 nm. Pharmacokinetic parameters of the vesicles were investigated in healthy mice i.p.

Unformulated artemisinin was quickly cleared from plasma and its detection after 1 h after administration was difficult. By contrast, when both liposomal formulations were administered, artemisinin was measurable after 3 and 24 hours for conventional and long circulating liposomes respectively. AUC (0–24 h) values were improved by ca. six times in both formulations, when compared with unformulated artemisinin. Its half-life was improved more than five-fold when PEGylated liposomes were administered (Isacchi et al. 2011b).

The therapeutic efficacy of the PEGylated liposomes loaded with artemisinin or artemisinin plus curcumin were investigated *in vivo* for the antimalarial activity. Formulations were evaluated in *Plasmodium berghei* NK-65 infected mice treated with artemisinin at the dose of 50 mg/kg/days alone or artemisinin plus curcumin, administered at the dosage of 100 mg/kg/days. When unformulated artemisinin was administered to infected mice, the parasitaemia decreased only 7 days after the beginning of the treatment. By contrast, treatments with artemisinin-loaded in conventional liposomes, artemisinin-curcumin loaded in conventional liposomes, artemisinin loaded in PEGylated liposomes, artemisinin-curcumin loaded in PEGylated liposomes had an immediate antimalarial effect. Nanoencapsulated artemisinin and artemisinin plus curcumin formulations cured all infected mice. Additionally, all formulations had a reduced variability of artemisinin plasma concentrations. In particular, artemisinin loaded in PEGylated liposomes gave the most noticeable and statistically significant antimalarial activity (Isacchi et al. 2012).

Dihydroartemisinin, a metabolite of artemisinin, is reported as an effective anticancer compound, but the molecule is scarcely water-soluble with poor bioavailability and short half-life (34–90 min). Accordingly, conventional and PEGylated liposomes were loaded with dihydroartemisinin and tested in cancer cells. Encapsulation efficiency was high for both conventional liposomes (71%) and PEGylated ones (69%). Physical and chemical stabilities were good under storage conditions or when added to albumin. It was found that cellular uptake of PEGylated liposomes was less than the uptake of conventional liposomes, probably due to hydrophilic steric barrier of PEG chains. Cytotoxicity studies were evaluated in the breast cancer MCF-7 cell line, which confirmed the lack of toxicity in blank formulations (Righeschi et al. 2014).

Leto et al. (2016) have tested artemisinin loaded in transferrin-conjugated liposomes, comparing their performance with artemisinin loaded in PEGylated liposomes. Both cell uptake and cytotoxicity studies of the nanoformulations were investigated in the human colon carcinoma (HCT-8) cell line, because of transferrin receptor overexpression. The studies established the improved delivery of artemisinin at the human colon carcinoma (HCT-8) cell line from the carrier decorated with transferrin compared with long-circulating liposomes performance. The benefits of using transferrin are that it is very biocompatible due to non-immunogenicity, it can be easily conjugated to diverse NPs. Consequently, artemisinin was loaded in transferrin conjugated liposomes, and stealth liposomes. Cell uptake and cytotoxicity of both liposomes were tested in the human colon carcinoma (HCT-8) cell line, selected among several cell lines because of the overexpression of transferrin receptor. The study revealed an increased delivery of artemisinin from transferrin conjugated liposomes in comparison with stealth liposomes. An improved cytotoxicity was also evidenced as a result of the presence of iron ions which gave a synergistic activity (Leto et al. 2016).

Chitosan-coated liposomes containing curcumin well as curcumin loaded anionic liposomes were also evaluated (Cuomo et al. 2018). Values of diameter, polydispersity index and surface charge for curcumin loaded anionic liposomes were 129 nm, 0.095 and −49 mV, respectively. Anionic liposomes and chitosan-coated liposomes, were tested for *in-vitro* digestion tests. The digestion model used simulates the food/beverages ingestion from mouth to the small intestinal phase. In every digestion compartment, the collective surface charge was measured. It was demonstrated that chitosan-coated liposomes in the mouth phase interact with mucin because mucin is negatively charged at the mouth pH. After chitosan-coating, diameter and polydispersity index remain unvaried while the surface charge gets positive. Slightly higher curcumin concentrations were found after the mouth and the stomach digestion phases when curcumin was loaded in anionic liposomes. On the contrary, after the intestinal phase, a higher percentage of curcumin was found when chitosan-coated liposomes were used as carrier, both in the raw digesta and in the bile salt micellar phase (Fig. 1.12). It was shown that the presence of a positively charged surface allows a better absorption of curcumin in the small intestine phase, which increases the overall curcumin bioavailability (Cuomo et al. 2018).

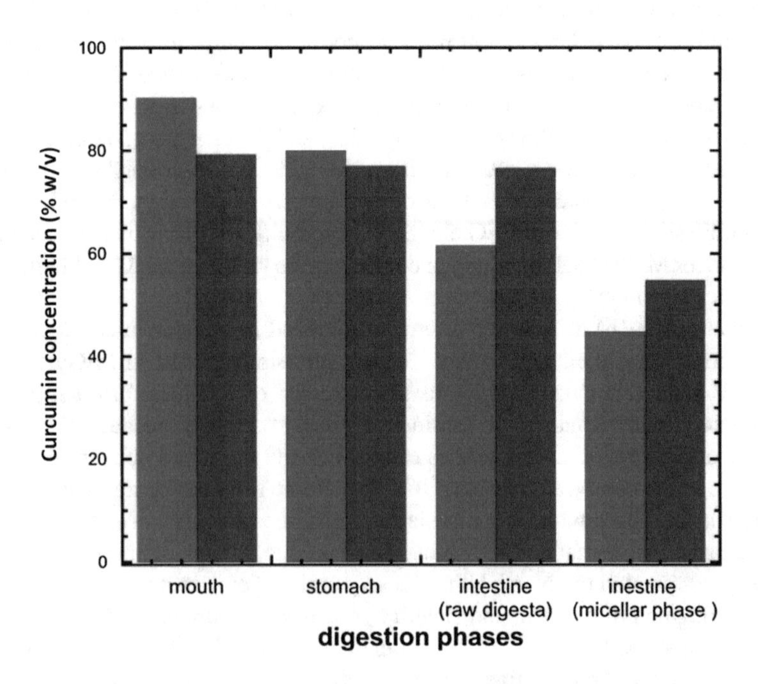

Fig. 1.12 Variation of curcumin content in liposomes (grey) and chitosan-coated liposomes (red) during the digestion phases. (Reproduced with permission from Cuomo et al. 2018 from Elsevier)

In a further study, different pluronics (F127, F87 and P85) were used as modifiers to improve the stability and bioaccessibility of curcumin liposomes (Li et al. 2018). The particle size and polydispersity index of liposomes containing pluronics was significantly lower than conventional liposomes. *In vitro* release studies demonstrated that 73.4%, 63.9%, 66.7% and 58.9% of curcumin was released from conventional liposomes, F127-liposomi, F87-liposomi and P85-liposomes, respectively. Stability studies indicated that pluronics modification could enhance their pH and thermal stability. *In vitro* simulated gastrointestinal tract studies suggested that pluronics modification could significantly improve the absorption of curcumin in liposomes. The pluronics modified curcumin liposomes exhibited superior bioaccessibility, defined as the fraction of the curcumin that is solubilized in the digesta and therefore available for absorption as reported in Fig. 1.13 (Li et al. 2018).

Resveratrol is a simple polyphenol with potential activity in the prevention of coronary disease and neurodegenerative pathologies, but the low bioavailability, low water solubility, and instability have compromised its wide array of activity (Wenzel and Somoza 2005) and numerous nanoformulations have been currently investigated to enhance its efficacy.

In a study, mitochondrial targeting resveratrol liposomes modified with a dequalinium polyethylene glycol-distearoylphosphatidyl ethanolamine conjugate were developed. The nanopartices induced apoptosis in both non-resistant and resistant cancer cells by the dissipating mitochondrial membrane potential. It also increased caspase-9 and caspase-3 activities. Significant antitumor efficacy was exerted by

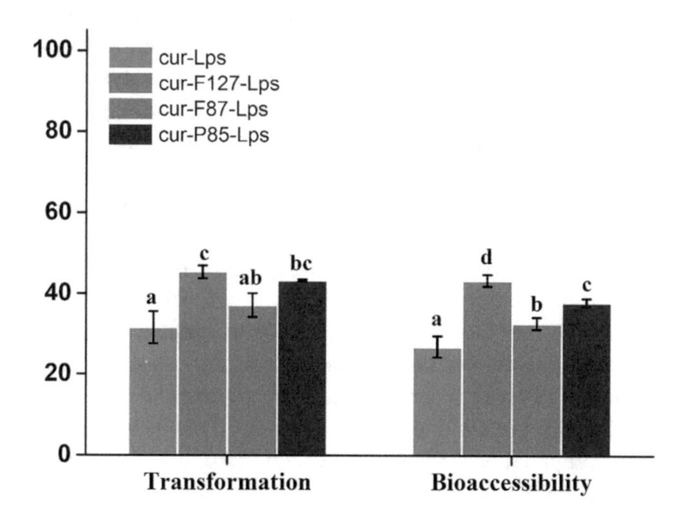

Fig. 1.13 Transformation and bioaccessibility of curcumin liposomes (cur-Lps), Pluronic F127 modified curcumin liposomes (cur-F127-Lps), Pluronic F87 modified curcumin liposomes (cur-F87-Lps) and Pluronic P85 modified curcumin liposomes (cur-P85-Lps). Different capital letters mean significant differences ($p < 0.05$). (Reproduced with permission from Li et al. 2018 from Elsevier)

resveratrol liposomes in xenografted resistant lung adenocarcinoma A549/cDDP cancers in nude mice and tumour spheroids by deep penetration (Wang et al. 2011a).

In another study, the viability of human embryonic kidney (HEK 293) cell line after treatment with liposomal resveratrol was tested. In addition the photoprotection after UV-B irradiation was investigated after treatment with free curcumin or the liposomal formulation. Interestingly, cell viability was found to be decreased at a 100 μM concentration, and cell proliferation increased at 10 μM, achieving the most effective photoprotection. This study showed the effectiveness of resveratrol at 10 μM and also toxicity at higher concentrations considering the changes in apoptotic features and cell shape and its detachment. Interestingly, liposomes prevented the cytotoxicity of resveratrol at high concentrations, even at 100microM, avoiding its immediate and massive intracellular distribution, and increased the ability of resveratrol to stimulate the proliferation of the cells and their ability to survive under stress conditions caused by UV-B light (Caddeo et al. 2008).

Resveratrol was also formulated in phosphatidylcholine and cholesterol liposomes obtaining an encapsulation efficiency of 96.5% ± 2.1. The formulation increased resveratrol solubility. Skin- parallel artificial membrane permeability assay proved an increased cutaneous permeability of resveratrol when loaded into liposomes maintaining the antioxidant capacity of resveratrol, as evidenced by 2,2-diphenyl-1-picrylhydrazyl (DPPH) test (Casamonti et al. 2019b).

The investigation of Manconi et al. (2018) reported the development of PEG-modified liposomes for the delivery of resveratrol via parental administration. Small, spherical, unilamellar vesicles were produced (size 86 ± 2.7, polydispersion index 0.20 ± 0.03, potential −21 ± 4.1, EE% 94.9 ± 2.9). The liposomes were stable, as demonstrated by their size basically unchanged after 2 months of storage (size 91.6 ± 3.2 nm, polydispersion index was 0.19 ± 0.02). The biocompatibility of the formulations was demonstrated in an *ex vivo* model of haemolysis in human erythrocytes. Further, the incorporation of resveratrol in polyethilenglycol-modified liposomes did not affect its intrinsic antioxidant activity, as 2,2-diphenyl-1-picrylhydrazyl radical was almost completely inhibited, and the vesicles were also able to ensure an optimal protection against oxidative stress in an *ex vivo* human erythrocytes-based model.

Ethemoglu et al. (2017) investigated the effects of resveratrol and resveratrol loaded in liposome on penicillin-induced epileptic activity in adult male Sprague–Dawley rats. Liposomes were the most effective anticonvulsant treatment on penicillin-induced epileptic seizures when compared to control, and immediately decreased the number of spikes per minute. S-transferase and superoxide dismutase activity, as well as the glutathione levels, were significantly increased in the liposomes group as compared with the control group. Also, the malondialdehyde levels were significantly higher in the liposomes compared to resveratrol and control groups (Ethemoglu et al. 2017).

In a further study, resveratrol was loaded into PEGylated liposomes and the liposome surface was modified with transferrin to obtain an active targeting against cancer cells (Jhaveri et al. 2018). Liposomes were stable, had a good drug-loading capacity, prolonged drug-release *in vitro* and were easily scalable. Flow cytometry and confocal microscopy were used to study the association with, and internalization of liposomes decorated with transferrin into human brain(glioblastoma astrocytoma) (U-87 MG) cells. These were significantly more cytotoxic and induced higher levels of apoptosis accompanied by activation of caspases 3/7 in glioblastoma cells when compared to free resveratrol or resveratrol loaded in PEGylated liposomes. The liposomes decorated with transferrin were more effective than other treatments in their ability to inhibit tumour growth and improve survival in mice (Jhaveri et al. 2018).

In an another study, chitosan-coated liposomes were investigated for use in enhanced transdermal delivery of resveratrol (Park et al. 2014). The particle size of uncoated liposomes was found to be 212.83 nm. After coating with chitosan solution, this increased to 279.85, 432.58, and 558.35 nm, for 0.1%, 0.3%, and 0.5% chitosan solutions, respectively. The zeta potential of the liposomes also followed the same trend, i.e., it changed from a negative value (−9.4 mV) for uncoated liposomes to increasingly positive values for the chitosan-coated ones (26.5, 34.5, and 39.2 mV, for 0.1%, 0.3%, and 0.5% chitosan solutions, respectively). The chitosan coating increased the stability of the liposomes by preventing their aggregation. The chitosan also increased skin-permeation efficiency (Park et al. 2014).

Two studies reported on the formulation of honokiol, 3′,5-di(2-propenyl)-1,1-biphenyl-2,4-diol, a constituent of *Magnolia officinalis* with several pharmacological effects, including a potent antitumor activity, but with a very limited bioavailability preventing vascular administration (Wang et al. 2011b, c). In the first study, stable liposomes containing an inclusion complex of honokiol in hydroxylpropil-β-cyclodextrin (HP-β-CD) used a molar ratio of honokiol/HP-β-CD/phosphatidilcholine 1:2: 2. The mean particle size was 123.5 nm, the zeta potential was −25.6 mV, and the EE was 91.09 ± 2.76%. The release profile *in vitro* demonstrated that honokiol was released from the liposome with a sustained and slow profile. A pharmacokinetic study revealed that honokiol-in-HP-β-CD-in-liposome significantly retarded the elimination of honokiol and prolonged the residence time in the circulating system. Honokiol-in-HP-β-CD-in-liposome had antiproliferative activity in adenocarcinomic human alveolar basal epithelial (A549) and human liver cancer (HepG2) tumour cells compared with free honokiol (Wang et al. 2011b). In the second study, PEGylated liposomal honokiol with a particle size of ca. 100 nm was developed, and both pharmacokinetic properties and human plasma protein binding ability were investigated. The pharmacokinetics properties were studied after intravenous administration in Balb/c mice. The results suggested that PEGylated liposome improved honokiol solubility, increased the drug concentration in plasma, and decreased the clearance (Wang et al. 2011c).

A study formulated and evaluated liposomes of andrographolide for brain-targeting. The surface of liposomes was modified by adding Tween 80 alone or in combination with didecyldimethylammonium bromide to enhance the penetration into the brain of AG and to understand the effect of the surface charge on the interaction with human brain endothelial cells (hCMEC/D3). The ability of LPs to increase the permeability of AG was evaluated by parallel artificial membrane permeability assay and human brain endothelial cells (hCMEC/D3). The cellular uptake mechanism was identified using fluorescence microscopy and HPLC. The study demonstrated that the presence of Tween 80 alone or in combination with didecyldimethylammonium bromide on the surface of liposomes enhanced solubility and cellular permeability of andrographolide loaded-liposomes (Fig. 1.14). The different surface charge did not significantly influence the endocytic pathway of liposomes: both formulations were mainly taken up by caveolate-mediated endocytosis (Piazzini et al. 2018b).

Lecithin-based liposomes loaded with silymarin were developed to increase its oral bioavailability and to make it targeted specific for the liver with enhanced hepatoprotection. The formulation showed maximum entrapment (55%) for a lecithin–cholesterol ratio of 6:1. Comparative release profile of formulation was better than unformulated silymarin at pH 1.2 and pH 7.4. *In vitro* studies showed a better hepato-protection efficacy for formulation (one and half times) and better prevention of ROS production (ten times) compared to silymarin. Paracetamol was used to obtain significant hepatotoxicity in Wistar rats, which were tested for the performance of liposomal silymarin. The formulation was found more efficacious than silymarin suspension in protecting the liver against paracetamol toxicity and the associated inflammatory conditions. The liposomal formulation yielded a three and

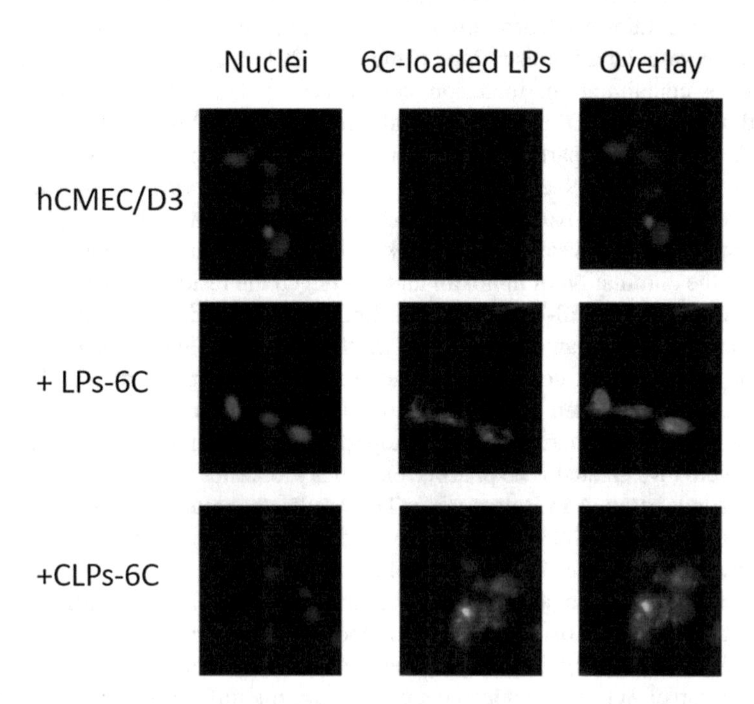

Fig. 1.14 Cellular uptake of Tween 80 liposomes loaded with 6-coumarin (LPs-6C) and Didecyldimethylammonium bromide liposomes loaded with 6-coumarin (CLPs-6C) by a model of the human blood brain barrier hCMEC/D3 cells after 2 h incubation at 37 °C. Images of nuclei stained with DAPI (blue), 6-Coumarin (green) and their overlay. Scale bar: 20 μm

half fold higher bioavailability of silymarin as compared with silymarin suspension (Kumar et al. 2014).

In a further study by the same researchers, four different liposomal formulations namely conventional, dicetyl phosphate, stearyl amine and PEGylated liposomes were prepared to improve the *in vitro* and *in vivo* hepatoprotection and increase silymarin oral bioavailability against alcohol intoxication. The conventional liposomes increased *in vitro* release profile at pH 1.2 and 6.8 and also showed better in vitro protection on Chang liver cells compared to silymarin alone.

Efficacious liposomes were selected for *in vivo* hepatoprotection study and bioavailability in alcohol intoxicated Wistar rats. Conventional and PEGylated liposomes showed better improvement in liver function, better efficacy in combating inflammatory conditions, better improvement in antioxidant levels and reversal of histological changes compared to unformulated silymarin. Conventional liposomes also showed an almost fourfold increase in area under the curve compared to silymarin suspension (Kumar et al. 2019).

An additional liposomal formulation loaded with silymarin was reported in the literature for targeting liver. It included hydrogenated soy phosphatidylcholine, cholesterol, distearoylphosphoethanolamine-(polyethyleneglycol)-2000 and various amounts of β-sitosterol-β-D-glucoside. Increasing the amount of β-sitosterol-β-D-glucoside in the liposomes gradually decreased drug encapsulation efficiencies from about 70% to about 60%. Addition of β-sitosterol-β-D-glucoside to non-PEGylated liposomes clearly affected their drug release profiles and plasma protein interactions, whereas no effect on these was seen for the PEGylated liposomes. Non-PEGylated liposomes with 0.17 M ratio of β-sitosterol-β-D-glucoside exhibited the highest cellular drug uptake of 37.5% for all of the studied liposome formulations. The highest cellular drug uptake in the case of PEGylated liposomes was 18%, which was achieved with 0.17 and 0.33 M ratio of added β-sitosterol-β-D-glucoside. The liposome formulations with the highest drug delivery efficacy showed hemolytic activities around 12.7% and were stable for at least 2 months upon storage in 20 mM HEPES buffer (pH 7.4) containing 1.5% polysorbate 80 at 4 °C and room temperature (Elmowafy et al. 2013).

Zhang et al. (2012b) developed a core-shell hybrid liposomal vesicles to improve the limited bioavailability and to enhance both protective effects and *in vivo* oral administration of *Panax notoginseng* saponins. Saponins were loaded in diverse conventional nanoparticles and in core-shell hybrid liposomal vesicles. Morphology, particle size, zeta potential, encapsulation efficiency (EE%), stability of all the nanocarriers and in vitro release studies were performed. The latter carriers had the best performance, they were stable for at least 12 months at 4 °C. Satisfactory improvements in the encapsulation efficacy of notoginsenoside R1, ginsenoside Rb1, and ginsenoside Rg1 were also found and the greatest controlled drug release profiles were exhibited from the core-shell hybrid liposomal vesicles. The effects on global cerebral ischemia/reperfusion and acute myocardial ischemia injury in rats was evaluated and the core-shell hybrid liposomal vesicles were able to significantly inhibit the oedema of brain and reduce the infarct volume better than the other formulations (Zhang et al. 2012b).

1.3.3 Nanocochleates

The therapeutic use of vesicles has some limitations, principally poor stability and availability under the harsh conditions typically presented in the gastrointestinal tract. A very limited number of studies report on the use of cochleates as an alternative platform to vesicles in order to overcome these limitations. Cochleates were first observed by Verkleij and coworkers using phosphatidylglycerol liposomes and later by Papahadjopoulos and coworkers, using phosphatidylserine liposomes in the presence of divalent metal cations (Me^{2+}). Cochleates can be produced as nano- and microstructures and they are extremely biocompatible, with excellent stability due to their unique compact structure. They present an elongated shape and a carpet roll-like morphology always accompanied by narrowly packed bilayers, through the interaction with Me^{2+} as bridging agents between the bilayers as reported in Fig. 1.8 (Bozó et al. 2017; Asprea et al. 2017, 2019).

A recent study reported the development of nanocochleates loaded with andrographolide and based on phosphatidylserine or phosphatidylcholine, cholesterol and calcium ions in order to overcome andrographolide low water solubility, its instability under alkaline conditions and its rapid metabolism in the intestine. The nanocochleates with the best performance were those obtained from phosphadityl choline nanoliposomes in terms of size and homogeneity. They had an extraordinary stability after lyophilisation without the use of lyoprotectants and after storage at room temperature. The encapsulation efficiency was 71%, while approximately 95% of andrographolide was released in PBS after 24 h, with kinetics according to the Hixson–Crowell model. The *in vitro* gastrointestinal stability and safety of the developed nanocochleates, both in macrophages and 3 T3 fibroblasts, were also assessed. Additionally, the nanocarriers had extraordinary uptake properties in macrophages (Asprea et al. 2019).

In a further study similar nanocochleates were loaded with 1 mg or 0.5 mg/100 ml of essential oil of thyme and fully characterized. Particle size of nanocochleates were ca. 250 nm and 210 nm, respectively, and all were negatively charged. Thymol and carvacrol encapsulation efficiencies, selected as characteristic constituents of the essential oil, were 46.3 ± 0.01 and 50.9 ± 0.02 respectively, using 1 mg/ml of essential oil. *In vitro* release studies of thymol and carvacrol indicated a steady release in the first h and after 6 h about 41% and 31% for thymol and carvacrol were released. Free essential oil showed a strong antioxidant activity (75.2%) at the concentration of 1 mg/ml and it was similar to that produced by nanocochleates (72.2%) (Asprea et al. 2017).

1.3.4 Solid Lipid Nanoparticles

Solid lipid nanoparticles (Fig. 1.8) are colloidal systems with a diameter between 50 and 1000 nm made of a lipid matrix solid at physiological temperature which is stabilized in aqueous solution by one or more surfactants (Mehnert 2001). Typically, the lipid core consists of glycerides, fatty acids, fatty alcohols, and waxes because they are solid at room temperature and most of them are approved as GRAS (Generally Recognized As Safe). The selection of the surfactant mainly depends on the chosen lipid and are represented by phospholipids, bile salts, polysorbates, poloxamers, polyoxyethylene ethers, polyvinyl alcohols. Solid lipid nanoparticles can be obtained through different methods including high shear homogenization and ultrasonication, high-pressure homogenization, hot homogenization, cold homogenization, solvent emulsification/evaporation, and microemulsion. Solid lipid nanoparticles are delivery systems for topical, nasal, oral, ocular and parenteral administration of drugs and are characterized by high biocompatibility, biodegradability, lower toxicity and higher stability respect to polymeric nanoparticles and vesicles. Other advantages of Solid lipid nanoparticles are the simple and economical large-scale production and the versatility. Lipophilic bioactive compounds are generally dispersed in the lipid matrix, whereas the hydrophilic bioactive compounds can be uploaded outside of the lipid matrix. Moreover, Solid lipid nanoparticles provide protection to the incorporated compound from chemical and physical processes of degradation with a prolonged release over time of the incorporated bioactive molecule. Finally, the surface characteristics of Solid lipid nanoparticles can be modified delivering the bioactives to specific targets and capable of crossing many biological barriers, improving plasma stability and biodistribution and consequently, the bioavailability of encapsulated drugs (Bilia et al. 2018; Uner and Yener 2007; Gastaldi et al. 2014; Bayon-Cordero et al. 2019). The scientific literature reports the use of Solid lipid nanoparticles to enhance the biopharmaceutical properties and the therapeutic effect of plant-derived molecules and herbal extracts.

Solid lipid nanoparticles loaded with curcumin were developed and investigated for their efficacy in an allergic rat model of asthma. The plasmatic and tissue concentrations, especially in lung and liver, of curcumin administered as solid lipid nanoparticles were significantly higher than those obtained with free curcumin. The formulation showed an average size of 190 nm with a zeta potential value of −20.7 mV and 75% drug entrapment efficiency. The release profile of curcumin-solid lipid nanoparticles was an initial burst followed by sustained release. Moreover, curcumin SLNs, suppressed airway hyper-responsiveness, inflammatory cell infiltration and significantly inhibited the expression of T-helper-2-type cytokines, such as interleukin-4 and interleukin-13, in broncho-alveolar lavage fluid compared to the asthma group and curcumin-treated group (Wang et al. 2012). In a further study, Singh et al. (2014) compared the efficacy of curcumin-solid lipid nanoparticles with that of unformulated curcumin and silymarin against carbon tetrachloride-induced hepatic injury in rats. Curcumin-solid lipid nanoparticles (12.5 mg/kg) significantly attenuated histopathological changes and oxidative stress, and also decreased the levels of alanine aminotransferase and aspartate aminotransferase and the induction of tumor necrosis factor alpha in comparison with free curcumin (100 mg/kg), silymarin (25 mg/kg), and self-recovery groups (Singh et al. 2014). Curcumin loaded in solid lipid nanoparticles was also developed for the treatment of oral mucosal infections. Gelucire 39/01, Gelucire 50/13, Precirol, Compritol, were used as excipients and poloxamer 407 as a surfactant. In an *ex-vivo* study mucoadhesion was assessed.

The nanoformulation exhibited higher antimicrobial activity when compared with unformulated curcumin showing interesting MIC (0.185, 0.09375, 0.75, 3, 1.5, and 0.1875 mg/mL) against *Staphylococcus aureus*, *Streptococcus mutans*, *Viridans streptococci*, *Escherichia coli*, *Lactobacillus acidophilus*, and *Candida albicans*, respectively (Hazzah et al. 2015). The performance of curcumin encapsulated into SLNs (10 and 30 mg/Kg) was evaluated in arthritic rats and it was compared with unformulated curcumin (10 and 30 mg/Kg). The rats showed a significant increase in blood leukocyte count, oxidative-nitrosative stress, tumour necrosis factor-alpha, C-reactive protein, cyclic citrullinated peptide antibody levels and radiological alterations in tibiotarsal joint. All these symptoms were significantly and dose dependently ameliorated by the nanoformulation, in particular the oxido-inflammatory and immunomodulatory cascade and preserved radiological alterations in joints of arthritic rats. (Arora et al. 2014). Righeschi et al. (2016) developed solid lipid nanoparticles loaded with curcumin suitable for oral administration employing generally recognized as safe ingredients. Parallel artificial membrane permeability assay was used to simulate the intestinal epithelium, showed a considerable increase of curcumin permeation when formulated into solid lipid nanoparticles (Righeschi et al. 2016). In a further study solid lipid nanoparticles loaded with curcumin were formulated and formulated in a topical Carbopol gel for the treatment of pigmentation and irritant contact dermatitis. *In vitro* tyrosinase inhibition assay indicates that the formulated gel had a substantial potential in skin depigmentation. The gel also confirmed proficient suppression of ear swelling and reduction in skin water content in the BALB/c mouse (Shrotriya et al. 2017).

A series of solid lipid nanoparticles loaded with silymarin have been reported. In a first paper the formulation enhanced the solubility and bioavailability of silymarin and the effect on D-galactosamine/tumour necrosis factor-alpha-induced liver damage in Balb/c mice was evaluated. The results evidenced that the nanoformulation significantly reduced D-galactosamine/tumour necrosis factor-alpha-induced hepatotoxicity compared to free silymarin (Cengiz et al. 2015). Solid lipid nanoparticles loaded with silymerin were also formulated as a sunscreen cream. The nanoparticles were prepared by micro-emulsion method using the glyceryl monostearate as lipid, and Tween 80 was used as an emulsifier. After assessing the stability of the sunscreen under accelerated conditions, *in vitro* and *in vivo* experiments showed that the sun protection factor was 13.8 and 14.1, respectively (Netto and Jose 2017). Chitosan-coated nanoparticles were also developed to enhance the solubility and intestinal absorption of silybin, the major flavonolignan of silymarin. The presence of chitosan improved the mucoadhesion properties of the nanoformulation. Moreover, the developed nanoparticles were stable in simulated gastrointestinal conditions and enhanced the permeation of silybin across parallel artificial membrane permeability assay and human epithelial colorectal adenocarcinoma cells layer (Piazzini et al. 2018c). Some studies have reported on the development of solid lipid nanoparticles to increase the solubility and stability of andrographolide and evaluated its antihyperlipidemic activity after oral administration. The bioavailability of andrographolide was increased to 241% by solid lipid nanoparticles as compared with the suspension. Moreover, SLNs influenced the transport mechanism of andrographolide across human epithelial colorectal adenocarcinoma cells layer, increased the stability in the intestine and improved the antihyperlipidemic effect of this diterpenoid (Yang et al. 2013). The anticancer potential of andrographolide encapsulated into solid lipid nanoparticles in Balb/c mice treated with Ehrlich's ascites carcinoma cells. Pharmacokinetic studies showed that solid lipid nanoparticles enhanced the bioavailability of andrographolide after oral administration. Furthermore, the life span of tumor bearing mice was increased in solid lipid nanoparticles -andrographolide-treated group (Parveen et al. 2013). In a further study solid lipid nanoparticles were prepared to deliver andrographolide into the brain. The developed SLNs were stable in presence of human serum albumin and after incubation in plasma. The ability of solid lipid nanoparticles to cross the blood-brain barrier was evaluated first *in vitro* by applying a permeation test with parallel artificial membrane permeability assay to predict passive and transcellular permeability through the blood-brain barrier, and then by using a model of the human blood brain barrier called hCMEC/D3 cells. *In vitro* results proved that SLNs improved permeability of andrographolide compared to the free molecule. Fluorescent nanoparticles were then prepared for *in vivo* tests in healthy rats. After intravenous administration, fluorescent solid lipid nanoparticles were detected in brain parenchyma outside the vascular bed, confirming their ability to overcome the blood-brain barrier (Fig. 1.15) (Graverini et al. 2018).

A series of solid lipid nanoparticles loaded with resveratrol were developed to increase stability and bioavailability. A study investigated the brain targeting ability of glyceryl behenate-based solid lipid nanoparticles as potential candidate for the treatment of neoplastic diseases. Tween 80 and polyvinyl alcohol were employed as surfactants. The authors demonstrated by *in vivo* biodistribution study using Wistar rats that solid lipid nanoparticles could significantly increase the brain concentration of resveratrol (17.28 ± 0.63 μg/g) as compared to free resveratrol (3.45 ± 0.39 μg/g) (Fig. 1.16) (Jose et al. 2014). In a study the surface of the solid lipid nanoparticles loaded with resveratrol made of Precirol ATO 5, PA, Gelucire 50/13, Tween 80 was modified adding N-trimethyl chitosan graft palmitic acid as mucoadhesive copolymer and to confer stability to the nanoparticles in stomach acidic pH conditions. The release of resveratrol was negligible in simulated gastric fluid and sustained in simulated intestinal conditions. Moreover, the relative bioavailability of resveratrol was found to be 3.8-fold higher from solid lipid nanoparticles than that from resveratrol suspension (Ramalingam and Ko 2016). Shrotriya and co-workers developed resveratrol-SLNs-engrossed gel for the treatment of chemically induced irritant contact dermatitis. Resveratrol-solid lipid nanoparticles made of Precirol ATO 5 and Tween 20 were incorporated into Carbopol gel. *Ex vivo* study of resveratrol- solid lipid nanoparticles -loaded gel exhibited controlled drug release up to 24 h; similarly, *in vitro* drug deposition studies showed potential of skin targeting with no skin irritation. Resveratrol-solid lipid nanoparticles gel confirmed competent suppression of ear swelling and reduction in skin water content in the BALB/c mouse model of irritant contact dermatitis when compared to marketed gel (Shrotriya et al. 2016).

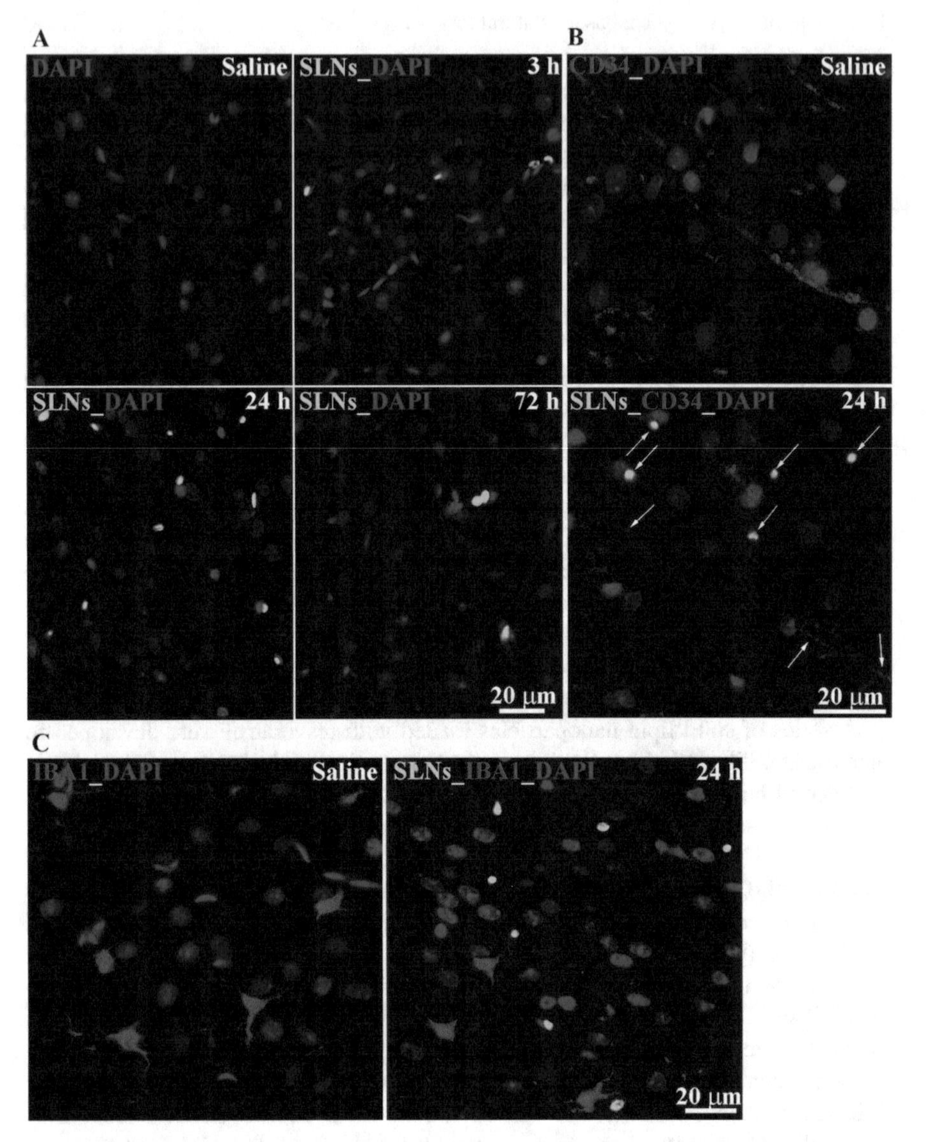

Fig. 1.15 (**a**) Intravenous administration of fluorescent solid lipid nanoparticles: representative photomicrographs showing the presence of solid lipid nanoparticles (green) in brain parenchyma of rats 3, 24 and 72 h after injection. Note the high presence of fluorescent solid lipid nanoparticles at 24 h after injection. Scale bar = 20 μm applies to all images. (**b**) Immunohistochemical detection of solid lipid nanoparticles (green) with the endothelial marker CD34 (red) in the brain of rats 24 h after injection. Solid lipid nanoparticles are detected both within blood vessels (white arrows) and brain parenchyma (yellow arrows). (**a**–**b**) No solid lipid nanoparticles are detected in the brain of rats injected with saline. Scale bar = 20 μm applies to both images. (**c**) Immunohistochemical detection of solid lipid nanoparticles (green) with the microglia antibody ionized calcium binding adaptor molecule 1 (red) in the brain of rats 24 h after injection. Solid lipid nanoparticles are not detected within microglial cells and no microglia activation is detected compared to saline injected rats. Scale bar = 20 μm applies to both images. A–C) 4′,6-Diamidino-2-Phenylindole is in blue. Reproduced with permission from Graverini et al. 2018 from Elsevier. *SLNs* solid lipid nanoparticles, *DAPI* 4′,6-Diamidino-2-Phenylindole, *IBA1* microglia antibody ionized calcium binding adaptor molecule

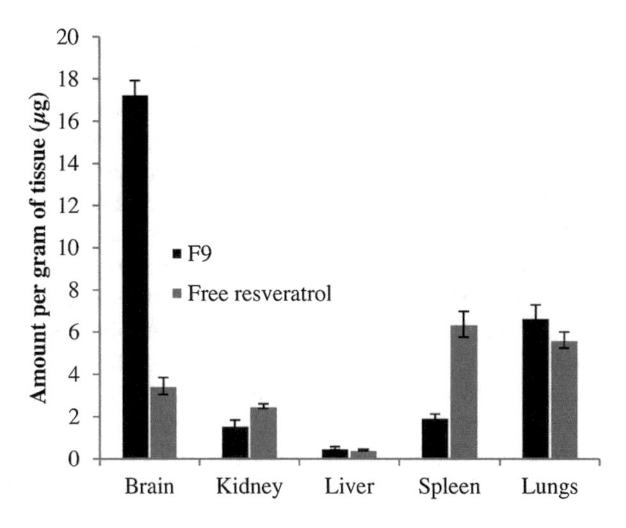

Fig. 1.16 Amount of drug per gram of brain, kidney, liver, spleen, and lungs at 90 min after intraperitoneal administration (5 mg/kg) of resveratrol loaded in solid lipid nanoparticles (F9) and bulk resveratrol (S); mean ± S.E.M (n = 6). (Reproduced with permission from Jose et al. 2014 from Elsevier)

Quercetin was also loaded in solid lipid nanoparticles to increase the oral bioavailability. Li et al. developed solid lipid nanoparticles having a size, average drug entrapment efficiency, drug loading and zeta potential of 155.3 nm, 91.1%, 13.2% and − 32.2 mV, respectively. *In vivo* studies in rats evidenced that the absorption percent of solid lipid nanoparticles loaded with quercetin in the stomach for 2 h was only 6.20%, the absorption process of intestine was first-process with passive diffusion mechanism, and the main absorptive segments were ileum and colon. Moreover, the relative bioavailability of quercetin-SLNs to quercetin suspension was 571.4% (Li et al. 2009). The anti-tumor activity of quercetin loaded in the solid lipid nanoparticles was also investigated. The nanoformulation induced a significant decrease in the viability and proliferation of breast cancer MCF-7 human breast cancer cells, compared to the free quercetin. In addition, it significantly increased reactive oxygen species level and malondialdehyde contents, significantly decreased antioxidant enzyme activity and reduced the expression of the B-cell lymphoma 2 protein in the breast cancer MCF-7 cell line. Furthermore, quercetin-solid lipid nanoparticles significantly increased apoptotic and necrotic indexes in these cells (Niazvand et al. 2019).

1.3.5 Nanostructured Lipid Carriers

Evolution of solid lipid nanoparticles to nanostructured lipid carriers (Fig. 1.8) is mainly due to limitations of the drug load in the solid lipid core, drug expulsion phenomenon, particle concentration in the aqueous dispersion ranging from about

1% to a maximum of only 30%. To overcome these shortcomings, lipid particles with a modified nanostructure containing a liquid lipid at room temperature were developed (Das and Chaudhury 2010). nanostructured lipid carriers are fabricated by the same methods of nanostructured lipid carriers employing solid lipids, liquid lipids, emulsifiers and water. In the nanostructured lipid carriers, solid and liquid lipids were blended to form the matrix; it has been reported that the high entrapment efficiency is attributed to the structural differences of the lipids which form a lot of imperfections resulting in higher space to accommodate drugs in molecular form and amorphous clusters. Moreover, the high entrapment efficiency is due to the higher solubility of drugs in liquid lipids in comparison to solid lipids. Furthermore, nanostructured lipid carriers can incorporate both hydrophilic and lipophilic molecules (Bilia et al. 2018a; Poonia et al. 2016). Among the liquid lipids, medium chain triglycerides and oleic acid are the most used. Medium chain triglycerides mainly composed of caprylic (C8:0; 50–80%) and capric (C10:0; 20–50%) fatty acids with a minor level of caproic (C6:0; ≤ 2%), lauric (C12:0; ≤ 3%) and myristic (C14:0; ≤ 1%) fatty acids, have been approved as GRAS by the US Food and Drug Administration. Oleic acid is used as emulsifying agent in many food and pharmaceutical formulations and its health promoting value is also well documented. Other edible oils such as vitamin E, squalene, soybean oil, corn oil and sunflower oil can also be used as liquid lipid for nanostructured lipid carriers fabrication (Tamjidi et al. 2013). Nanostructured lipid carriers have been applied to improve the therapeutic effect of different plant-derived molecules, such as curcumin, silymarin, quercetin and resveratrol.

The effects of curcumin loaded in nanostructured lipid carrier to the central nervous system was evaluated and compared with the free compound. Western blot analysis shows that intraperitoneal injection of curcumin-nanostructured lipid carriers (100 mg/kg) in mice induces a marked hypoacetylation of histone 4 at lysine 12 in the spinal cord compared with control group due to the inhibition of p300-specific histone acetyltransferase (Fig. 1.17). Notably, curcumin solution (100 mg/kg) did not change the acetylation level in the central nervous system (Puglia et al. 2012). In a further study, curcumin was encapsulated in nanostructured lipid carriers with a particle size of about 150 nm, a polydispersity index of 0.18, an entrapment efficiency of 90.86%, and the zeta potential of −21.4 mV. The researchers observed increased cytotoxicity of curcumin-nanostructured lipid carriers compared to free curcumin in astrocytoma-glioblastoma cell line (U373MG). Furthermore, biodistribution studies in Wistar rats showed higher drug concentration in brain after intranasal administration of nanostructured lipid carriers than plain drug suspension (Madane and Mahajan 2014). Nanostructured lipid carriers loaded with curcumin were developed and ip administered in mice. The plasmatic concentration of curcumin delivered by the nanoparticles was highly increased compared with the unformulated molecule. Moreover, curcumin nanostructured lipid carriers enhanced the targeting effect of curcumin to brain and tumor, which finally increased the inhibition efficiency of curcumin from 19.5% to 82.3%. The authors also demonstrated that the inhibition effect was related to apoptosis and not necrosis (Chen et al. 2015). In a further study, taurocholic acid was selected as a ligand for uptake

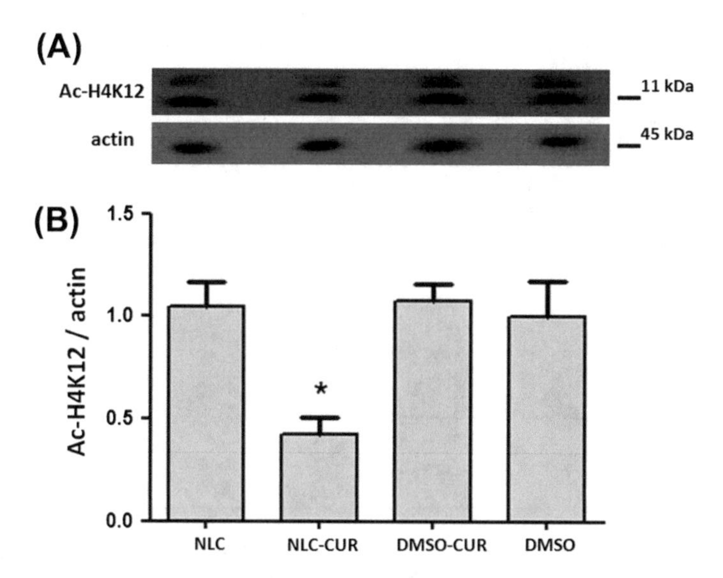

Fig. 1.17 Western blot analysis to examine acetylation of histone H4 at lysine 12 (H4K12) in the mouse spinal cord. Upper panel: representative western blot of H4K12 acetylation levels in mice intraperitoneally injected with different curcumin formulations and respective vehicles. Lower panel: densitometric analysis of acetylated H4K12 normalized by actin. Data are the means ± S.E.M. of five animals. $^{*}p < 0.05$ (One-way analysis of variance ± Fisher's protected least significant difference) versus values obtained in animals treated with the respective vehicle. Reproduced with permission from Puglia et al. 2012 from Elsevier. *Ac-H4K12* acetylation on lysine 12 of histone H4, *NLC* nanostructured lipid carriers, *NLC-CUR* curcumin-loaded nanostructured lipid carriers, *DMSO-CUR* curcumin solution in dimethyl sulfoxide, *DMSO* dimethyl sulfoxide

of nanostructured lipid carriers mediated by a bile-acid transporter to improve oral bioavailability of curcumin. The developed nanostructured lipid carriers exhibited a particle size less than 150 nm and encapsulation efficiency more than 90%; in situ intestinal perfusion studies demonstrated improved absorption rate and permeability coefficient of nanostructured lipid carriers loaded with curcumin and modified with taurocholic acid. Furthermore, the presence of taurocholic acid considerably increased the area under the curve of curcumin after oral administration in rats as compared with unmodified nanostructured lipid carriers (Tian et al. 2017).

Nanostructured lipid carriers loaded with silymarin and made of glyceryl monostearate and oleic acid were developed using using a Box-Behnken design. *In vivo* studies indicated that silymarin loaded in the nanoformulation was absorbed after oral administration through the lymphatic pathway. Moreover, the relative bioavailability of silymarin was two-fold higher when loaded into the nanoformulation when compared with the suspension of silymarin. Finally, a significant amount of silymarin reached the liver 2 h after administration, whereas the concentration of the silymarin suspension in the other organs was negligible as reported in Fig. 1.18 (Chaudhary et al. 2015). In a further study silymarin was formulated in nanostructured lipid carriers and incorporated into Carbopol gel to increase its permeation and anticancer activity. Photostability of the nanoformulation was assessed. The

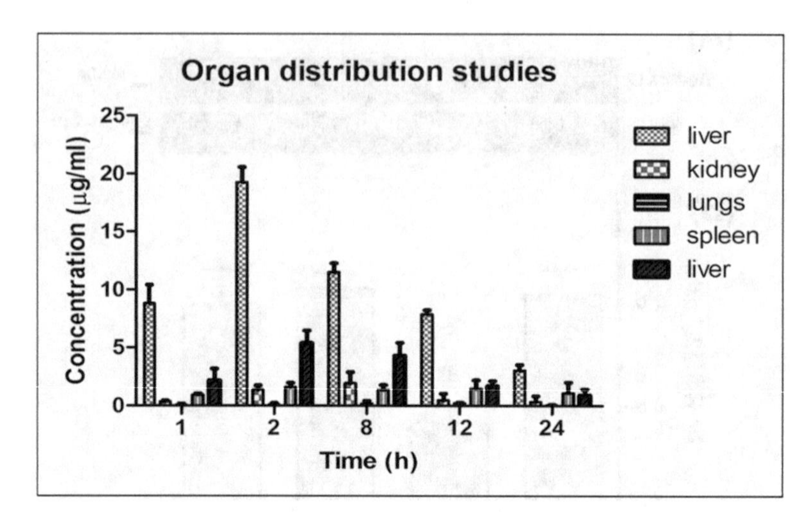

Fig. 1.18 Organ distribution studies of silymarin loaded in nanostructured lipid carriers. (Reproduced with permission from Chaudhary et al. 2015 from Elsevier)

permeation parameters of the nanostructured lipid carriers loaded with silymarin was significantly higher when compared to a marketed phytosome formulation. Furthermore, the developed nanoformulation showed anticancer activity in a dose-dependent manner (10–80 µM) and induced apoptosis at 80 µM in melanoma SK-MEL 2 cancer cells (Singh et al. 2016). Nanostructured lipid nanocarriers made of stearic acid and Capryol 90 as lipid mixture and polyoxyethylene (20) stearyl ether (Brij S20) as stealth agent were also developed for oral administration. Nanoformulation was stable after incubation in simulated gastrointestinal fluids and showed sustained and prolonged release of silymarin in physiological pH conditions. Both parallel artificial membrane permeability assay and human epithelial colorectal adenocarcinoma cells experiments demonstrated the potential of nanostructured lipid carriers to increase the intestinal permeation of silymarin. Finally, uptake studies indicated that nanostructured lipid carriers entered into Caco-2 cells via active transport mechanisms (Piazzini et al. 2018d).

Several studies reported on the nanostructured lipid carrier formulation of quercetin. A study developed nanostructured lipid carriers as topical delivery systems. The average particle size was 215 nm and the encapsulation efficiency was about 90%. *In vitro* drug permeation studies through excised mouse skin and *in vivo* drug distribution experiments in epidermis and dermis of mice showed that nanoformulation could promote the permeation of quercetin, increase the amount of quercetin retention in epidermis and dermis, and enhance the effect of anti-oxidation and anti-inflammation exerted by quercetin (Chen-Yu et al. 2012). The biodistribution *in vivo* of quercetin-cationic-nanostructured lipid carriers after oral administration was also evaluated. Cationic nanoparticles were formulated because they can interact with negatively charged intestinal mucosa and increase their residence time in the gastrointestinal tract. Nanostructured lipid carriers exhibited particle size of

126.6 nm and zeta potential value of 40.5 mV due to the presence of didodecyldimethylammonium bromide. Nanostructured lipid carriers significantly increase the concentration of quercetin in lung, liver and kidney compared with quercetin suspension following oral administration and may be an attractive strategy for therapy of kidney, liver and lung diseases (Liu et al. 2014).

A study by Montenegro et al. (2017) compared the performance of solid lipid nanoparticles, nanostructured lipid carriers and nanoemulsions loaded with resveratrol and evaluated the relationship between *in vitro* occlusion factor and *in vivo* skin hydration. All these nanocarriers were prepared using as solid lipid cetyl palmitate, a synthetic wax of generally regarded as safe status, and as liquid lipid isopropyl myristate, a synthetic oil commonly used in pharmaceutical and cosmetic formulations. Before *in vitro* and *in vivo* experiments, gel formulations were prepared by dispersing Carbopol Ultrez 21 (0.8% *w/w*) in the colloidal dispersions. *In vitro* occlusion factor was in the order solid lipid nanoparticles>nanostructured lipid carriers>nanoemulsions. Gels containing unloaded or resveratrol-loaded nanocarriers were applied on the back of a hand of 12 healthy volunteers twice a day for 1 week, recording skin hydration changes using the instrument Soft Plus. An increase of skin hydration was observed for all lipid nanocarriers (solid lipid nanoparticles>nanostructured lipid carriers>nanoemulsions). Furthermore, a linear relationship (r^2 = 0.969) was observed between occlusion factor and *in vivo* increase of skin hydration (Montenegro et al. 2017). Nanostructured lipid carriers loaded with resveratrol were developed to explore the ability to inhibit the tyrosinase enzyme, as a potential application as a skin-lightning agent and for the treatment of skin disorders associated with hyperpigmentation and melanogenesis. The authors prepared NLCs with different compositions for resveratrol delivery, based on glyceryl behenate or polyoxyethylene 40 stearate and then characterized the formulations, investigated the cytotoxicity, and evaluated the tyrosinase inhibition activity. Compared with glyceryl behenate, polyoxyethylene 40 stearate produced nanocarriers with smaller particle size and polydispersity, higher drug encapsulation and more sustained release, due the superior emulsifying ability of polyoxyethylene 40. The cytotoxicity of resveratrol against a keratinocyte cell line was not enhanced when loaded into nanostructured lipid carriers. Finally, resveratrol loaded into NLCs prepared with polyoxyethylene 40 stearate resulted on greater tyrosinase inhibition than resveratrol solution and formulation containing glyceryl behenate, equivalent to 1.31 and 1.83 times higher, respectively, demonstrating that the incorporation of resveratrol into nanostructured lipid carriers allowed an enhanced tyrosinase inhibitory activity (Fachinetti et al. 2018). Finally, the permeation properties of unformulated resveratrol, resveratrol-loaded in nanostructured lipid carriers and resveratrol encapsulated into liposomes was investigated using skin- parallel artificial membrane permeability assay. Both nanoformulations enhanced considerably the solubility of resveratrol and the encapsulation efficiency was more than 85% for both formulations. *In vitro* release studies of both formulations at pH 5.5 and 7.4 showed a prolonged release of resveratrol. Moreover, the antioxidant capacity of resveratrol was not altered by the nanoformulations. Finally, skin-parallel artificial membrane permeability assay proved an increased cutaneous permeation of resveratrol when loaded into both nanoformulations (Casamonti et al. 2019b).

1.4 Conclusions

The chapter has discussed the current status on nanotechnology applications for natural products delivery. Extensive research is reported in the literature dealing on natural products and herbal extracts loaded into nanosized drug delivery systems, mainly polymeric and lipid nanoparticles, vesicles, nanosized emulsions. The application of nanotechnology to drug delivery has already had a significant impact on many areas of medicine, and this approach appears imperative to develop proper therapeutic treatments of many drugs, principally antitumor and antiparasistic agents. Nanosized drug delivery systems loaded with natural products such as artemisinin, curcumin, resveratrol, honokiol, salvianolic acid B, andrographolide, green tea catechins, silymarin and other extracts, could represent valuable therapeutic approaches for a close future. The nanoencapsulation of natural products has led to very encouraging landscapes. Generally, nanosized drug delivery systems have better stability, improved solubility, protection from physical and chemical degradation, sustained delivery and release of the drug. Consequently, drug bioavailability is improved as well as the tissue macrophage distribution and the pharmacological activity optimized using much lower doses, increasing long-term safety of these constituents.

Acknowledgments The authors acknowledge the funding received from the Department of Chemistry.

References

Arora R, Kuhad A, Kaur I, Chopra K (2014) Curcumin loaded solid lipid nanoparticles ameliorate adjuvant-induced arthritis in rats. Eur J Pain 19:940–952. https://doi.org/10.1002/ejp.620

Artiga-Artigas M, Lanjari-Pérez Y, Martín-Belloso O (2018) Curcumin-loaded nanoemulsions stability as affected by the nature and concentration of surfactant. Food Chem 266:466–474. https://doi.org/10.1016/j.foodchem.2018.06.043

Asprea M, Leto I, Bergonzi MC, Bilia AR (2017) Thyme essential oil loaded in nanocochleates: encapsulation efficiency, in vitro release study and antioxidant activity. LWT 77:497–502. https://doi.org/10.1016/j.lwt.2016.12.006

Asprea M, Tatini F, Piazzini V, Rossi F, Bergonzi MC, Bilia AR (2019) Stable, monodisperse, and highly cell-permeating nanocochleates from natural soy lecithin liposomes. Pharmaceutics 11(1):pii: E34. https://doi.org/10.3390/pharmaceutics11010034

Bayón-Cordero L, Alkorta I, Arana L (2019) Application of solid lipid nanoparticles to improve the efficiency of anticancer drugs. Nano 9:474. https://doi.org/10.3390/nano9030474

Bergonzi MC, Hamdouch R, Mazzacuva F, Isacchi B, Bilia AR (2014) Optimization, characterization and in vitro evaluation of curcumin microemulsions. LWT-Food Sciences and Technology 59:148–155. https://doi.org/10.1016/j.lwt.2014.06.009

Bergonzi MC, Guccione C, Grossi C, Piazzini V, Torracchi A, Luccarini I, Casamenti F, Bilia AR (2016) Albumin nanoparticles for brain delivery: a comparison of chemical versus thermal methods and in vivo behaviour. ChemMedChem 11(16):1840–1849. https://doi.org/10.1002/cmdc.201600080

Bilia AR, Isacchi B, Righeschi C, Guccione C, Bergonzi MC (2014a) Flavonoids loaded in nanocarriers: an opportunity to increase oral bioavailability and bioefficacy. Food Nutr Sciences 5:1212–1227. https://doi.org/10.4236/fns.2014.513132
Bilia AR, Guccione C, Isacchi B, Righeschi C, Firenzuoli F, Bergonzi MC (2014b) Essential oils loaded in nanosystems: a developing strategy for a successful therapeutic approach. Evid Based Complement Alternat Med 2014:651593. https://doi.org/10.1155/2014/651593
Bilia AR, Piazzini V, Guccione C, Risaliti L, Asprea M, Capecchi G, Bergonzi MC (2017) Improving on nature: the role of nanomedicine in the development of clinical natural drugs. Planta Med 83:366–381. https://doi.org/10.1055/s-0043-102949
Bilia AR, Piazzini V, Risaliti L, Vanti G, Casamonti M, Wang M, Bergonzi MC (2018a) Nanocarriers: a successful tool to increase solubility, stability and optimise bioefficacy of natural constituents. Curr Med Chem 25:1–24. https://doi.org/10.2174/0929867325666181101110050
Bilia AR, Piazzini V, Asprea M, Risaliti L, Vanti G, Bergonzi MC (2018b) Plants extracts loaded in nanocarriers: an emergent formulating approach. NPC 13:1157–1160
Bilia AR, Nardiello P, Piazzini V, Leri M, Bergonzi MC, Bucciantini M, Casamenti F (2019) Successful brain delivery of andrographolide loaded in human albumin nanoparticles to TgCRND8 mice, an Alzheimer's disease mouse model. Submitted to Front Pharmacology
Bonaccini L, Karioti K, Bergonzi MC, Bilia AR (2015) Effects of Salvia miltiorrhiza on CNS neuronal injury and degeneration: a plausible complementary role of tanshinones and depsides. Planta Med 81:1003–1016. https://doi.org/10.1055/s-0035-1546196
Bozó T, Wacha A, Mihály J, Bóta A, Kellermayer MSZ (2017) Dispersion and stabilization of cochleate nanoparticles. Eur J Pharm Biopharm 117:270–275. https://doi.org/10.1016/j.ejpb.2017.04.030
Bozzuto G, Molinari A (2015) Liposomes as nanomedical devices. Int J Nanomedicine 10:975–999. https://doi.org/10.2147/IJN.S68861
Caddeo C, Teskac K, Sinico C, Kristl J (2008) Effect of resveratrol incorporated in liposomes on proliferation and UVB protection of cells. Int J Pharmac 363:183–191. https://doi.org/10.1016/j.ijpharm.2008.07.024
Cameron A, Ewen M, Auton M, Abegunde D (2011) The world medicines situation 2011: medicines prices, availability and affordability. WHO (World Health Organization). Geneva
Casamonti M, Risaliti L, Vanti G, Piazzini V, Bergonzi MC, Bilia AR (2019a) Andrographolide loaded in micro- and Nano-formulations: improved bioavailability, target-tissue distribution, and efficacy of the "king of bitters". Engineering 5:69–75
Casamonti M, Piazzini V, Bilia AR, Bergonzi MC (2019b) Evaluation of skin permeability of resveratrol loaded liposomes and nanostructured lipid carriers using a skin mimic artificial membrane (skin-PAMPA). Drug Deliv Lett 9:134–145. https://doi.org/10.2174/2210303109666190207152927
Cengiz M, Kutlu HM, Burukoglu DD, Ayhancı A (2015) A comparative study on the therapeutic effects of silymarin and silymarin-loaded solid lipid nanoparticles on D-GaIN/TNF-α-induced liver damage in Balb/c mice. Food Chem Toxicol 77:93–100. https://doi.org/10.1016/j.fct.2014.12.011
Chaudhary S, Garg T, Murthy R et al (2015) Development, optimization and evaluation of long chain nanolipid carrier for hepatic delivery of silymarin through lymphatic transport pathway. Int J Pharm 485:108–121. https://doi.org/10.1016/j.ijpharm.2015.02.070
Chen Y, Pan L, Jiang M, Li D, Jin L (2015) Nanostructured lipid carriers enhance the bioavailability and brain cancer inhibitory efficacy of curcumin both *in vitro* and *in vivo*. Drug Deliv:1–10. https://doi.org/10.3109/10717544.2015.1049719
Chen Z, Shu G, Taarji N, Barrow CJ, Nakajima M, Khalid N, Neves MA (2018) Gypenosides as natural emulsifiers for oil-in-water nanoemulsions loaded with astaxanthin: insights of formulation, stability and release properties. Food Chem 261:322–328. https://doi.org/10.1016/j.foodchem.2018.04.054

Chen-Yu G, Chun-Fen Y, Qi-Lu L, Qi T, Yan-wei X, Wei-na L, Guang-xi Z (2012) Development of a quercetin-loaded nanostructured lipid carrier formulation for topical delivery. Int J Pharm 430:292–298. https://doi.org/10.1016/j.ijpharm.2012.03.042

Coimbra M, Isacchi B, van Bloois L, Torano JS, Ket A, Wu X, Broere F, Metselaar JM, Rijcken CJ, Storm G, Bilia R, Schiffelers RM (2011) Improving solubility and chemical stability of natural compounds for medicinal use by incorporation into liposomes. Int J Pharm 416:433–442. https://doi.org/10.1016/j.ijpharm.2011.01.056

Cuomo F, Cofelice M, Venditti F, Ceglie A, Miguel M, Lindman B, Lopez F (2018) *In-vitro* digestion of curcumin loaded chitosan-coated liposomes. Colloids Surf B Biointerf 168:29–34. https://doi.org/10.1016/j.colsurfb.2017.11.047

Das S, Chaudhury A (2010) Recent advances in lipid nanoparticle formulations with solid matrix for Oral drug delivery. AAPS PharmSciTech 12:62–76. https://doi.org/10.1208/s12249-010-9563-0

Elmowafy M, Viitala T, Ibrahim HM, Abu-Elyazid SK, Samy A, Kassem A, Yliperttul M (2013) Silymarin loaded liposomes for hepatic targeting: in vitro evaluation and HepG2 drug uptake. Eur J Pharmaceut Sciences 50:161–171. https://doi.org/10.1016/j.ejps.2013.06.012

Ethemoglu MS, Seker FB, Akkaya H, Kilic E, Aslan I, Erdogan CS, Yilmaz B (2017) Anticonvulsivant activity of resveratrol-loaded liposomes *in vivo*. Neuroscience 357:12–19. https://doi.org/10.1016/j.neuroscience.2017.05.026

Fachinetti N, Rigon RB, Eloy JO, Sato MR, Dos Santos KC, Chorilli M (2018) Comparative study of glyceryl behenate or polyoxyethylene 40 stearate-based lipid carriers for trans-resveratrol delivery: development, characterization and evaluation of the *in vitro* tyrosinase inhibition. AAPS PharmSciTech 19:1401–1409. https://doi.org/10.1208/s12249-018-0961-z

Falconieri MC, Adamo M, Monasterolo C, Bergonzi MC, Coronnello M, Bilia AR (2017) New dendrimer-based nanoparticles enhance curcumin solubility. Planta Med 83(5):420–425. https://doi.org/10.1055/s-0042-103161.

Gastaldi L, Battaglia L, Peira E, Chirio D, Muntoni E, Solazzi I, Gallarate M, Dosio F (2014) Solid lipid nanoparticles as vehicles of drugs to the brain: current state of the art. Eur J Pharm Biopharm 87:433–444. https://doi.org/10.1016/j.ejpb.2014.05.004

Ghiasia Z, Esmaelia F, Aghajania M, Ghazi-Khansarib M, Faramarzic MA, Aman A (2019) Enhancing analgesic and anti-inflammatory effects of capsaicin when loaded into olive oil nanoemulsion: an *in vivo* study. Int J Pharm 559:341–347. https://doi.org/10.1016/j.ijpharm.2019.01.043

Graverini G, Piazzini V, Landucci E, Pantano D, Nardiello P, Casamenti F, Pellegrini-Giampietro DE, Bilia AR, Bergonzi MC (2018) Solid lipid nanoparticles for delivery of andrographolide across the blood-brain barrier: in vitro and in vivo evaluation. Colloids Surf B Biointerfaces 161:302–313. https://doi.org/10.1016/j.colsurfb.2017.10.062

Grossi C, Guccione C, Isacchi B, Bergonzi MC, Luccarini I, Casamenti F, Bilia AR (2017) Development of blood-brain barrier permeable nanoparticles as potential carriers for salvianolic acid B to CNS. Planta Med 83(5):382–391. https://doi.org/10.1055/s-0042-101945.

Guccione C, Oufir M, Piazzini V, Eigenmann DE, Jähne EA, Zabela V, Faleschini MT, Bergonzi MC, Smiesko M, Hamburger M, Bilia AR (2017) Andrographolide-loaded nanoparticles for brain delivery: formulation, characterisation and *in vitro* permeability using hCMEC/D3 cell line. Eur J Pharm Biopharm 119:253–263. https://doi.org/10.1016/j.ejpb.2017.06.018

Guccione C, Bergonzi MC, Awada KM, Piazzini V, Bilia AR (2018) Lipid nanocarriers for oral delivery of *Serenoa repens* CO2 extract: a study of microemulsion and self-microemulsifying drug delivery systems. Planta Med 84:736–742. https://doi.org/10.1055/a-0589-0474

Gupta S, Bansal R, Ali J, Gabrani R, Dang S (2014) Development and characterization of Polyphenon 60 and caffeine microemulsion for enhanced antibacterial activity. BioMed Res Intl:Article ID 932017, 7 pages. https://doi.org/10.1155/2014/932017

Han L, Fu Y, Cole AJ, Liu J, Wang J (2012) Co-encapsulation and sustained-release of four components in ginkgo terpenes from injectable PELGE nanoparticles. Fitoterapia 83:721–731. https://doi.org/10.1016/j.fitote.2012.02.014

Hazzah HA, Farid RM, Nasra MM et al (2015) Gelucire-based nanoparticles for curcumin targeting to oral mucosa: preparation, characterization, and antimicrobial activity assessment. J Pharm Sci 104:3913–3924. https://doi.org/10.1002/jps.24590

Isacchi B, Fabbri V, Galeotti N, Bergonzi MC, Karioti A, Ghelardini C, Vannucchi MG, Bilia AR (2011a) Salvianolic acid B and its liposomal formulations: anti-hyperalgesic activity in the treatment of neuropathic pain. Eur J Pharm Sci 44:552–558. https://doi.org/10.1016/j.ejps.2011.09.019

Isacchi B, Arrigucci S, la Marca G, Bergonzi MC, Vannucchi MG, Novelli A, Bilia AR (2011b) Conventional and long-circulating liposomes of artemisinin: preparation, characterization, and pharmacokinetic profile in mice. J Liposome Res 21:237–244

Isacchi B, Bergonzi MC, Grazioso M, Righeschi C, Pietretti A, Severini C, Bilia AR (2012) Artemisinin and artemisinin plus curcumin liposomal formulations: enhanced antimalarial efficacy against Plasmodium bergheiinfected mice. Eur J Pharm Biopharm 80:528–534. https://doi.org/10.1016/j.ejpb.2011.11.015

Isacchi B, Bergonzi MC, Iacopi R, Ghelardini C, Galeotti N, Bilia AR (2016) Liposomal formulation to increase stability and prolong antineuropathic activity of verbascoside. Planta Med 83:412–419. https://doi.org/10.1055/s-0042-106650

Jain S, Jain AK, Pohekar M, Thanki K (2013) Novel self-emulsifying formulation of quercetin for improved *in vivo* antioxidant potential: implications for drug-induced cardiotoxicity and nephrotoxicity. Free Radic Biol Med 65:117–130. https://doi.org/10.1016/j.freeradbiomed.2013.05.041

Jhaveri A, Deshpande P, Pattni B, Torchilin V (2018) Transferrin-targeted, resveratrol-loaded liposomes for the treatment of glioblastoma. J Control Release 277:89–101. https://doi.org/10.1016/j.jconrel.2018.03.006

Jose S, Anju S, Cinu T et al (2014) In vivo pharmacokinetics and biodistribution of resveratrol-loaded solid lipid nanoparticles for brain delivery. Int J Pharm 474:6–13. https://doi.org/10.1016/j.ijpharm.2014.08.003

Kang BK, Lee JS, Chon SK, Jeong SY, Yuk SH, Khang G, Lee HB, Cho SH (2004) Development of self-microemulsifying drug delivery systems (SMEDDS) for oral bioavailability enhancement of simvastatin in beagle dogs. Int J Pharm 274:65–73. https://doi.org/10.1016/j.ijpharm.2003.12.028

Kumar N, Rai A, Reddy ND, Raj PV, Jain P, Deshpande P, Mathew G, Kutty NG, Udupa N, Mallikarjuna Rao CM (2014) Silymarin liposomes improves oral bioavailability of silybin besides targeting hepatocytes, and immune cells. Pharmacol Rep 66:788–798. https://doi.org/10.1016/j.pharep.2014.04.007

Kumar N, Rai A, Reddy ND, Shenoy RR, Mudgal J, Bansal P, Mudgal PP, Arumugam K, Udupa N, Sharma N, Rao CM (2019) Improved *in vitro* and *in vivo* hepatoprotective effects of liposomal silymarin in alcohol-induced hepatotoxicity in Wistar rats. Pharmacol Rep. https://doi.org/10.1016/j.pharep.2019.03.013

Kumari A, Yadav SK, Pakade YB, Singh B, Yadav SC (2010) Development of biodegradable nanoparticles for delivery of quercetin. Colloids Surf B Biointerfaces 80(2):184–192. https://doi.org/10.1016/j.colsurfb.2010.06.002

Leto I, Coronnello M, Righeschi C, Bergonzi MC, Mini E, Bilia AR (2016) Enhanced efficacy of artemisinin loaded in transferrin-conjugated liposomes *versus* stealth liposomes against HCT-8 colon cancer cells. ChemMedChem 11(16):1745–1751. https://doi.org/10.1002/cmdc.201500586

Li H, Zhao X, Ma Y et al (2009) Enhancement of gastrointestinal absorption of quercetin by solid lipid nanoparticles. J Control Release 133:238–244. https://doi.org/10.1016/j.jconrel.2008.10.002

Li Z-L, Penga S-F, Chena X, Zhua Y-Q, Zoua L-Q, Liua W, Liu C-M (2018) Pluronics modified liposomes for curcumin encapsulation: sustained release, stability and bioaccessibility. Food Res Int 108:246–253. https://doi.org/10.1016/j.foodres.2018.03.048

Liu W, Tian R, Hu W, Jia Y, Jiang H, Zhang J, Zhang L (2012) Preparation and evaluation of self-microemulsifying drug delivery system of baicalein. Fitoterapia 83:1532–1539. https://doi.org/10.1016/j.fitote.2012.08.021

Liu L, Tang Y, Gao C et al (2014) Characterization and biodistribution in vivo of quercetin-loaded cationic nanostructured lipid carriers. Colloids Surf B Biointerfaces 115:125–131. https://doi.org/10.1016/j.colsurfb.2013.11.029

Machado FC, Prandini Adum de Matosa R, Primo FL, Tedesco AC, Rahal P, Calmon MF (2019) Effect of curcumin-nanoemulsion associated with photodynamic therapy in breast adenocarcinoma cell line. Bioorg Med Chem 27:1882–1890. https://doi.org/10.1016/j.bmc.2019.03.044

Madane RG, Mahajan HS (2014) Curcumin-loaded nanostructured lipid carriers (NLCs) for nasal administration: design, characterization, and in vivo study. Drug Deliv:1–9. https://doi.org/10.3109/10717544.2014.975382

McClements DJ (2012) Nanoemulsions versus microemulsions: terminology, differences, and similarities. Soft Matter 8:1719–1729. https://doi.org/10.1039/c2sm06903b

Mehnert W (2001) Solid lipid nanoparticles production, characterization and applications. Adv Drug Deliv Rev 47:165–196. https://doi.org/10.1016/s0169-409x(01)00105-3

Meng Q, Long P, Zhou J, Ho C-T, Zou X, Chen B, Zhang L (2019) Improved absorption of β-carotene by encapsulation in an oil-in-water nanoemulsion containing tea polyphenols in the aqueous phase. Food Res Int 116:731–736. https://doi.org/10.1016/j.foodres.2018.09.004

Montenegro L, Parenti C, Turnaturi R, Pasquinucci L (2017) Resveratrol-loaded lipid nanocarriers: correlation between in vitro occlusion factor and in vivo skin hydrating effect. Pharmaceutics 9:58. https://doi.org/10.3390/pharmaceutics9040058

Netto G, Jose J (2017) Development, characterization, and evaluation of sunscreen cream containing solid lipid nanoparticles of silymarin. J Cosmet Dermatol 17:1073–1083. https://doi.org/10.1111/jocd.12470

Newman DJ, Cragg GM (2012) Natural products as sources of new drugs over the 30 years from 1981 to 2010. J Nat Prod 75:311–335. https://doi.org/10.1021/np200906s

Niazvand F, Orazizadeh M, Khorsandi L et al (2019) Effects of quercetin-loaded nanoparticles on MCF-7 human breast cancer cells. Medicina 55:114. https://doi.org/10.3390/medicina55040114

Nicolas J, Mura S, Brambilla D, Mackiewicz N, Couvreur P (2013) Design, functionalization strategies and biomedical applications of targeted biodegradable/biocompatible polymer-based nanocarriers for drug delivery. Chem Society Reviews 42:1147–1235. https://doi.org/10.1039/C2CS35265F

Park SN, Jo NR, Jeon SH (2014) Chitosan-coated liposomes for enhanced skin permeation of resveratrol. J Ind Engin Chem 20:1481–1485. https://doi.org/10.1016/j.jiec.2013.07.035

Parveen R, Ahmad FJ, Iqbal Z, Samim M, Ahmad S (2013) Solid lipid nanoparticles of anticancer drug andrographolide: formulation, *in vitro* and *in vivo* studies. Drug Dev Ind Pharm 40:1206–1212. https://doi.org/10.3109/03639045.2013.810636

Piazzini V, Monteforte E, Luceri C, Bigagli E, Bilia AR, Bergonzi MC (2017a) Nanoemulsion for improving solubility and permeability of *Vitex agnus-castus* extract: formulation and *in vitro* evaluation using PAMPA and Caco-2 approaches. Drug Deliv 24:380–390. https://doi.org/10.1080/10717544.2016.1256002

Piazzini V, Rosseti C, Bigagli E, Luceri C, Bilia AR, Bergonzi MC (2017b) Prediction of permeation and cellular transport of *Silybum marianum* extract formulated in a nanoemulsion by using PAMPA and Caco-2 cell models. Planta Med 83:1184–1193. https://doi.org/10.1055/s-0043-110042

Piazzini V, Bigagli E, Luceri C, Bilia AR, Bergonzi MC (2018a) Enhanced solubility and permeability of Salicis cortex extract by formulating as a microemulsion. Planta Med 84:976–984. https://doi.org/10.1055/a-0611-6203

Piazzini V, Landucci E, Graverini G, Pellegrini-Giampietro D, Bilia AR, Bergonzi MC (2018b) Stealth and cationic nanoliposomes as drug delivery systems to increase andrographolide BBB permeability. Pharmaceutics 10:128. https://doi.org/10.3390/pharmaceutics10030128

Piazzini V, Cinci L, Dambrosio M, , Luceri C, Bilia AR, Bergonzi MC (2018c) Solid lipid nanoparticles and chitosan-coated solid lipid nanoparticles as promising tool for silybin delivery: formulation, characterization, and in vitro evaluation. Curr Drug Deliv 16:142–152. doi: https://doi.org/10.2174/1567201815666181008153602

Piazzini V, Lemmi B, D'Ambrosio M, Cinci L, Luceri C, Bilia AR, Bergonzi MC (2018d) Nanostructured lipid carriers as promising delivery systems for plant extracts: the case of silymarin. Appl Sci 8:1163. https://doi.org/10.3390/app8071163

Piazzini V, Landucci E, D'Ambrosio M, Tiozzo Fasiolo L, Cinci L, Colombo G, Pellegrini-Giampietro DE, Bilia AR, Luceri C, Bergonzi MC (2019a) Chitosan coated human serum albumin nanoparticles: a promising strategy for nose-to-brain drug delivery. Int J Biol Macromol 129:267–280. https://doi.org/10.1016/j.ijbiomac.2019.02.005

Piazzini V, D'Ambrosio M, Luceri C, Cinci L, Landucci E, Bilia AR, Bergonzi MC (2019b) Formulation of nanomicelles to improve the solubility and the oral absorption of silymarin. Molecules 24(9):pii: E1688. https://doi.org/10.3390/molecules24091688

Pleguezuelos-Villa M, Nácher A, Hernández MJ, Ofelia Vila Buso MA, Sauri AR, Díez-Sales O (2019) Mangiferin nanoemulsions in treatment of inflammatory disorders and skin regeneration. Int J Pharmac 564:299–307. https://doi.org/10.1016/j.ijpharm.2019.04.056

Poonia N, Kharb R, Lather V, Pandita D (2016) Nanostructured lipid carriers: versatile oral delivery vehicle. Future Sci OA 2:FSO135. https://doi.org/10.4155/fsoa-2016-0030

Puglia C, Frasca G, Musumeci T, Rizza L, Puglisi G, Bonina F, Chiechio S (2012) Curcumin loaded NLC induces histone hypoacetylation in the CNS after intraperitoneal administration in mice. Eur J Pharm Biopharm 81:288–293. https://doi.org/10.1016/j.ejpb.2012.03.015

Ramalingam P, Ko YT (2016) Improved oral delivery of resveratrol from N-trimethyl chitosan-g-palmitic acid surface-modified solid lipid nanoparticles. Colloids Surf B Biointerfaces 139:52–61. https://doi.org/10.1016/j.colsurfb.2015.11.050

Righeschi C, Coronnello M, Mastrantoni A, Isacchi B, Bergonzi MC, Mini E, Bilia AR (2014) Strategy to provide a useful solution to effective delivery of dihydroartemisinin: development, characterization and *in vitro* studies of liposomal formulations. Colloids Surf B Biointerfaces 116:121–127. https://doi.org/10.1016/j.colsurfb.2013.12.019

Righeschi C, Bergonzi MC, Isacchi B, Bazzicalupi C, Gratteri P, Bilia AR (2016) Enhanced curcumin permeability by SLN formulation: the PAMPA approach. LWT Food Sci Technol 66:475–483. https://doi.org/10.1016/j.lwt.2015.11.008

Shrotriya SN, Ranpise NS, Vidhate BV (2016) Skin targeting of resveratrol utilizing solid lipid nanoparticle-engrossed gel for chemically induced irritant contact dermatitis. Drug Deliv Transl Res 7:37–52. https://doi.org/10.1007/s13346-016-0350-7

Shrotriya S, Ranpise N, Satpute P, Vidhate B (2017) Skin targeting of curcumin solid lipid nanoparticles-engrossed topical gel for the treatment of pigmentation and irritant contact dermatitis. Artif Cells Nanomed Biotechnol 46:1471–1482. https://doi.org/10.1080/21691401.2017.1373659

Shu G, Khalid N, Chen Z, Neves MA, Barrow CJ, Nakajim M (2018) Formulation and characterization of astaxanthin-enriched nanoemulsions stabilized using ginseng saponins as natural emulsifiers. Food Chem 255:67–74. https://doi.org/10.1016/j.foodchem.2018.02.062

Singh N, Khullar N, Kakkar V, Kaur IP (2014) Attenuation of carbon tetrachloride-induced hepatic injury with curcumin-loaded solid lipid nanoparticles. BioDrugs 28:297–312. https://doi.org/10.1007/s40259-014-0086-1

Singh P, Singh M, Kanoujia J, Arya M, Saraf SK, Saraf SA (2016) Process optimization and photostability of silymarin nanostructured lipid carriers: effect on UV-irradiated rat skin and SK-MEL 2 cell line. Drug Deliv Transl Res 6:597–609. https://doi.org/10.1007/s13346-016-0317-8

Song Q, Wang X, Hu Q, Huang M, Yao L, Qi H, Qiu Y, Jiang X, Chen J, Chen H, Gao X (2013) Cellular internalization pathway and transcellular transport of pegylated polyester nanoparticles in Caco-2 cells. Int J Pharm 445(1–2):58–68. https://doi.org/10.1016/j.ijpharm.2013.01.060

Tahmasebi Mirgani M, Isacchi B, Sadeghizadeh M, Marra F, Bilia AR, Mowla SJ, Najafi F, Babaei E (2014) Dendrosomal curcumin nanoformulation downregulates pluripotency genes via miR-145 activation in U87MG glioblastoma cells. Int J Nanomedicine 9:403–417. https://doi.org/10.2147/IJN.S48136.

Tamjidi F, Shahedi M, Varshosaz J, Nasirpour A (2013) Nanostructured lipid carriers (NLC): a potential delivery system for bioactive food molecules. Innov Food Sci Emerg Technol 19:29–43. https://doi.org/10.1016/j.ifset.2013.03.002

Tian C, Asghar S, Wu Y, Chen Z, Jin X, Yin L, Huang L, Ping Q, Xiao Y (2017) Improving intestinal absorption and oral bioavailability of curcumin via taurocholic acid-modified nanostructured lipid carriers. Int J Nanomedicine 12:7897–7911. https://doi.org/10.2147/ijn.s145988

Uner M, Yener G (2007) Importance of solid lipid nanoparticles (SLN) in various administration routes and future perspectives. Int J Nanomedicine 2:289–300

Wang XX, Li YB, Yao HJ, Ju RJ, Zhang Y, Li RJ, Yu Y, Zhang L, Lu WL (2011a) The use of mitochondrial targeting resveratrol liposomes modified with a dequalinium polyethylene glycol distearoylphosphatidyl ethanolamine conjugate to induce apoptosis in resistant lung cancer cells. Biomaterials 32:5673–5687. https://doi.org/10.1016/j.biomaterials.2011.04.029

Wang X, Deng L, Cai L, Zhang X, Zheng H, Deng C, Duan X, Zhao X, Wei Y, Chen L (2011b) Preparation, characterization, pharmacokinetics, and bioactivity of honokiol-in-hydroxypropyl-β-cyclodextrin-in-liposome. J Pharm Sci 100:3357–3364. https://doi.org/10.1002/jps.22534

Wang XH, Cai LL, Zhang XY, Deng LY, Zheng H, Deng CY, Wen JL, Zhao X, Wei YQ, Chen LJ (2011c) Improved solubility and pharmacokinetics of PEGylated liposomal honokiol and human plasma protein binding ability of honokiol. Int J Pharm 410:169–174. https://doi.org/10.1016/j.ijpharm.2011.03.003

Wang W, Zhu X, Li A, Xiao Y, Li K, Liu H, Cui D, Chen Y, Wang S (2012) Enhanced bioavailability and efficiency of curcumin for the treatment of asthma by its formulation in solid lipid nanoparticles. Int J Nanomed:3667–3677. https://doi.org/10.2147/ijn.s30428

Wang Y, Wang S, Firempong CK, Zhang H, Wang M, Zhang Y, Zhu Y, Yu J, Xu X (2017) Enhanced solubility and bioavailability of naringenin via liposomal nanoformulation: preparation and *in vitro* and *in vivo* evaluations. AAPS PharmSciTech 18:586–594. https://doi.org/10.1208/s12249-016-0537-8

Wenzel E, Somoza V (2005) Metabolism and bioavailability of trans-resveratrol. Molec Nutrit Food Res 49:472–481. https://doi.org/10.1002/mnfr.200500010

Xie X, Tao Q, Zou Y, Zhang F, Guo M, Wang Y, Wang H, Zhou Q, Yu S (2011) PLGA nanoparticles improve the oral bioavailability of curcumin in rats: characterizations and mechanisms. J Agric Food Chem 59(17):9280–9289. https://doi.org/10.1021/jf202135j

Yang T, Sheng H-H, Feng N-P, Wei H, Wang ZT, Wang CH (2013) Preparation of andrographolide-loaded solid lipid nanoparticles and their *in vitro* and *in vivo* evaluations: characteristics, release, absorption, transports, pharmacokinetics, and antihyperlipidemic activity. J Pharm Sci 102:4414–4425. https://doi.org/10.1002/jps.23758

Zhang Y, Wang R, Wu J, Shen Q (2012a) Characterization and evaluation of self-microemulsifying sustained-release pellet formulation of puerarin for oral delivery. Int J Pharmac 427:337–344. https://doi.org/10.1016/j.ijpharm.2012.02.013

Zhang J, Han X, Li X, Luo Y, Zhao H, Yang M, Ni B, Liao Z (2012b) Core-shell hybrid liposomal vesicles loaded with *Panax notoginseng*: preparation, characterization and protective effects on global cerebral ischemia/reperfusion injury and acute myocardial ischemia in rats. Int J Nanomedicine 7:4299–4310. https://doi.org/10.2147/IJN.S32385

Chapter 2
Nanotechnology-Inspired Bionanosystems for Valorization of Natural Origin Extracts

Ana Catarina Sousa Gonçalves, Ana Sofia Mendes Ferreira, Alberto Dias, Marisa P. Sárria, and Andreia Castro Gomes

Abstract Medicinal plants are the richest bioresource of drugs on traditional systems of medicine and their use in treating diseases by ancestral societies has caught the attention of the scientific community. Application of these plants in daily diet, cosmetic and pharmaceutical industries is widely implemented for many years. However, with the increasing need for more sustainable and environmentally friendly techniques, substituting chemical processes by plants in the production and enrichment of nanomaterials is certainly a very appealing alternative.

Studies have shown that among many examples of green synthesized drug delivery systems, those that receive the most attention include nanometallic particles, polymers and biological materials. Nanotechnology has enabled the creation of new drug delivery systems with the ability to increase the efficacy and improve the bioavailability of plant-derived bioactive compounds, promoting their release in a controlled manner, requiring a reduced dose, and reducing side effects while potentiating their activity. This review highlights the use of biosynthesized nanomaterials as a viable alternative to conventional techniques, and values plant extracts as a source of new nanomedicines, acting as an ally or alternative to existing therapies.

Authors Ana Catarina Sousa Gonçalves and Ana Sofia Mendes Ferreira have equally contributed to this chapter.

A. C. S. Gonçalves · A. S. M. Ferreira
CBMA – Centre of Molecular and Environmental Biology, Department of Biology, University of Minho, Campus of Gualtar, Braga, Portugal

A. Dias · A. C. Gomes (✉)
CBMA – Centre of Molecular and Environmental Biology, Department of Biology, University of Minho, Campus of Gualtar, Braga, Portugal

IB-S Institute of Science and Innovation for Sustainability, University of Minho, Campus of Gualtar, Braga, Portugal
e-mail: agomes@bio.uminho.pt

M. P. Sárria (✉)
INL – International Iberian Nanotechnology Laboratory, Braga, Portugal
e-mail: marisa.passos@inl.int

A. Saneja et al. (eds.), *Sustainable Agriculture Reviews 44*, Sustainable Agriculture Reviews 44, https://doi.org/10.1007/978-3-030-41842-7_2

Keywords Plant extracts · Nanotechnology · Plant-based nanostructures · Green synthesis, eco-friendly nanotherapeutics

2.1 Introduction

Nanotechnology has been one of the trendiest technologies in the last decades with potential applications at all industrial sectors, from medicine to clean water and energy (Fig. 2.1). Namely, nanotechnology has had tremendous impact in medicine in the synthesis and development of various types of nanomaterials and nanoparticles, used as vehicles of antioxidants, antimicrobials and anticancer agents for therapeutic and diagnostic purposes, and as components of nanosensors (Allafchian et al. 2018). These new nanotechnology tools for medical research and practice are more effective and enable faster response to new diseases, allow continuous patient health control, besides increasing the accuracy of examinations and surgeries (Ocsoy et al. 2018).

Knowledge and use of medicinal plants to treat diseases has had immense relevance along mankind's history. Actually, a large segment of the population in many countries continues to use them solely or as complementary practices to conventional medicines. Although many pharmaceutical products have been implemented in the clinic, the chemical composition of medicinal plants and their popular uses have become focus of intense research by the scientific community, pursuing

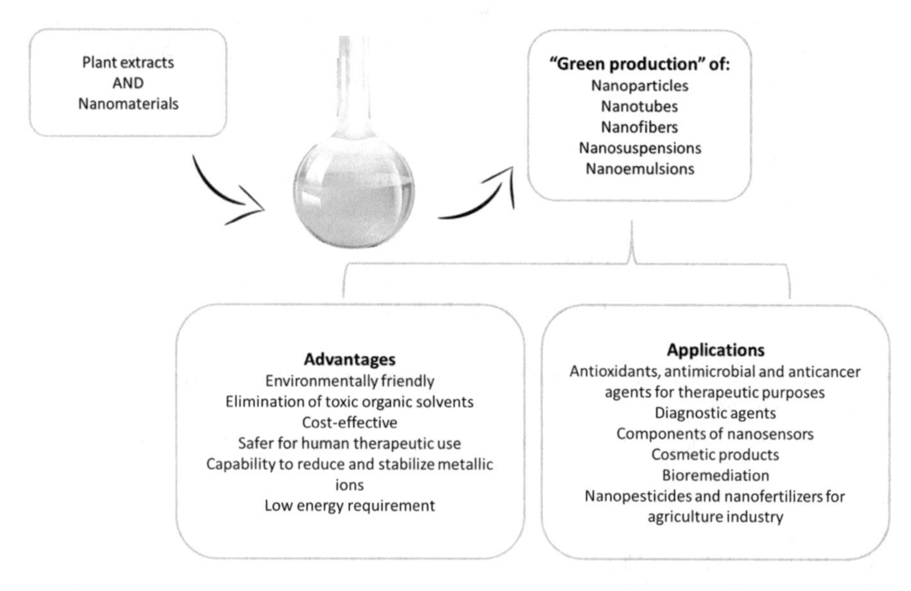

Fig. 2.1 Nanotechnology approach for exploration of plant extracts for biomedical and environmental applications. Use of plant extracts in the production and/or their incorporation in nanotechnology materials has advantages that can be applied to various areas

increasingly innovative products, with fewer side effects, compared to those associated with existing drugs (Islam et al. 2015).

Many bioactive compounds present in plant extracts, such as some alkaloids and phenolic compounds, are highly water soluble and therefore unable to cross lipid membranes. Whereas others, like many terpenoids, are poorly water soluble, making them poorly bioavailable and effective (Afrin et al. 2018). However, several nanotechnological strategies have been developed, such as polymer and solid lipid nanoparticles, liposomes, nanoemulsions, nanosuspensions, nanofibers and nanotubes to try to overcome this limitative feature. These discoveries have revolutionized drug delivery by developing new targeted systems with the ability to increase the efficacy and bioavailability of bioactive compounds, promoting their release in a controlled manner, requiring a reduced dose, reducing side effects and potentiating the activity of the vegetal extracts more efficiently (Afrin et al. 2018).

Multiple strategies of applying nanotechnology to plant extracts have been widely cited in the literature as "green nanoparticles" and "green synthesis". Many original research and review articles focusing on green synthesis of metal nanoparticles, as gold, silver, zinc, copper, iron and others metal oxides report the use of plants extracts rich in secondary metabolites due to their capability to reduce and stabilize metallic ions (Selvam et al. 2017). Furthermore, green synthesis avoid the use of toxic organic solvents and severe reaction conditions, making it an eco-friendly alternative, cost-effective, sustainable and safer for human therapeutic use (Thatoi et al. 2016).

2.2 Methodology

The goal of this review is to provide an overview of the major developments in this field, summarizing the multitude of plant extracts applications in nanotechnology, showing how relevant these natural compounds can be and how it is becoming a trend in nanotherapeutics research.

Literature searching strategy started by pre-selecting keywords related to nanotechnology-inspired bionanosystems using plant extracts and inserting these as search topic into the selected electronic databases, namely PubMed, Scopus and Web of Science.

Two searches were carried out: (1) at PubMed using the following keywords with 205 results: ((plant extract [Title/Abstract]) AND (nanotechnology [Title/Abstract])) and these were being adapted to the syntax and subject headings of the databases; (2) at the three databases, using the keywords: ((plant extract [Title/Abstract] AND (nanoparticles [Title/Abstract])), ((plant extract [Title/Abstract] AND (nanostructures [Title/Abstract])), ((plant extract [Title/Abstract] AND [nanomaterials [Title/Abstract])), and ((plant extract [Title/Abstract] AND (nanocomposites [Title/Abstract])). Results from both searches were merged, resulting in 1060 publications, and after duplicate elimination, just 634 remained.

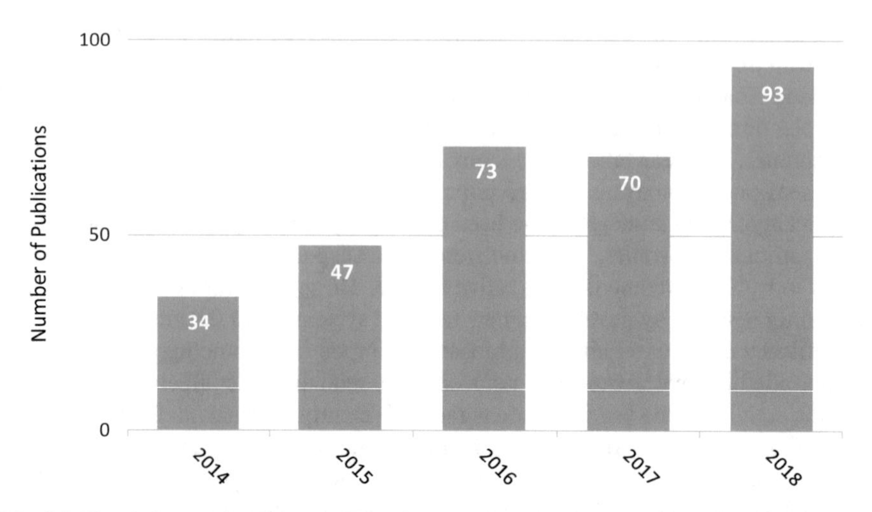

Fig. 2.2 Trend of research articles published from 2014 to 2018, by searching on PubMed, Scopus and Web of Science databases for "plant extract" AND "nanotechnology", "plant extract" AND "nanocomposites", "plant extract" AND "nanostructure", "plant extract" AND "nanoparticles" and "plant extract" AND "nanomaterials", after removing duplicates and reviews

Of these, 317 publications were rejected given that keywords selected were not present in the title or abstract, as or were not written in English language, or were in the format of reviews or book chapters. References in reviews were, however, manually consulted to identify citations that could have been missed in the original database searches. After this step, the collected publications were organized by years, as indicated in Fig. 2.2, demonstrating the increasing trend in this field for the last 5 years. Figure 2.3 shows the intersection between the keywords selected.

The full text of the remaining 182 studies was carefully examined. From this analysis, 51 studies were rejected because either these did not present conclusive results, or had similar or contradictory results.

From this screening and eligibility process, 107 published studies were eligible to be included in this review -see flowchart diagram at Fig. 2.4. The last online search was performed on 13th April 2019.

2.3 Medicinal Plants and Nanotechnology

Medicinal use of plants by mankind is known since ancient times in different contexts. Medicinal plants have historically proved their value as a natural source of molecules with therapeutic potential, and currently represent an important resource for the development of new drugs (Atanasov et al. 2015). Nowadays, the use of natural-derived products as agents for drug discovery has gained considerable attention by the pharmaceutical industry, and is expected to be among the most important sources of new drugs. However, despite their immense potential and considerable availability, only a few medicinal plants have been marketed. This highlights the complexity of

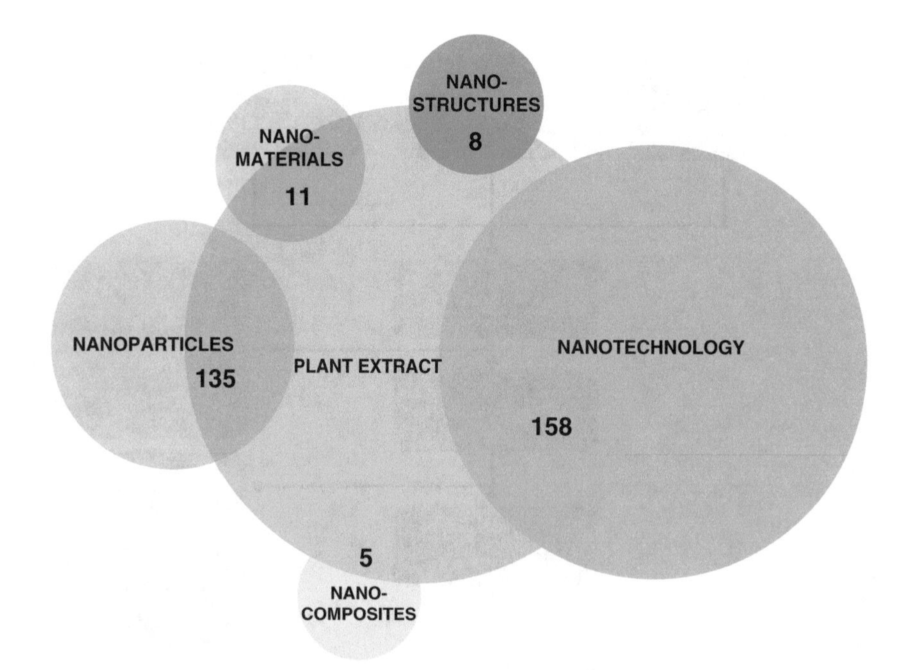

Fig. 2.3 Venn diagram with keywords used in PubMed, Scopus and Web of Science databases, after removing duplicates, reviews and book chapters. The numbers indicate the quantity of publications with the combination of two keywords selected

drug discovery, and the need for an interdisciplinary approach, as well as the valence of considering nanobiotechnology as an ally for the development of new products, that can be used in the treatment of various diseases (Atanasov et al. 2015).

This review aims to recognize how medicinal plants, among of those referred/selected at published articles identified, are distributed throughout the planet. Worldwide distribution of medicinal plants is given in Fig. 2.5. Results show that 45% of the medicinal plants studied are originary from Asia. Countries like China and India have a long tradition in using plants as medicines, apart from being a source of food (Dzoyem et al. 2013). In recent decades, the interest in these plants has increased, with the recognition of their extraordinary pharmacological properties (Dzoyem et al. 2013). Within the search results, 20% of the plants studied were found in Europe, 16% in Africa, 13% in the American continent, and 6% in Oceania. However, it is important to consider that these values are not only related to the number of plants with potential *per* continent, but are also related to the investment that had been made in investigating them.

Plants produce a vast and diverse assortment of organic compounds, such as terpenoids, alkaloids, amino acids, phenolic compounds, glutathiones, polysaccharides, organic acids and quinones, with a vast array of pharmacologic activities like antioxidants, anti-inflammatory and antimicrobial. These substances, traditionally referred as "secondary metabolites" are derived from various parts of plants as leaves, stems, roots shoots, flowers, barks and seeds (Vijayaraghavan and Ashokkumar 2017). Recently, many studies have proven that plant extracts can act as a potential precursor for the

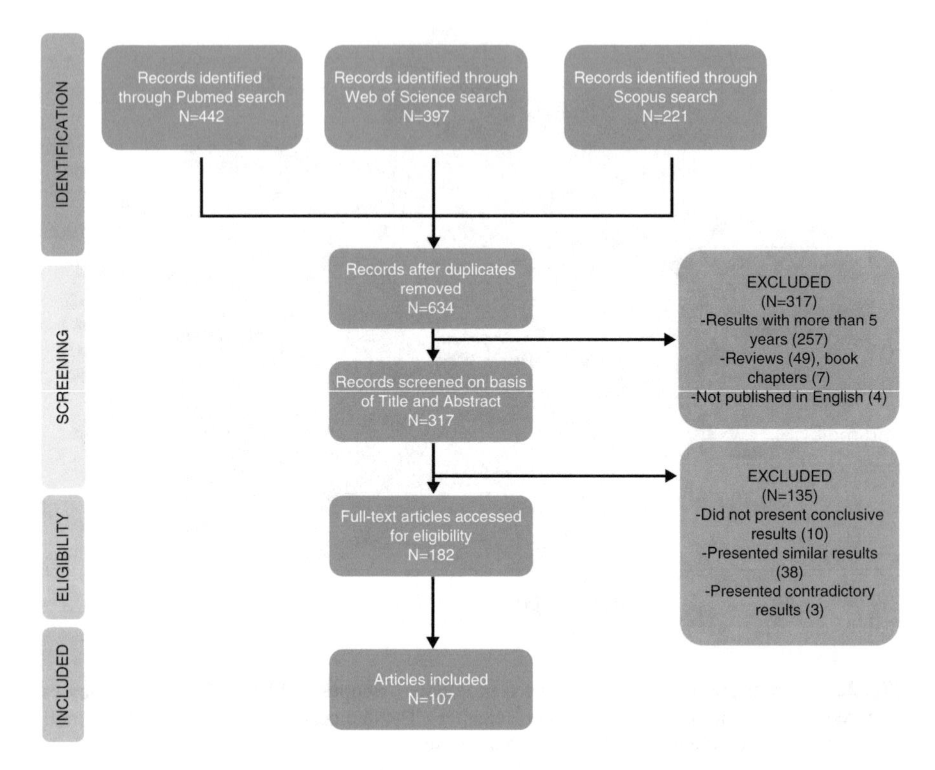

Fig. 2.4 Flowchart diagram of the study selection for this systematic review, from the selection of the keywords in Pubmed, Scopus and Web of Science databases, to the application of filters until the final set of articles

Fig. 2.5 Percentage of worldwide distribution of plant species studied in selected articles. The largest percentage of plants in the studies considered are derivated from Asia, where use and consumption of medicinal plants is significant, as due to their more extensive study

synthesis of nanomaterials in non-hazardous ways, through a process designated green synthesis (Chintamani et al. 2018; Mahendra et al. 2017; Marslin et al. 2015). Balamurugan and colleagues have tested the use of *Eucalyptus globulus* in the biosynthesis of iron oxide nanoparticles, resulting in a reproducibly high yield of about 83% (Balamurugan et al. 2014). Also, Klekotko and his team, in a study of the cytotoxicity of gold nanoparticles found that biological synthesis of these using the extract of *Mentha Piperita* revealed a promising lower cytotoxic profile than those synthesized chemically (Klekotko et al. 2015). Other studies with leaves of lemon and leaves of coriander have been employed for the extracellular biosynthesis of silver nanoparticles. Green leaves were preferably selected for synthesis of these metal-based nanoparticles given that these are photosynthetic organs of the plants, and therefore, H^+ ions to reduce the metal ions are more available (Amooaghaie et al. 2015).

The incorporation of natural products into different forms of nano-scale particles and materials, or the synthesis of these nanosystems with plant extracts has also been explored in nanotechnology. The low insolubility of organic compounds, low stability, poor absorption and rapid systemic elimination causes inherent limitations, impeding the use of these natural phytobioactives *per se* (Marslin et al. 2015). Emergent developments in nanotechnology contribute to surmount some of these obstacles. One of these strategies to circumvent undesirable effects is the incorporation of the natural bioactive agent in nanoparticles, to make its beneficial pharmacological properties more accessible and efficient (Dai et al. 2017).

In addition to the use of plant-based compounds incorporated into nanoparticles, their use for the synthesis of nanomaterials is a fast, economical and easily scalable synthetic route technique, that can enhance the biological activity of the compounds, showed in Fig. 2.6 (Mahendra et al. 2017).

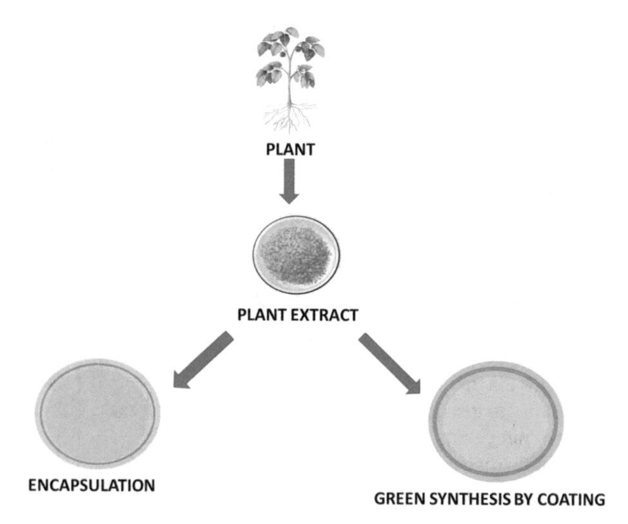

Fig. 2.6 Use of plant extracts in nanotechnology: encapsulation of plant extracts to enhance efficiency for a certain therapeutic target, represented as green core, in the left; and green synthesis of nanoparticles, at which plant derived components nucleate and stabilize nanoparticles formed from a colloidal solution

2.4 Nanomaterials Used in Combination with Plants Extracts

In recent years, nanotechnology has been increasingly used for the synthesis, engineering and design of various nanomaterials. Although the green synthesis is more applied to nanoparticles production, plant extracts have been used to fabricate nanofibers, nanotubes, nanosuspensions and nanoemulsions, as presented in Table 2.1.

Table 2.1 Types of nanostructures used in combination with plant extracts, their composition and advantages

Types of nanomaterials	Composition	Advantages	References
Nanoemulsions	Colloidal particulate system composed of a mixture of water and oil along with a suitable stabilizing surfactant	High biocompatibility, long-time circulation, low toxicity, efficient drug loading efficiency, targeted and sustained drug-release system	Periasamy et al. (2016), Hoscheid et al. (2015), Gumus et al. (2015) and Balestrin et al. (2016)
Nanosuspensions	Sub-micron colloidal dispersions of pure particles of drug, which are stabilized by surfactants	High solubility and bioavailability, drug safety and efficacy; long time circulation	Li et al. (2018), Jahan et al. (2016) and Periasamy et al. (2016)
Nanofibers	Ultra-fine solid fibers with small diameters (lower than 100 nm)	Improvement of drug release, drug loading and drug stability	Ahn et al. (2018), Ajoumshariati et al. (2016), Youse et al. (2017) and Miftahul (2018)
Nanotubes	Nanotubes are cylindrical structures with small diameters (lower than 100 nm)	High biocompatibility, mechanical strength, electrical and thermal conductivity	André et al. (2014), Foo et al. (2018), Tripathi et al. (2017) and Tostado-plascencia et al. (2018)
Nanoparticles	Particles that exist on a nanometer scale, under 100 nm. Nanoparticles types are divided in two main groups: organic and inorganic nanoparticles	High stability, high carrier capacity, feasibility of incorporation of hydrophilic and hydrophobic substances, and variable routes of administration	Riaz et al. (2018), Hassanien et al. (2018), Murad et al. (2018) and Sahni et al. (2015)

2.4.1 Nanoemulsions

Nanoemulsions are dispersions of two immiscible liquids, stabilized by a surfactant or surfactant mixture (Zorzi et al. 2015). Previous studies have suggested that nanoemulsions can be used as carriers of drugs, and may confer long-term validity on carrier drugs for antimicrobial, anticancer, larvicidal and intestinal activities (Zorzi et al. 2015). These can be administered via oral, ocular and intravenous to reduce side effects and potentiate pharmacological effects. A variety of essential oils derived from plants has been used to prepare nanoemulsions, including those derived from neem, basil, lemongrass, clove and kalojeere (Periasamy et al. 2016). Spontaneous emulsification and high pressure homogenization are the most used techniques when associating plant extracts to nanoemulsions (Zorzi et al. 2015). Main advantages of these are high biocompatibility, long-time circulation, low toxicity and efficient drug loading efficiency, targeted and sustained drug-release. Combining these advantages of nanoemulsions with the use of plant extracts, it can be obtained a product with exceptional biological properties that can be used for the production of hydrogels, creams and other products that can be applied at the pharmaceutical industry (Balestrin et al. 2016).

2.4.2 Nanosuspensions

Nanosuspensions are defined as unique liquid submicron colloidal dispersions of nanosized pure drug particles, stabilized by a suitable polymer and/or surfactant (Zorzi et al. 2015). Nanosuspensions represent a potential method for improving bioavailability of hydrophobic drugs (Odei-Addo et al. 2017). In most cases, nanoparticles need to be functionalized by various coating agents before application. Plant extracts act as capping agents and stabilize these (Chung et al. 2016). Nanosuspensions and nanoemulsions have common advantages, just differing between the two types of nanostructures. Nanoemulsions are composed of a mixture of water and oil, together with a suitable stabilizing surfactant, while nanosuspensions are mainly a dispersion of pure particles of a drug, which are stabilized by surfactants (Zorzi et al. 2015). These nanostructures in combination with plant extracts as agents for nanoparticle synthesis, or as a nanoparticle-incorporating bioactive compounds, or even the use at nanoemulsions, have attracted increasing attention by researchers working on targeted and sustained drug-release systems (Li et al. 2018).

2.4.3 Nanofibers

Fibers with diameter size at nanometric range are known as nanofibers. These can be generated from different polymers, and therefore, can have different physical properties and potential applications. Some examples of natural polymers include collagen, silk fibroin, keratin and gelatin (Youse et al. 2017). In addition to these, polymers of vegetable origin such as cellulose and alginate have also been used frequently (Youse et al. 2017). Electrospinning is the most commonly used method for their production (Ghayempour and Montazer 2018). It is a simple, versatile and efficient process to produce ultrafine fibers, as it can be applicable to different polymers (Ghayempour and Montazer 2018). The great advantage of nanofibers lies in their large surface area for volume ratio, as well as their porous structure, that favors cell adhesion, proliferation and differentiation *in vitro*. Functionalized nanofibers with antibacterial and healing properties have been extensively developed (Youse et al. 2017). There has been a growing use of nanofibers containing medicinal compounds from plant extracts for biomedical applications, particularly for wound dressings (Youse et al. 2017).

2.4.4 Nanotubes

Since the discovery and synthesis of carbon nanotubes, these materials have been highly investigated for their peculiar properties. Carbon nanotubes are cylinders made up of carbon atoms that have extraordinary mechanical, electrical and thermal properties, and have been the basis for various applications (Tostado-plascencia et al. 2018). Carbon nanotubes can be made by only one of these cylinders, being classified as single wall nanotubes, and as multi-walled nanotubes, which are formed by several cylinders that are concentrically coiled (Foo et al. 2018). In the field of biomedical applications, because of their extremely small size and lightness, nanotubes can reach the interior of a cell, and therefore, can be used as biosensors on medical diagnostics and therapies (Tostado-plascencia et al. 2018). However, a major inconvenience is though their uncontrolled cytotoxicity. To prevent it, some authors proposed a coating strategy by using a synthetic polymer capable of mimicking cell surface receptors. Although the use of plant extracts is not as widely used in nanotubes, recent studies have shown that their functionalization with plant extracts enhances their biological activity (Foo et al. 2018; Tostado-plascencia et al. 2018).

2.4.5 Nanoparticles

Nanoparticles are the most used nanostructures in combination with plant extracts (P. Sathishkumar et al. 2016). With a particle size between 1 and 100 nm these are composed of synthetic or semi-synthetic polymers with nano- or subnano-sized

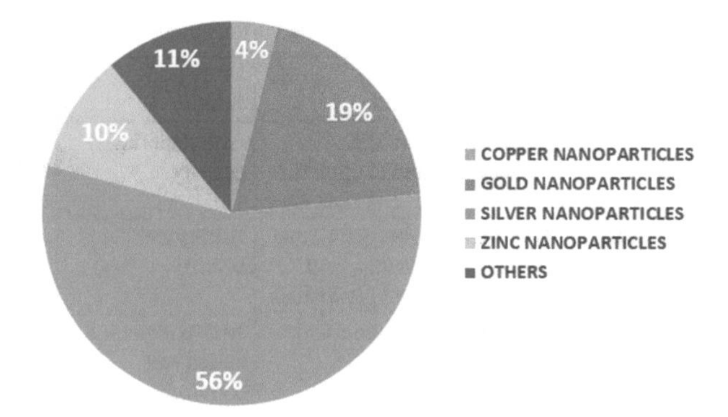

Fig. 2.7 Percentage of different types of nanomaterials used as object of study in articles, for the last 5 years, showing the trend of using metals in nanoparticles fabrication

structures (Zorzi et al. 2015). Encapsulation of plant extracts in nanoparticles is an effective way used to protect compounds against deterioration or interaction with other factors. Besides, several other advantages are of reference, namely increased solubility, efficacy and bioavailability, reduced dosage and improved drug absorption (Zorzi et al. 2015). On the other hand, plant extracts are also used for the production of nanoparticles, being the extract simply mixed with a salt metal solution at room temperature, to promote a reaction that is complete in a few minutes. Most conventional methods for producing nanoparticles are expensive, and environmentally hazardous due to the application of chemicals with high biological risk. (Mittal et al. 2013)

Green synthesis processes have been developed to overcome these disadvantages and are more acceptable for medical applications, due to their high biocompatibility, biodegradability, non-toxicity and green nature of the agents, as well as their cost-effectiveness (P. Sathishkumar et al. 2016). Plant extracts are used for the bioreduction of metal ions to form nanoparticles, and in support of their subsequent stability (Sathishkumar et al. 2016). The most common metal nanoparticles in (bio) medical applications are nanomaterials based on silver and gold, given their biocide effect, showed in Fig. 2.7 (Parveen et al. 2016).

Generally, nanoparticles are prepared by physical and chemical methods. Most of these protocols involve the use of toxic reagents that are potentially hazardous to the environment and relatively expensive (Sathishkumar et al. 2016). Hence, there was a need to develop nanoparticles synthesis techniques that were economically sustainable, environmentally friendly and simple. Secondary metabolites present in plants act as reducing agents towards stabilization of nanoparticles (Lediga et al. 2018). Green synthesis of nanoparticles has demonstrated advantage over chemically synthesized nanoparticles (Pirtarighat et al. 2019). Methods applied do not require any high temperature processing, or relevant toxic chemicals. Instead, these are eco-friendly, time affordable and cost effective procedures. More importantly, these are free of hazardous materials on their surface, and can be coated with bioorganic compounds that make them more biocompatible (Pirtarighat et al. 2019).

Table 2.2 Green-synthesized metal nanoparticles using various plant extracts

Metal nanoparticles	Active compound	Biological activity	References
Silver nanoparticles using *Malva sylvestris* extract	Quercetin, malvidin 3-glucoside and scopoletin	Antibacterial activity	Mahmoodi Esfanddarani et al. (2018)
Silver nanoparticles using *Gymnema sylvestre* extract	Gymnemagenin, gymnemic acids, gymnemanol, and β-amyrin-related glycosides	Anticancer activity	Arunachalam (2014)
Gold nanoparticles and silver nanoparticles using *Stereospermum suaveolens* extract	n-triacontanol, p-coumaric acid, beta-sitosterol, lapachol and cycloolivil	Antioxidant and anticancer activity	Francis et al. (2018)
Magnesium nanoparticles using *Artemisia abrotanum* extract	Aglycones and glycosylates, and hydroxycinnamic derivatives	Catalytic and antioxidant activity	Dobrucka (2018)
Copper nanoparticles using *Punica* granatum extract	Ellagic acid, punicalagins A and B, ellagitannins, gallic acid and gallotannins	Photocatalytic activity	Nazar et al. (2018)

Metallic nanoparticles are of great scientific interest, mainly for biomedical applications, with many of these being used as antibacterial nanoagents (Ocsoy et al. 2018). Several articles have reported the synthesis of metal-based nanoparticles using plant extracts. Some of related examples are listed in Table 2.2.

2.5 Properties and Applications of Plant-Based Nanomaterials

Nanomaterials properties can be very distinct. Their constituents used, as well as their modification, shape and size can give rise to a wide range of useful properties (Al-Huqail et al. 2018; Balalakshmi et al. 2017; Hussain et al. 2018). Also, to standardize production of nanomaterials using plant extracts it is necessary to change the reaction parameters, such as reaction time and temperature, pH, metal ion and plant extract concentrations (Devi et al. 2017). Final products can be characterized by a series of methods as UV-visible spectrophotometer, to validate the reduction of ions; TEM analysis to analyze morphology and size distribution; and X-ray diffraction for crystallographic characterization (Abbasi and Anjum 2016; Judith Vijaya et al. 2017; Thatoi et al. 2016; Yadi et al. 2018).

One of the most used and diverse type of nanoparticles, in terms of properties and applications, are metallic-based nanoparticles, as shown in Fig. 2.7. These are prepared by using various kinds of metals as silver, zinc, iron, copper, gold and titanium (Botha et al. 2019; Saravanan et al. 2016).

There are innumerous reports documenting that these metals are known to have multifunctional bio-applications as antibacterial, antimicrobial, antiproliferative, antifungal, antioxidant, anti-inflammatory, antivirus, antimosquitoes, larvicidal activity and others, as presented in Table 2.3 and Fig. 2.8.

Table 2.3 Applications of metal-based nanomaterials synthesized from plant extract

Metal-based nanomaterials	Applications	References
Silver	Antibacterial; anticancer; antioxidant; antimicrobial; anti-inflammatory; antifungal	Ocsoy et al. (2017), Kotakadi et al. (2014), Amooaghaie et al. (2015) and Rasheed et al. (2017)
Gold	Anticancer; antioxidant; antimicrobial	Rijo et al. (2016), Clemente et al. (2017), Yallappa et al. (2015) and Nambiar et al. (2018)
Copper	Antibacterial; anticancer; antimicrobial; antifungal; antioxidant	Hassanien et al. (2018), Saran et al. (2018), Pansambal et al. (2017) and Yugandhar et al. (2017)
Zinc	Antibacterial; antioxidant; antimicrobial; anticancer	Mahendra et al. (2017), Jafarirad et al. (2016), Suresh et al. (2015) and Vijayakumar et al. (2018)
Nickel	Anticancer; antibacterial; antimicrobial	Ezhilarasi et al. (2016, 2018), Amiri et al. (2018) and Saleem et al. (2017)
Titanium	Antioxidant; antibacterial	Santhoshkumar et al. (2014) and Nadeem et al. (2018)
Magnesium	Antioxidant; antibacterial	Sushma et al. (2016), Dobrucka (2018) and Jamal et al. (2016)

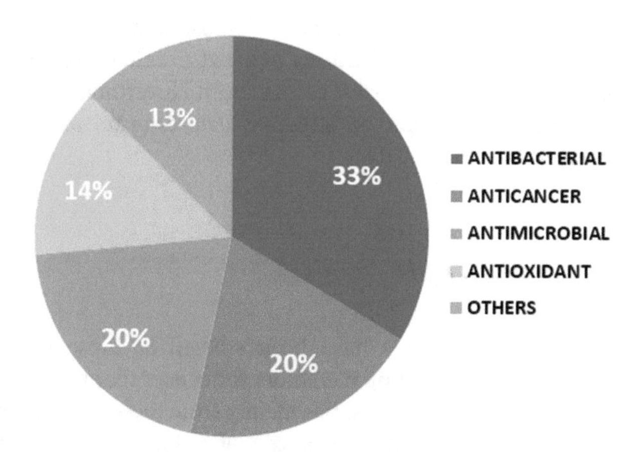

Fig. 2.8 Percentage of biomedical applications of nanomaterials synthesized from plant extracts considering data from the selected articles. It can be concluded that antibacterial, anticancer and antimicrobial are currently their main applications

2.5.1 Silver Nanomaterials

Noble metal nanomaterials have attracted attention because of their structures, as these exhibit better physical, chemical, biological properties and functionalities, due to their ultra-small sizes (Ali et al. 2017). Among the various nanomaterials, silver nanoparticles have been the most studied, given mostly because of their high electrical and chemical conductivity, chemical stability, high catalytic activity and antimicrobial, anticancer and larvicidal activities (Allafchian et al. 2018; Doddapaneni et al. 2018; Zhou and Tang 2018).

Silver nanoparticles can cause cell lyses, inhibit cell transduction, induce changes in cell membrane permeability, affect the microbial genome and cause DNA fragmentation (Ouda Sahar 2014). Oh et al. suggested that the eco-friendly silver nanoparticles prepared with *C. sinensis* were effective as antimicrobial agent against pathogenic *E. coli* and *S. aureus* (Oh et al. 2018).

The unique properties of silver nanoparticles are mainly advantageous, since these also led to an enhanced chemotherapeutic efficacy with minimal toxicity (Adil et al. 2016; Doddapaneni et al. 2018; Jena et al. 2016). Nowadays, nanoparticles-based combinatorial therapies using those with anticancer activity in combination with a chemotherapeutic agent are being translated into nanomedicine, in order to counteract resistance to the standard treatments (Banu et al. 2018; Dehghanizade et al. 2017).

In developing countries, vector-borne diseases are a major public health problem. Suman et al. (2018) indicates that silver nanoparticles synthesized with *H. enneaspermus* are highly stable and had significant mosquito larvicidal activity, having the potential to be used as an ideal ecofriendly approach for vector control programs (Suman et al. 2016).

Less common properties were reported by Sathishkumar et al. wherein silver nanoparticles using *C. sativum* leaf extract showed excellent anti-acne and anti-dandruff against *P. acnes* and *M. furfur*, respectively. Nevertheless, further investigation is still needed to turn it into an effective nano-drug for *in vivo* therapeutic application (Sathishkumar et al. 2016).

2.5.2 Gold Nanomaterials

Biocompatible gold nanomaterials have been studied as biosensors, diagnostic agents, cell targeting vectors, heating mediators for cancer thermotherapy and drug delivery (Yallappa et al. 2015). These stems mainly from their less cytotoxicity, high stability, ability to bind to biomolecules and also their visible light behavior (Guruviah Karthiga Devi and Sathishkumar 2016; Nambiar et al. 2018; Yallappa et al. 2015). Gold nanoparticles have been successfully prepared with *M. foetida* leaves extract and conjugated with folic acid and doxorubicin. Such complex showed low toxicity towards normal epithelial cells, and high toxicity to human cancer cells

(Yallappa et al. 2015). Also, investigation on combined use of *P. emodi* and gold nanoparticles demonstrated enhanced potential for treatment and management of diseases, as is the case of atherosclerosis and myocardial infarction (Ibrar et al. 2018).

Antimicrobial activity of gold nanoparticles has also been widely reported. Oh et al. suggested that gold nanoparticles produced with *C. sinensis,* besides being effective against breast cancer cells, were successful antimicrobial agents against pathogenic *E. coli* and *S. aureus* (Oh et al. 2018).

In addition, the use of gold nanoparticles as catalyst in photocatalytic degradation, and reduction of pollutant, has become a prominent approach for environmental solutions. Choudhary et al. studied a green synthesis approach of gold nanoparticles mediated by *L. speciosa* leaf extract. Obtained nanoparticles showed strong photocatalytic behavior, adequate to be used for treatment of wastewater containing toxic dyes and other organic pollutants (Choudhary et al. 2017).

2.5.3 *Zinc Nanomaterials*

Zinc nanomaterials have drawn attention for their exclusive optical and chemical behaviors, which are strongly related to their size and morphology (Ambika and Sundrarajan 2015). Zinc nanoparticles combined with *V. negundo* leaves extract were demonstrated to inhibit the growth of *S. aureus* and *E. coli* bacterial cells (Ambika and Sundrarajan 2015). Furthermore, zinc nanoparticles have highly promising prospective on biological functions for sensing, as drug delivery system, or as anticancer, antibacterial, antilarvicidal, antioxidant and antidiabetic agents (Jafarirad et al. 2016; Mahendra et al. 2017; Ovais 2017; Surface et al. 2018; Thatoi et al. 2016; Vijayakumar et al. 2018; Yadi et al. 2018). Zinc nanoparticles produced in combination with *V. Arctostaphylos* have been referred as nano-antidiabetic drugs (Bayrami et al. 2017).

2.5.4 *Iron Nanomaterials*

Due to the biodegrability, biocompatibility and easy synthesis nature of iron oxide nanomaterials, these have a great potential to be used as treatment agent towards different types of cancer (Zhao et al. 2018). *In vitro* studies revealed that green synthesis of iron nanoparticles with *C. sativum* can be used to enhance cytotoxic effects on HeLa cancer cells, for example (Sathya et al. 2018).

Moreover, iron has been the first choice metal for groundwater treatment due to its high intrinsic reactivity, low toxicity, biodegrability, low cost, abundance and magnetic properties, being considered a universal material for water treatment (Nasiri et al. 2019). It has been used to remediate contaminants such as heavy metals, azo dyes and organic compounds (Carvalho and Carvalho 2017; Nasiri et al. 2019).

Also, iron nanoparticles produced in combination with *C. guianensis* possess a great bactericidal effect against different bacterial human pathogens (Sathishkumar et al. 2018).

2.5.5 Nanofibers

Nanofibers have been mainly recognized for their applications in drug delivery (Charernsriwilaiwat et al. 2013). In this context, these nanomaterials present very large surface area to volume ratios having potential to improve drug release significantly (Ghayempour and Montazer 2018). Furthermore, the small dimension of fibers combined with their microporous structure provided by the polymer, mimics a protective shield, resulting in increased drug loading and stability (Charernsriwilaiwat et al. 2013). Recently, some studies have reported successful development of *Mangosteen pericarp* extract nanofibers intended for various purposes. For instance, it was loaded onto polyvinyl alcohol (PVA) nanofibers for dermal delivery purpose and it was spun in a mixture of chitosan/EDTA/PVA for wound healing (Charernsriwilaiwat et al. 2013; Ghayempour and Montazer 2018).

2.5.6 Nanoemulsions

Main advantages of nanoemulsions compared to conventional emulsions include improved physical stability and absorption/penetration, as high bioavailability, properties that are attributed to their reduced droplet size and great surface area (da Silva Gündel et al. 2018). Besides, these permit the dispersion of lipophilic molecules in aqueous media and can be administered by several vias, from topic, oral, nasal and ocular to intravenous (Zorzi et al. 2015).

Jaboticaba is a fruit of Jaboticabeira tree (*Myrciaria spp.*, *Myrtaceae* family) and highly appreciated in the Brazilian culinary because of its medicinal properties. Pharmacological activities reported include antioxidant, anti-inflammatory, antimicrobial, hypoglycemic and antiproliferative effects (Mazzarino et al. 2017). In a recent study, Mazzarino and colleagues reported that this plant nanoemulsions can be considered a potential antioxidant agent for cosmetic and pharmaceutical applications (Mazzarino et al. 2017).

In another study, microemulsions were fabricated to improve oral bioavailabitilty of the total flavones extracted from *Hippophae rhamnoides*, that exhibits therapeutically significant effects on cardiovascular and cerebrovascular diseases, and showed protective applications in aging, radiation, oxidative injury, myocardial ischemia and cerebral thrombus (Guo and Guo 2017).

2.6 Natural Products from Plants Approved for Therapeutic Use

Over the years, some plant compounds have been integrated into the pharmaceutical industry as drugs for the treatment of various diseases (Afrin et al. 2018; David et al. 2015; Newman and Cragg 2016). Table 2.4 presents some chemical entities derived from plants that are currently used in various therapies.

Table 2.4 Plant-derived natural products approved for therapeutic use

Generic name and chemical structure	Plant species	Commercial name and year of introduction	Therapeutic use
Galantamine 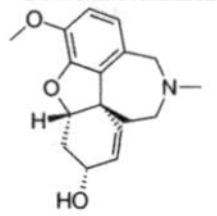	*Galanthus woronowii Losinsk.* 	Razadyne, 2001	Early-onset Alzheimer's disease
Nitisinone 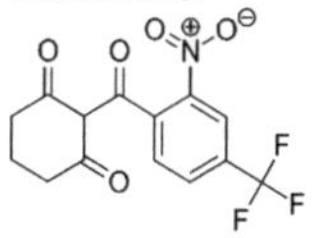	*Callistemon citrinus* 	Orfadin, 2002	Hereditary tyrosinemia type 1
Varenicline 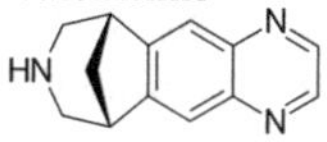	*Cytisus laburnum L.* 	Chantix, 2006	Tobacco dependence
Paclitaxel 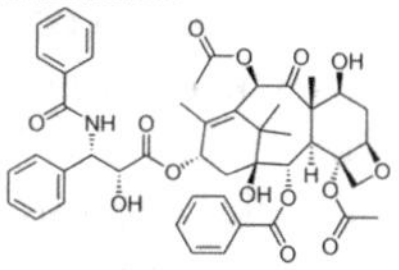	*Taxus brevifolia Nutt.* 	Nanoxelc, 2007	Cancer chemotherapy
Capsaicin 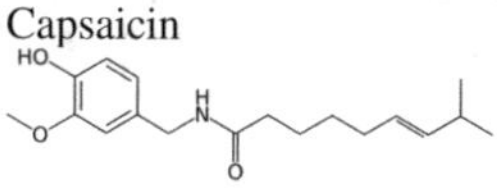	*Capsicum annum L.* 	Qutenza, 2010	Peripheral neuropathic pain
Omacetaxine mepesuccinate 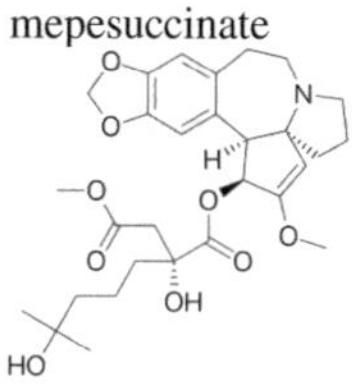	*Cephalotaxus harringtonia* 	Synribo, 2012	Oncology

The process of selection and study of bioactive compounds of plants with potential for the development of new drugs is time consuming, since much information is needed about its habitat, abundance, use by population, besides their phytochemical composition that often is quite complex. However, with the development of new technologies and improved instruments, such as high-performance liquid chromatography (HPLC), coupled to mass spectrometry (MS)/MS (liquid chromatography, LC-MS), higher magnetic field-strength nuclear magnetic resonance (NMR) instruments, and robotics to automate high-throughput bioassays, yield study of plants extractives easier and faster (Atanasov et al. 2015).

Several herbal medicines have been released on the market in recent years, and many plant-derived compounds are currently undergoing clinical trials for the potential treatment of various diseases (Atanasov et al. 2015). As there are still many plants that have not been studied and tested as potential sources of bioactive compounds, the discovery of plant-based drugs continues surely to be essential for the release of new and highly competent phytopharmaceuticals, and even more with the development of more sensitive and more versatile analytical methods.

2.7 Usage of Curcumin in Nanoparticles for Periodontal Disease Therapy

Curcumin is a hydrophobic polyphenolic compound derived from the rhizomes of *Curcuma longa*. This natural compound has a long history of use as curry (turmeric) in East Asian countries. India is the largest turmeric producer worldwide, being turmeric a major driver of Indian economy. Data from 2016 estimates that India share in exportations was 65.5% of world exportation, valued at US$186.46 million and estimated to grow yearly above 10% (Tridge. 2019). Demand is driven by increasing awareness of its beneficial characteristics in cosmetics, food and medicine (Sadegh et al. 2019). It is fairly intuitive that, if nanotechnology is to eliminate one of the biggest limitations of curcumin's use in medicine by improving its solubility, and thus bioavailability, this will have a tremendous economical impact on the market and its players, creating opportunities for more business and increased margins of profit (Hatamipour et al. 2019; Sadegh et al. 2019).

Curcumin is "generally recognized as safe" (GRAS) by the Food and Drug Administration (FDA), and is characterized by a wide range of antibacterial, antifungal, antiviral, antioxidative, anti-inflammatory, anticancer and antiproliferative activities (Dai et al. 2017; Loo et al. 2016; Nambiar et al. 2018; Yallapu et al. 2013).

Periodontal disease is a chronic inflammatory condition affecting the tissues that support and protect the teeth, and may lead to the destruction of alveolar bone and periodontal ligament (Preshaw 2015). Its prevalence and severity are highly variable in different populations, but it is estimated that 15–20% of adult individuals are affected by the more severe forms of the disease, while the less severe forms affect 35–60% of the global population (Preshaw 2015).

At literature, there is considerable evidence from both *in vitro* and *in vivo* studies indicating that the anti-inflammatory properties of curcumin attenuate the response of immune cells to periodontal disease-associated bacterial antigens and inhibit periodontal tissue destruction (Zambrano et al. 2018). However, most of these used a systemic route of administration, and their results may be limited by the poor pharmacodynamic properties of curcumin, as well as its high hydrophobicity that conducts to low absorption rates at gastrointestinal tract, besides having extremely short plasma half-life (Pr et al. 2016; Zhou et al. 2013).

Zambrano et al. (2018) reported that local administration of curcumin-loaded nanoparticles effectively inhibited inflammation and bone re-absorption associated with experimental periodontal disease (Zambrano et al. 2018).

A search on U.S. National Institutes of Health Public online database of clinical trials (http://www.clinicaltrials.gov) in May 2019, using the keyword "curcumin", returned 198 registered trials on various clinical conditions such as asthma, multiple myeloma and various other types of cancer, diabetes, Alzheimer's disease, schizophrenia and inflammatory intestinal diseases. Interestingly, this search returned 9 phase IV clinical trials, including 1 on topical use of curcumin as adjunct in treatment of periodontal disease. The primary outcome of this single study was the antioxidant activity in saliva of periodontitis patients, but no results were posted.

2.8 Conclusion

Nanotechnology is in great evolution and has allowed the development of nanomaterials with applications in the most diverse areas, especially in the biomedical field. The variety of existing nanostructures with different properties added different possibilities of treatments that had not existed before, or did not work so effectively. The use of plant extracts allied to nanoparticles, nanoemulsions, nanosuspensions, nanofibers and nanotubes has been shown to improve the efficacy and enhance the biological properties of the phyto-extracts. Although there are many studies of green synthesized nanomaterials and its properties, it is still necessary to continue investigating these derived formulations to validate them as potential agents in new nanotherapies. These technological developments will certainly lead to great economical impact in the plant derived products markets, creating opportunities for more business and increased margins of profit in value added products.

Acknowledgements This work was supported by the strategic programme UID/BIA/04050/2019 funded by national funds through the FCT I.P.

References

Abbasi B, Anjum S (2016) Thidiazuron-enhanced biosynthesis and antimicrobial efficacy of silver nanoparticles via improving phytochemical reducing potential in callus culture of *Linum usitatissimum* L. Int J Nanomedicine 11:715

Adil S et al (2016) Apoptosis inducing ability of silver decorated highly reduced graphene oxide nanocomposites in A549 lung cancer. Int J Nanomedicine 11:873

Afrin S, Jahan I, Nazmul Hasan AHM, Deepa KN (2018) Novel approaches of herbal drug delivery. J Pharm Res Int 21(5):1–11

Ahn S et al (2018) Soy protein/cellulose nanofiber scaffolds mimicking skin extracellular matrix for enhanced wound healing. Adv Healthc Mater 7(9):1–13. https://doi.org/10.1002/adhm.201701175

Ajoumshariati S, Eyedramin P, Eyedeh S, Imia K, Avari Y, Ohammad M, Hokrgozar ALIS (2016) Physical and biological modification of polycaprolactone electrospun nanofiber by *Panax ginseng* extract for bone tissue engineering application. Ann Biomed Eng 44(5):1808–1820

Al-Huqail AA, Hatata MM, AL-Huqail AA, Ibrahim MM (2018) Preparation, characterization of silver phyto nanoparticles and their impact on growth potential of *Lupinus termis* L. seedlings. Saudi J Biol Sci 25(2):313–319

Ali MS, Altaf M, Al-Lohedan HA (2017) Green synthesis of biogenic silver nanoparticles using *Solanum tuberosum* extract and their interaction with human serum albumin: evidence of 'Corona' formation through a multi-spectroscopic and molecular docking analysis. J Photochem Photobiol B Biol 173:108–119

Allafchian AR et al (2018) Green synthesis of silver nanoparticles using *Glaucium corniculatum* (L.) curtis extract and evaluation of its antibacterial activity. IET Nanobiotechnol 12(5):574–578

Ambika S, Sundrarajan M (2015) Antibacterial behaviour of Vitex negundo extract assisted ZnO nanoparticles against pathogenic bacteria. J Photochem Photobiol B Biol 146:52–57

Amiri M, Pardakhti A, Ahmadi-zeidabadi M, Akbari A (2018) Magnetic nickel ferrite nanoparticles: green synthesis by urtica and therapeutic effect of frequency magnetic field on creating cytotoxic response in neural cell lines. Colloids Surf B: Biointerfaces 172:244–253. https://doi.org/10.1016/j.colsurfb.2018.08.049

Amooaghaie R, Reza M, Azizi M (2015) Synthesis, characterization and biocompatibility of silver nanoparticles synthesized from *Nigella sativa* leaf extract in comparison with chemical silver nanoparticles. Ecotoxicol Environ Saf 120:400–408. https://doi.org/10.1016/j.ecoenv.2015.06.025

André C, Kapustikova I, Lethier L (2014) A particulate biochromatographic support for the research of arginase inhibitors doped with nanomaterials: differences observed between carbon and boron nitride nanotubes. Application to three plant extracts. Chromatographia 77(21–22):1521–1527

Arunachalam K, Arun L, Annamalai S, Arunachalam A (2014) Potential anticancer properties of bioactive compounds of *Gymnema sylvestre* and its biofunctionalized silver nanoparticles. Int J Nanomedicine 10:31–41. http://www.embase.com/search/results?subaction=viewrecord&from=export&id=L600969558%5Cn. https://doi.org/10.2147/IJN.S71182

Atanasov AG et al (2015) Discovery and resupply of pharmacologically active plant-derived natural products : a review. Biotechnol Adv 33(8):1582–1614. https://doi.org/10.1016/j.biotechadv.2015.08.001

Balalakshmi C et al (2017) Green synthesis of gold nanoparticles using a cheap *Sphaeranthus indicus* extract: impact on plant cells and the aquatic crustacean *Artemia* nauplii. J Photochem Photobiol B Biol 173:598–605

Balamurugan M, Saravanan S, Soga T (2014) Synthesis of Iron oxide nanoparticles by using *Eucalyptus globulus* plant extract. J Surf Sci Nanotechnol 12:363–367

Balestrin LA et al (2016) Protective effect of a hydrogel containing *Achyrocline satureioides* extract-loaded nanoemulsion against UV-induced skin damage. J Photochem Photobiol B Biol 163:269–276. https://doi.org/10.1016/j.jphotobiol.2016.08.039

Banu H et al (2018) Gold and silver nanoparticles biomimetically synthesized using date palm pollen extract-induce apoptosis and regulate P53 and Bcl-2 expression in human breast adenocarcinoma cells. Biol Trace Elem Res 186(1):122–134

Bayrami A, Parvinroo S, Habibi-Yangjeh A, Rahim Pouran S et al (2017) Bio-extract-mediated ZnO nanoparticles: microwave-assisted synthesis, characterization and antidiabetic activity evaluation. Artif Cells Nanomed Biotechnol: 1–10. https://doi.org/10.1080/21691401.2017.1337025

Botha TL et al (2019) Cytotoxicity of Ag, Au and Ag-Au bimetallic nanoparticles prepared using golden rod (*Solidago canadensis*) plant extract. Sci Rep 9(1):1–8

Carvalho SSF, Carvalho NMF (2017) Dye degradation by green heterogeneous Fenton catalysts prepared in presence of *Camellia sinensis*. J Environ Manag 187:82–88

Charernsriwilaiwat N et al (2013) Electrospun chitosan-based nanofiber mats loaded with *Garcinia mangostana* extracts. Int J Pharm 452(1–2):333–343

Chintamani RB, Salunkhe KS, Chavan MJ (2018) Emerging use of green synthesis silver nanoparticle: an updated review. Int J Pharm Sci Res 9(10):4029–4055

Choudhary BC et al (2017) Photocatalytic reduction of organic pollutant under visible light by green route synthesized gold nanoparticles. J Environ Sci (China) 55:236–246

Chung IM et al (2016) Plant-mediated synthesis of silver nanoparticles: their characteristic properties and therapeutic applications. Nanoscale Res Lett 11(1):1–14. https://doi.org/10.1186/s11671-016-1257-4

Clemente I, Ristori S, Pierucci F, Muniz-miranda M (2017) Gold nanoparticles from vegetable extracts using different plants from the market: a study on stability, shape and toxicity. ChemistrySelect 2(30):9777–9782

da Silva Gündel S et al (2018) Nanoemulsions containing *Cymbopogon flexuosus* essential oil: development, characterization, stability study and evaluation of antimicrobial and antibiofilm activities. Microb Pathog 118:268–276

Dai X et al (2017) Nano-formulated curcumin accelerates acute wound healing through Dkk-1-mediated fibroblast mobilization and MCP-1-mediated anti-inflammation. NPG Asia Mater 9(3):368

David B, Wolfender JL, Dias DA (2015) The pharmaceutical industry and natural products: historical status and new trends. Phytochem Rev 14:299–315

Dehghanizade S, Arasteh J, Mirzaie A (2017) Green synthesis of silver nanoparticles using *Anthemis atropatana* extract: characterization and *in vitro* biological activities. Artif Cells Nanomed Biotechnol:1–9. https://doi.org/10.1080/21691401.2017.1304402

Devi GK, Sathishkumar K (2016) Synthesis of gold and silver nanoparticles using *Mukia maderaspatna* plant extract and its anticancer activity. IET Nanobiotechnol 11(2):143–151

Devi GK, Kumar KS, Parthiban R, Kalishwaralal K (2017) An insight study on HPTLC fingerprinting of *Mukia maderaspatna*: mechanism of bioactive constituents in metal nanoparticle synthesis and its activity against human pathogens. Microb Pathog 102:120–132

Dobrucka R (2018) Synthesis of MgO nanoparticles using *Artemisia abrotanum* herba extract and their antioxidant and photocatalytic properties. Iran J Sci Technol Trans A Sci 42(2):547–555

Doddapaneni SJDS et al (2018) Antimicrobial and anticancer activity of AgNPs coated with Alphonsea sclerocarpa extract. 3 Biotech 8(3):1–9. https://doi.org/10.1007/s13205-018-1155-9

Dzoyem JP, Tshikalange E, Kuete V (2013) Medicinal plant research in Africa. In: Medicinal plants market and industry in Africa: pharmacology and chemistry. Elsevier Inc, London. https://doi.org/10.1016/B978-0-12-405927-6.00024-2

Ezhilarasi A et al (2016) Green synthesis of NiO nanoparticles using *Moringa oleifera* extract and their biomedical applications: cytotoxicity effect of nanoparticles against HT-29 cancer cells. J Photochem Photobiol B Biol 164:352–360. https://doi.org/10.1016/j.jphotobiol.2016.10.003

Ezhilarasi AA et al (2018) Green synthesis of NiO nanoparticles using *Aegle marmelos* leaf extract for the evaluation of *in vitro* cytotoxicity, antibacterial and photocatalytic properties. J Photochem Photobiol B Biol 180:39–50. https://doi.org/10.1016/j.jphotobiol.2018.01.023

Foo ME, Gyun C, Radi A, Yaakub W (2018) Antimicrobial activity of functionalized single – walled carbon nanotube with herbal extract of *Hempedu bumi*. Surf Interface Anal 2018:354–361

Francis S, Koshy EP, Mathew B (2018) Green synthesis of *Stereospermum suaveolens* capped silver and gold nanoparticles and assessment of their innate antioxidant, antimicrobial and antiproliferative activities. Bioprocess Biosyst Eng 41(7):939–951. https://doi.org/10.1007/s00449-018-1925-0

Ghayempour S, Montazer M (2018) A modified microemulsion method for fabrication of hydrogel tragacanth nanofibers. Int J Biol Macromol 115:317–323

Gumus ZP et al (2015) Herbal infusions of black seed and wheat germ oil: their chemical profiles, *in vitro* bio-investigations and effective formulations as phyto-nanoemulsions. Colloids Surf B: Biointerfaces 133:73–80. https://doi.org/10.1016/j.colsurfb.2015.05.044

Guo R, Guo X (2017) Fabrication and optimization of self-microemulsions to improve the oral bioavailability of total flavones of Hippopha ë Rhamnoides L. J Food Sci 82(7):1–9

Hassanien R, Husein DZ, Al-hakkani MF (2018) Biosynthesis of copper nanoparticles using aqueous *Tilia* extract: antimicrobial and anticancer activities. Heliyon 4(12):e01077. https://doi.org/10.1016/j.heliyon.2018.e01077

Hatamipour M et al (2019) Novel nanomicelle formulation to enhance bioavailability and stability of curcuminoids. Iran J Basic Med Sci 22(3):282–289. (11)

Hoscheid J et al (2015) Development and characterization of *Pterodon pubescens* oil nanoemulsions as a possible delivery system for the treatment of rheumatoid arthritis. Colloids Surf A Physicochem Eng Asp 484:19–27. https://doi.org/10.1016/j.colsurfa.2015.07.040

Hussain M et al (2018) Green synthesis and characterisation of silver nanoparticles and their effects on antimicrobial efficacy and biochemical profiling in *Citrus reticulata*. IET Nanobiotechnol 12(4):514–519

Ibrar M, Khan MA, Abdullah, Imran M (2018) Evaluation of Paeonia emodi and its gold nanoparticles for cardioprotective and antihyperlipidemic potentials. J Photochem Photobiol B Biol 189:5–13

Islam NU et al (2015) Antinociceptive, muscle relaxant and sedative activities of gold nanoparticles generated by methanolic extract of *Euphorbia milii*. BMC Complement Altern Med 15:160

Jafarirad S, Mehrabi M, Divband B, Kosari-nasab M (2016) Biofabrication of zinc oxide nanoparticles using fruit extract of *Rosa canina* and their toxic potential against bacteria: a mechanistic approach. Mater Sci Eng C 59:296–302. https://doi.org/10.1016/j.msec.2015.09.089

Jahan N et al (2016) Formulation and characterisation of nanosuspension of herbal extracts for enhanced antiradical potential. J Exp Nanosci 11(1):72–80. https://doi.org/10.1080/17458080.2015.1025303

Jamal L, Umaralikhan M, Jaffar M (2016) Green synthesis of MgO nanoparticles and it antibacterial activity. Iran J Sci Technol Trans A Sci 12:1–9

Jena S et al (2016) Photo-bioreduction of Ag+ ions towards the generation of multifunctional silver nanoparticles: mechanistic perspective and therapeutic potential. J Photochem Photobiol B Biol 164:306–313

Judith Vijaya J et al (2017) Bioreduction potentials of dried root of *Zingiber officinale* for a simple green synthesis of silver nanoparticles: antibacterial studies. J Photochem Photobiol B Biol 177:62–68

Klekotko M et al (2015) Bio-mediated synthesis, characterization and cytotoxicity of gold nanoparticles. Phys Chem Chem Phys 17(43):29014–29019

Kotakadi VS et al (2014) Biofabrication of silver nanoparticles using *Andrographis paniculata*. Eur J Med Chem 73:135–140. https://doi.org/10.1016/j.ejmech.2013.12.004

Lediga ME et al (2018) Biosynthesis and characterisation of antimicrobial silver nanoparticles from a selection of fever-reducing medicinal plants of South Africa. S Afr J Bot 119:172–180. https://doi.org/10.1016/j.sajb.2018.08.022

Li H et al (2018) Folate-targeting annonaceous acetogenins nanosuspensions: significantly enhanced antitumor efficacy in HeLa tumor-bearing mice. Drug Deliv 25(1):880–887. https://doi.org/10.1080/10717544.2018.1455761

Loo C-y et al (2016) Combination of silver nanoparticles and curcumin nanoparticles for enhanced anti-bio film activities. J Agric Food Chem 64(12):2513–2522

Mahendra C et al (2017) Antibacterial and antimitotic potential of bio-fabricated zinc oxide nanoparticles of *Cochlospermum religiosum* (L.). Microb Pathog 110:620–629. https://doi.org/10.1016/j.micpath.2017.07.051

Mahmoodi Esfanddarani H, Abbasi Kajani A, Bordbar A-K (2018) Green synthesis of silver nanoparticles using flower extract of *Malva sylvestris* and investigation of their antibacterial activity. IET Nanobiotechnol 12(4):412–416

Marslin G et al (2015) Antimicrobial activity of cream incorporated with silver nanoparticles biosynthesized from *Withania somnifera*. Int J Nanomedicine 10:5955–5963

Mazzarino L et al (2017) Jaboticaba (*Plinia peruviana*) extract nanoemulsions: development, stability, and *in vitro* antioxidant activity. Drug Dev Ind Pharm 44(4):1–27

Miftahul M (2018) *Mangosteen pericarp* extract embedded in electrospun PVP nanofiber mats: physicochemical properties and release mechanism of α -mangostin. Int J Nanomedicine 13:4927–4941

Mittal AK, Chisti Y, Banerjee UC (2013) Synthesis of metallic nanoparticles using plant extracts. Biotechnol Adv 31(2):346–356. https://doi.org/10.1016/j.biotechadv.2013.01.003

Murad U et al (2018) Synthesis of silver and gold nanoparticles from leaf of *Litchi chinensis* and its biological activities. Asian Pac J Trop Biomed 8(3):142–149

Nadeem M et al (2018) The current trends in the green syntheses of titanium oxide nanoparticles and their applications. Green Chem Lett Rev 11(4):492–502. https://doi.org/10.1080/17518253

Nambiar S et al (2018) Synthesis of curcumin – functionalized gold nanoparticles and cytotoxicity studies in human prostate cancer cell line. Appl Nanosci 8(3):347–357. https://doi.org/10.1007/s13204-018-0728-6

Nasiri J, Motamedi E, Naghavi MR, Ghafoori M (2019) Removal of crystal violet from water using B-cyclodextrin functionalized biogenic zero-valent iron nanoadsorbents synthesized via aqueous root extracts of Ferula persica. J Hazard Mater 367:325–338

Nazar N et al (2018) Cu nanoparticles synthesis using biological molecule of *P. Granatum* seeds extract as reducing and capping agent: growth mechanism and photo-catalytic activity. Int J Biol Macromol 106:1203–1210. https://doi.org/10.1016/j.ijbiomac.2017.08.126

Newman DJ, Cragg GM (2016) Natural products as sources of new drugs from 1981 to 2014. J Nat Prod 79(3):629–661

Ocsoy I et al (2017) A green approach for formation of silver nanoparticles on magnetic graphene oxide and highly effective antimicrobial activity and reusability. J Mol Liq 227:147–152. https://doi.org/10.1016/j.molliq.2016.12.015

Ocsoy I, Tasdemir D, Mazicioglu S, Tan W (2018) Nanotechnology in plants. Adv Biochem Eng Biotechnol 164:263–275. https://doi.org/10.1007/10_2017_53

Odei-Addo F et al (2017) Nanoformulation of Leonotis leonurus to improve its bioavailability as a potential antidiabetic drug. 3 Biotech 7(5):344

Oh KH et al (2018) Biosynthesized gold and silver nanoparticles by aqueous fruit extract of *Chaenomeles sinensis* and screening of their biomedical activities. Artif Cells Nanomed Biotechnol 46(3):599–606

Ouda Sahar M (2014) Some nanoparticles effects on *Proteus Sp.* and *KLebsiella* Sp. isolated from water. Am J Infect Dis Microbiol 2(1):4–10

Ovais M (2017) Sageretia thea (Osbeck.) mediated synthesis of zinc oxide nanoparticles and its biological applications. Nanomedicine 12(15):1767–1789

Pansambal S et al (2017) Phytosynthesis and biological activities of fluorescent CuO nanoparticles using *Acanthospermum hispidum* L. extract. J Nanostruct 7(3):165–174

Parveen K, Banse V, Ledwani L (2016) Green synthesis of nanoparticles: their advantages and disadvantages. In: AIP conference proceedings 1724

Periasamy VS, Athinarayanan J, Alshatwi AA (2016) Anticancer activity of an ultrasonic nanoemulsion formulation of *Nigella sativa* L. essential oil on human breast cancer cells. Ultrason Sonochem 31:449–455. https://doi.org/10.1016/j.ultsonch.2016.01.035

Pires PR et al (2016) Systemic treatment with resveratrol and/or curcumin reduces the progression of experimental periodontitis in rats. J Periodont Res 55(11). https://doi.org/10.1111/jre.12382

Pirtarighat S, Ghannadnia M, Baghshahi S (2019) Green synthesis of silver nanoparticles using the plant extract of *Salvia spinosa* grown *in vitro* and their antibacterial activity assessment. J Nanostruct Chem 9(1):1–9. https://doi.org/10.1007/s40097-018-0291-4

Preshaw PM (2015) Detection and diagnosis of periodontal conditions amenable to prevention. BMC Oral Health 15(Suppl 1):S5

Rasheed T, Bilal M, Iqbal HMN, Li C (2017) Green biosynthesis of silver nanoparticles using leaves extract of *Artemisia vulgaris* and their potential biomedical applications. Colloids Surf B: Biointerfaces 158:408–415. https://doi.org/10.1016/j.colsurfb.2017.07.020

Riaz M et al (2018) Biogenic synthesis of AgNPs with *Saussurea lappa* C.B. Clarke and studies on their biochemical properties. J Nanosci Nanotechnol 18(12):8392–8398

Rijo P, Ascensão L, Roberto A, Sofia A (2016) Bioproduction of gold nanoparticles for photothermal therapy. Ther Deliv 7:287–304

Sadegh S et al (2019) Neuroprotective potential of curcumin- loaded nanostructured lipid carrier in an animal model of Alzheimer' s disease : behavioral and biochemical evidence. J Alzheimers Dis 69:671–686

Sahni G, Panwar A, Kaur B (2015) Controlled green synthesis of silver nanoparticles by Allium cepa and *Musa acuminata* with strong antimicrobial activity. Int Nano Lett 5:93–100. https://doi.org/10.1007/s40089-015-0142-y

Saleem S, Ahmed B, Saghir M, Al-shaeri M (2017) Inhibition of growth and bio film formation of clinical bacterial isolates by NiO nanoparticles synthesized from *Eucalyptus globulus* plants. Microb Pathog 111:375–387. https://doi.org/10.1016/j.micpath.2017.09.019

Santhoshkumar T et al (2014) Green synthesis of titanium dioxide nanoparticles using *Psidium guajava* extract and its antibacterial and antioxidant properties. Asian Pac J Trop Med 7(12):968–976. https://doi.org/10.1016/S1995-7645(14)60171-1

Saran M, Vyas S, Mathur M, Bagaria A (2018) Green synthesis and characterisation of CuNPs : insights into their potential bioactivity. IET Nanobiotechnol 12(3):357–364

Saravanan A, Kumar PS, Karthiga Devi G, Arumugam T (2016) Synthesis and characterization of metallic nanoparticles impregnated onto activated carbon using leaf extract of *Mukia maderasapatna*: evaluation of antimicrobial activities. Microb Pathog 97:198–203

Sathishkumar P et al (2016) Anti-acne, anti-dandruff and anti-breast cancer efficacy of green synthesised silver nanoparticles using *Coriandrum sativum* leaf extract. J Photochem Photobiol B Biol 163:69–76. https://doi.org/10.1016/j.jphotobiol.2016.08.005

Sathishkumar G et al (2018) Green synthesis of magnetic Fe 3 O 4 nanoparticles using *Couroupita guianensis* Aubl. fruit extract for their antibacterial and cytotoxicity activities. Artif Cells Nanomed Biotechnol 46(3):589–598

Sathya K, Saravanathamizhan R, Baskar G (2018) Ultrasonic assisted green synthesis of Fe and Fe/Zn bimetallic nanoparticles for *in vitro* cytotoxicity study against HeLa cancer cell line. Mol Biol Rep 45(5):1397–1404

Selvam K, Sudhakar C, Govarthanan M (2017) Eco-friendly biosynthesis and characterization of silver nanoparticles using *Tinospora cordifolia* (Thunb.) miers and evaluate its antibacterial, antioxidant potential. J Radiat Res Appl Sci 10(1):6–12. https://doi.org/10.1016/j.jrras.2016.02.005

Suman TY et al (2016) GC-MS analysis of bioactive components and biosynthesis of silver nanoparticles using *Hybanthus enneaspermus* at room temperature evaluation of their stability and its larvicidal activity. Environ Sci Pollut Res 23(3):2705–2714

Suman TY et al (2018) Toxicity of biogenic gold nanoparticles fabricated by *Hybanthus enneaspermus* aqueous extract against *Anopheles stephensi* and *Culex tritaeniorhynchus*. Res J Biotechnol 13(9):26–34

Suresh D, Nethravathi PC, Rajanaika H (2015) Green synthesis of multifunctional zinc oxide (ZnO) nanoparticles using *Cassia fistula* plant extract and their photodegradative, antioxidant and antibacterial activities. Mater Sci Semicond Process 31:446–454. https://doi.org/10.1016/j.mssp.2014.12.023

Surface, OF, Modified Zinc, Oxide Nanoparticles, and Human Pathogenic Bacteria (2018) Exploiting *in vitro* potential and characterization of surface modified zinc oxide nanoparticles of isodon rugosus extract: their clinical potential towards hepg2 cell line and human pathogenic bacteria. EXCLI J 17:671–687

Sushma NJ, Swathi DPG, Deva TMB, Raju P (2016) Facile approach to synthesize magnesium oxide nanoparticles by using *Clitoria ternatea* – characterization and *in vitro* antioxidant studies. Appl Nanosci 6(3):437–444

Thatoi P et al (2016) Photo-mediated green synthesis of silver and zinc oxide nanoparticles using aqueous extracts of two mangrove plant species, *Heritiera fomes* and *Sonneratia apetala* and investigation of their biomedica. J Photochem Photobiol B 163:311–318. https://doi.org/10.1016/j.jphotobiol.2016.07.029

Tostado-plascencia MM, Sanchez-tizapa M, Zamudio-ojeda A (2018) Synthesis and characterization of multiwalled carbon nanotubes functionalized with chlorophyll-derivatives compounds extracted from *Hibiscus tiliaceus*. Diam Relat Mater 89:151–162. https://doi.org/10.1016/j.diamond.2018.09.004

Tripathi N, Pavelyev V, Islam SS (2017) Synthesis of carbon nanotubes using green plant extract as catalyst: unconventional concept and its realization. Appl Nanosci 7(8):557–566

Vijayakumar S et al (2018) Green synthesis of zinc oxide nanoparticles using *Atalantia monophylla* leaf extracts : characterization and antimicrobial analysis. Mater Sci Semicond Process 82:39–45. https://doi.org/10.1016/j.mssp.2018.03.017

Vijayaraghavan K, Ashokkumar T (2017) Plant-mediated biosynthesis of metallic nanoparticles: a review of literature, factors affecting synthesis, characterization techniques and applications. J Environ Chem Eng 5(5):4866–4883. https://doi.org/10.1016/j.jece.2017.09.026

Yadi M et al (2018) Current developments in green synthesis of metallic nanoparticles using plant extracts: a review. Artif Cells Nanomed Biotechnol 46(Suppl 3):S336–S343

Yallappa S et al (2015) Phytosynthesis of gold nanoparticles using *Mappia foetida* leaves extract and their conjugation with folic acid for delivery of doxorubicin to cancer cells. J Mater Sci Mater Med 26(9):1–12

Yallapu MM, Jaggi M, Chauhan SC (2013) Curcumin nanomedicine: a road to cancer therapeutics. Curr Pharm Des 19(11):1994–2010

Youse I, Pakravan M, Rahimi H, Bahador A (2017) An investigation of *Electrospun henna* leaves extract-loaded chitosan based nano fibrous mats for skin tissue engineering. Mater Sci Eng C Mater Biol Appl 75:433–444

Yugandhar P, Vasavi T, Uma P, Devi M (2017) Bioinspired green synthesis of copper oxide nanoparticles from *Syzygium alternifolium* (Wt.) Walp : characterization and evaluation of its synergistic antimicrobial and anticancer activity. Appl Nanosci 7(7):417–427

Zambrano LMG et al (2018) Local administration of curcumin- loaded nanoparticles effectively inhibits inflammation and bone resorption associated with experimental periodontal disease. Sci Rep 8(1):1–11

Zhao X, Wang J, Song Y, Chen X (2018) Synthesis of nanomedicines by nanohybrids conjugating ginsenosides with auto-targeting and enhanced MRI contrast for liver cancer therapy. Drug Dev Ind Pharm 44(8):1307–1316

Zhou Y, Tang RC (2018) Facile and eco-friendly fabrication of AgNPs coated silk for antibacterial and antioxidant textiles using honeysuckle extract. J Photochem Photobiol B Biol 178:463–471

Zhou TE et al (2013) Curcumin inhibits in Fl ammatory response and bone loss during experimental periodontitis in rats. Acta Odontol Scand 71(2):349–356

Zorzi GK, Carvalho ELS, Von Poser GL, Teixeira HF (2015) On the use of nanotechnology-based strategies for association of complex matrices from plant extracts. Braz J Pharm 25(4):426–436. https://doi.org/10.1016/j.bjp.2015.07.015

Chapter 3
Nanocarriers as Tools for Delivery of Nature Derived Compounds and Extracts with Therapeutic Activity

Raju Saka and Naveen Chella

Abstract Natural medicine found its use over a long period of time. These were effectively used in treating various ailments since ancient times. Initially simple extracts were used for treatment followed by isolated compounds. Natural compounds formed integral part of drug discovery programmes for most of the time. However, with the introduction of combinatorial chemistry the use of natural compounds reduced gradually, nevertheless, natural compounds were still under extensive evaluation and newer products were being approved. Many natural compounds have potential to treat various life threatening diseases for which synthetic medicine doesn't have specific cure. However, natural compounds have limitations owing to their poor solubility, stability and permeability. This led to their reduced usage in clinical stage.

Nanocarriers due to their unique properties, offers successful delivery of various synthetic drugs by improving/modifying the properties such as solubility, pharmacokinetics and specificity. These advantages increased the interest towards development of various nanocarriers for the effective delivery of natural molecules. This chapter explores the various delivery options of different bioactive natural molecules and extracts using different nano carriers made of polymers, lipids and inorganic materials. Various natural molecules used in different disease conditions such as Alzheimer's, diabetes, microbial infections, inflammatory condition and psoriasis were selected and their limitation in reaching clinical stage was discussed. The role of nanocarriers in overcoming these limitations and improving their clinical promise was highlighted with relevant case studies or examples.

Keywords Nanotechnology · Antibiotics · Flavonoids · Plant extracts · Nanoparticles · Liposomes

---

R. Saka · N. Chella (✉)
Department of Pharmaceutics, National Institute of Pharmaceutical Education & Research (NIPER), Hyderabad, India
e-mail: naveen.niperhyd@gov.in

A. Saneja et al. (eds.), *Sustainable Agriculture Reviews 44*, Sustainable Agriculture Reviews 44, https://doi.org/10.1007/978-3-030-41842-7_3

Abbreviations

AUC Area under the curve
EGCG Epigallocatechin-3-gallate
NLC Nanostructured lipid carrier
PLGA Poly lactic-co-glycolic

3.1 Introduction

Nature has provided mankind with many wonderful gifts, one of which is wide variety of natural molecules with medicinal value. Natural products were used over a long time since the beginning of mankind. Various ancient literatures have mentioned the efficient use of natural medicine in the treatment of various ailments. Natural products were the principal source of medicine to humans before the advent of western/synthetic medicine. These ranged from plant extracts, fungi to inorganic chemicals. The use of natural products by man dates back to 2900 BC. Various literary works like Ebers Papyrus, Chinese Materia Medica, Shennong, and the Tang herbal and Charaka Samhita have compiled about the use of natural molecules from different sources as medicines to treat various ailments (Ghalioungui 1987; Patwardhan et al. 2005; Yuan et al. 2016). Before the synthetic era, majority of the available medical preparations are of natural origin. Initially, the majority of natural medicines were plant derived. With the diversification of scientific exploration, new sources like microbes and marine organisms were also identified which gave highly potent compounds to medical sciences.

Natural medicine dominated the world healthcare system for a very long time. When western medicine was introduced more emphasis was put upon the synthetic chemistry. Due to this, interest in natural medicine decreased gradually. The integration of natural products with chemistry was initiated in the early 1800s when Serturner discovered and isolated morphine from opium poppy (*Papaver somniferum*) (Klockgether-Radke 2002). This was followed by emetine, colchicine, atropine, quinine, and aspirin which dominated until early 1900s. The major breakthrough in natural product development was when Sir Alexander Fleming isolated the first antibiotic “Penicillin” from fungus *Penicillium notatum* which was termed as one of the biggest achievements in modern medicine (Dias et al. 2012).

Interest in natural products continued as newer and more potent antibiotics were identified and isolated. Eventually natural medicine diversified into other fields like anti-inflammation, anaesthesia and oncology. One of the first effective antineoplastic agents was nature derived. Paclitaxel, one of the widely used anti-cancer agents was first isolated from the bark of *Taxus brevifolia* (Cragg 1998). This was followed by other drugs like doxorubicin, asparagine, etoposide, teneposide, vincristine and vinblastine. These dominated the healthcare system for longer period (Shoeb 2006).

Advances in chemistry have shifted the interest towards synthetic drugs, as these were easy to synthesize in large amounts enough to carry out a full-fledged clinical trial which was not feasible with natural products. The introduction of high-throughput screening and combinatorial chemistry where, thousands of molecules can be evaluated in a short period has led to decreased interest in natural products (Broach and Thorner 1996; Von Nussbaum et al. 2006). Though this trend is being continued, still a significant proportion of newly approved molecules are of natural origin (Table 3.1). Apart from being directly used as a drug, natural compounds are also effectively used as a template to synthesize new lead molecules with higher efficacy. This approach led to the launch of numerous drug compounds that are derived from components of a natural product (Harvey 2008). A recent example is Eribulin mesylate from marine sponges which has been approved by the United States Food and Drug Administration for soft tissue sarcomas (Huyck et al. 2011).

The major advantages of natural products include (Siddiqui et al. 2014):

- Wider structural diversity compared to synthetic chemical libraries
- Safer than synthetic alternatives
- Bioactive by nature
- Obey Lipinski's "Rule of Five"
- Have better absorption than complex synthetic molecules
- Has long history of usage with wider public acceptance
- Can be used as scaffolds for the development of more effective semi-synthetic and synthetic molecules

Though the natural products have various advantages their use has been drastically reduced over time due to the following reasons (Siddiqui et al. 2014):

- Crude extracts obtained from natural sources are not compatible with high-throughput screening methods of lead identification
- The structure of natural compounds will be highly complex

Table 3.1 Representative examples of recently approved natural products along with their therapeutic indications

Year	Name	Therapeutic indication	Source	Company
2010	Fingolimod	Multiple sclerosis	*Mycelia sterilia*	Novartis Int. AG
2012	Ingenol mebutate	Actinic keratosis	*Euphorbia peplus*	LEO Pharma A/S
2011	Spinosad	Human head lice	*Saccharopolyspora spinosa*	ParaPRO, LLC
2013	Canagliflozin	Type-2 diabetes	Apple bark	Janssen Pharmaceuticals, Inc.
2016	Eribulin mesylate	Breast cancer and liposarcoma	*Halichondria* species	Eisai Co.
2018	Cannabidiol	Dravet syndrome and Lennox–Gastaut syndrome	*Cannabis* species	GW Pharmaceuticals

- Difficulty in isolating a pure compound from a mixture and the high cost involved in the extraction procedure
- Large scale supply is an issue if the isolated molecule is a hit
- Patenting issues of natural products
- Diversion of resources to combinatorial chemistry and high-throughput screening

3.2 Problems Associated with Natural Compounds

3.2.1 Solubility

Solubility is the intrinsic property and most important physicochemical property of the drug that plays critical role at an early stage of drug development. It is mostly dependent on the structure and physical nature of the drug (Yalkowsky 1999). Poor solubility will always be a problem during the early stage of development as it is essential for the drug to be soluble in sufficient quantities for absorption. The majority of complex natural compounds have poor water solubility (Bilia et al. 2017) which is a major hurdle for their development into clinical candidates. Conventional approaches like pH adjustment, co-solvency, and surfactant solubilization don't improve the intended outcomes as there is a possibility for precipitation after dilution. Hence nanotechnology can be applied to improve the solubility of the natural compounds. The major advantage is that, there will be no effect of dilution on the nanoformulation. Nanoformulations used to improve the solubility of various natural products were discussed in later sections of the chapter.

3.2.2 Permeability

Permeability is the second most important property after solubility, as the molecule need to cross the biological barriers to show efficacy. The permeability of the compound depends on physicochemical properties of the molecule like, molecular weight, terminal functional groups and partition coefficient (Log P) (Hansch and Clayton 1973). Many natural compounds have unfavourable Log P values that hinders their partition into the biological membrane followed by absorption (Watkins et al. 2015). This issue can be solved using permeation enhancers. Some drugs despite of having sufficient Log P values also show less absorption as they are expelled by efflux transporters present on the membrane (Chan et al. 2004). Permeation enhancers not always improve the permeation especially, if the drug is absorbed through carrier-mediated transport. Instead, carrier-mediated transport is dependent on transporter saturation. In such cases, permeation enhancers will have little effect. To overcome such problems various nanocarriers have been developed.

Nanocarriers prevent efflux and poor permeability by getting engulfed into the biological membrane thereby minimizing the transporter effect (Hillaireau and Couvreur 2009). Various nanocarriers have been discussed for separate natural molecules to improve their transport through biological membranes in later sections.

3.2.3 Stability

Chemical stability is one of the parameters of a compound which is usually neglected at initial stages of drug development. The stability of a compound is an important parameter as it governs the later developmental stages (Group 2003). Degraded compounds either do not show efficacy or produce toxic effects, both of which are undesirable at the clinical stage. Most of the drugs degrade either due to pH change, oxidation, hydrolysis, reduction, and light under *in vitro* and due to acidic or alkaline environment and enzymes *in vivo* (Blessy et al. 2014; Crowley and Martini 2001). Light-induced degradation can be avoided by using a light-protective container closure system. However, the prevention of degradation from acidic/basic conditions of biological environment is very difficult as the direct pH change of the external media may pose severe ramifications to the delivered drug molecules. In such cases, encapsulation of drug molecule protects the drug from the external environment and also improves cellular transport of drug (Sharma and Garg 2010). Apart from chemical stability, metabolic stability also plays a vital role in compound bioavailability (Masimirembwa et al. 2003). Many natural compounds (E.g. Artemisinin) to reach systemic circulation in sufficient concentrations to provide a therapeutic effect due to extensive metabolism by gut or hepatic enzymes (Efferth et al. 2016). Nanoencapsulation protects the compound from metabolic degradation as the free drug is not available for metabolism thus improving bioavailability. These approaches are explained in later sections of the chapter.

3.3 Nanotechnology in Medicine

Nanotechnology is defined as the science of design and production of nanoscale (1–100 nm) structures or devices that offer unparalleled advantages in terms of physical, chemical and material properties (Boisseau and Loubaton 2011; Freitas Jr 2005). Since these materials have similar dimensions to that of biological molecules, these can be used for medical applications. This is of special importance in cancer therapy where the effective treatment is based on targeting the disease at a molecular level which is possible only with nanoscale materials. Nanomaterials have unique structural properties that are different from macro-level materials in enormous magnitudes. These structural properties can be tailored and exploited for various medical applications such as diagnosis, imaging and treatment of various diseases (Logothetidis 2006; Surendiran et al. 2009).

Nanotechnology has been researched since long time, but only in the 1990's these gained interest of pharmaceutical industry with the approval of first nanomedicine based product by United States Food and Drug Administration (Barenholz 2012). This was followed by the approval of various nanotechnology-based products for multiple applications (Bajwa et al. 2017). Unlike macro systems, the nanoparticle has a high surface to volume ratio which is the most requisite characteristic property that paves way for higher drug loading, surface functionality tuning which is useful for targeted drug delivery (Merisko-Liversidge and Liversidge 2008). First generation nanomedicine works by passive targeting to the diseased site based on enhanced permeation and retention effect, where nanocarriers infiltrate the disease site through narrow fenestration of blood capillaries at the affected site (Iyer et al. 2006). Apart from this mechanism nanocarriers also infiltrate tough biological barriers with ease because of their lower particle size and unique surface properties. Newer generation nanocarriers make use of the higher surface to volume ratio thereby tuning their surface with targeting agents (ligands) and triggered release polymers so that the drug will be released accurately at the target site in response to specific receptors or physiological or pathological condition (Mout et al. 2012).

Over the time, various nanocarriers have been developed for multiple applications in medicine. The nanotechnology field started with metal nanoparticles followed by micelles, polymeric nanoparticles, and liposomes that dominated for a longer period. Recently, newer systems were introduced such as lipoplexes/polyplexes, quantum dots, nanofibres, dendrimers and carbon nanotubes. Various nanocarrier based products both approved and currently in development are presented in Table 3.2 (Caster et al. 2017).

Nonetheless, nanocarrier systems have various advantages like improved solubility, permeability due to smaller size, increased stability due to encapsulation, improved drug delivery and reduced toxicity due to targeted delivery by hindering the drug exposure to healthy tissues. There are disadvantages such as toxicity due to non-degradable materials used in the synthesis, not suitable for less potent drugs where the dose required increases and are costlier compared to the conventional dosage forms.

3.4 Development of Nanocarrier Systems

Formulation scientists consider various parameters while developing a formulation for preclinical and clinical studies. Most emphasis will be upon solubility rather than permeability and stability. Conventional approaches are prioritized initially rather than using nanotechnological approaches. Various authors described the formulation development of new hits for preclinical testing. They explained the sequential development of formulation using a conventional and nanotechnological approach (Fig. 3.1). Usually, the novel formulation development starts with nanonization approaches like nanocrystals and nanosuspension formation. Once the intended solubility values have been attained the formulation will be carried

Table 3.2 Representative examples of recently approved nanoformulations

Brand name	Therapeutic indication	Drug	Year/ Clinical trial phase	Formulation	Company
Krystexxa®	Multiple sclerosis	Pegloticase	2010	Polymer-protein conjugate	Horizon
Marqibo®	Acute Lymphoblastic Leukemia	Vincristine	2012	Liposomes	Onco TCS
Invega® Sustenna®	Schizophrenia Schizoaffective Disorder	Paliperidone Palmitate	2014	Nanocrystals	Janssen Pharmaceuticals, Inc.
Adynovate®	Hemophilia	PEGylated factor VIII	2015	Protein conjugate	Baxalta
Plegridy®	relapsing remitting multiple sclerosis (RRMS)	PEG-IFN-β-1a	2014	Drug-polymer conjugate	Biogen
Zilretta®	Knee osteoarthritis	Triamcinolone acetonide	2017	poly lactic-co-glycolic acid (PLGA) nanoparticles hydrogel	Flexion Therapeutics
ThermoDox®	Hepatobiliary Tumors	Doxorubicin	III	Liposomes	Celsion Corporation
Nanoplatin®	Advanced solid tumour malignancies	Cisplatin	II/III	Polymeric micelle	NanoCarrier Co., Ltd.
Promitil®	Advanced solid tumour malignancies	Mitomycin-C	I	Liposomes	Lipomedix
Paclical®	Gynaecological Malignancies	Paclitaxel	III	Polymeric micelle	Oasmia Pharmaceutical AB

forward for pharmacokinetic/pharmacodynamic studies. If the formulation approach fails, the compound is further developed into micro- and nanoemulsions and tested for pharmacokinetic and pharmacodynamic parameters. If this fails, it will be further developed into nano solid dispersions, polymeric nanoparticles, solid lipid nanoparticles (SLN) and nanostructured lipid carriers (NLC). Liposomes are considered as last resort formulations in preclinical development and thus used when all the above approaches fail to improve solubility or meet the desired specifications (Bittner and Mountfield 2002; Maas et al. 2007).

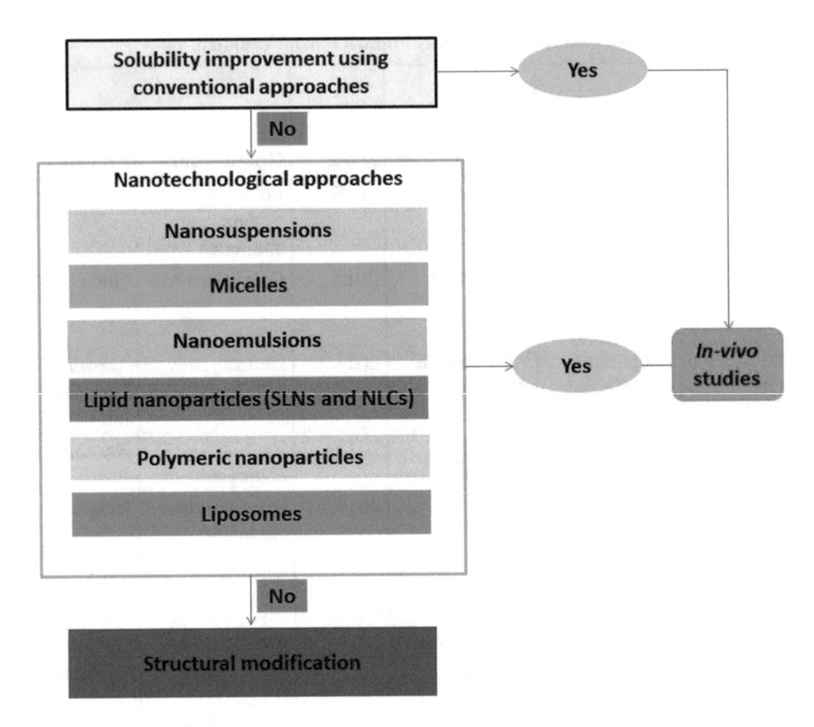

Fig. 3.1 Sequential development of nanoformulations for solubility improvement of new chemical entity

3.5 Nanotechnology in Delivery of Natural Products

Nanotechnology has greatly improved the delivery of complex synthetic molecules. Though the natural substances are safer than their synthetic counterparts they suffer from drawbacks owing to poor physicochemical properties. Nanotechnology is believed to address these issues and can effectively deliver the natural compounds to the target area. Unlike synthetic molecules where the diversity of nanomaterials is very high, natural products are mostly nano-processed into lipidic and polymeric systems. Various natural products have been effectively delivered to their target site by the use of nanocarrier systems (Devi et al. 2010; Gopi et al. 2016; Watkins et al. 2015). Since there are various compounds currently being investigated, the nanotechnological applications for various natural compounds are individually described in the following sections (Table 3.3).

Table 3.3 Various nanocarriers exploited for natural products delivery and their advantages

Nanocarrier	Structural description	Advantages
Nanocrystals	Nanosized drug crystals produced from particle size reduction approaches with sizes below 1 μm (Keck and Müller 2006)	Improved solubility Better dissolution profile Good safety as it is devoid of any excipients Can be suitable for high dose administration
Polymeric nanoparticles	Mesoscale particles comprised of polymer matrix or capsules loaded with drugs (Soppimath et al. 2001)	Suitable for various routes of administration Higher cell uptake Tunable surface
Solid lipid nanoparticles and nanostructured lipid carriers	Solid lipid nanoparticles are nanoparticles comprising solid lipid matrix while Nanostructured lipid carriers are nanoparticles comprised of irregularly distributed solid and liquid lipids. (Pardeike et al. 2009)	High scalability Improved stability of sensitive drugs High elasticity of the particles High drug loading
Magnetic nanoparticles	Inorganic magnetic core coated with various functional agents (McBain et al. 2008)	Highly tunable surface Has diagnostic and therapeutic purpose Gene delivery
Liposomes	Bilayerd lipid vesicles (Allen and Cullis 2013)	Improved stability of sensitive molecules Encapsulation any type of compound Effective for gene delivery More market presence
Micelles	Spherical supramolecular assemblies formed from amphiphilic copolymer molecules consisting hydrophobic core (Sutton et al. 2007)	Can be used for water insoluble drugs Protects drug from degradation First choice of nanoformulations for systemic delivery
Mesoporous silica nanoparticles	Solid mesoporous nanosized silica particles with large surface area for entrapment (Slowing et al. 2008)	Can be simultaneously used for diagnosis and treatment Bio-sensing Good stability
Dendrimers	Synthetic branched tree shaped macromolecules with monodispersed 3D structure (Gillies and Frechet 2005)	Can entrap multiple compounds Definitive tunable structure Uniform particle morphology Highly tunable terminal functional groups

(continued)

Table 3.3 (continued)

Nanocarrier	Structural description	Advantages
Carbon nanotubes	Needle like carbon based hollow carriers with various geometrical shapes and patterns (Bianco et al. 2005)	Can be used as implantable systems for longer duration of time Can be used as scaffolds to improve tissue growth apart from drug delivery High mechanical strength
Quantum dots	Nanosized semiconductor crystals with inherent fluorescent emission control (Qi and Gao 2008)	Low particle size Improved cellular uptake High tunability
Niosomes	Non-ionic surfactant vesicles (Uchegbu and Vyas 1998)	Can target reticulo-endothelial system Low cost excipients High drug loading

Farjadian et al. (2018)

3.5.1 Resveratrol

Resveratrol (3,4,5′-trihydroxystillbene) is a stillbenoid extracted and isolated from the peel of grapevine (*Vitis vinifera*) (Frémont 2000). It exists in both *cis* and *trans* forms of which later has more biological activity. The compound has both cosmetic and therapeutic applications like acne *vulgaris*, psoriasis, neurological disorders, wound healing, skin cancer, and diabetes (Baur and Sinclair 2006). Various research publications have proven that the compound has a huge potential to be a viable clinical candidate for these diseases. Despite these favourable factors resveratrol suffers from many shortcomings such as poor aqueous solubility, poor chemical and photo stability, short biological half-life and rapid clearance from the body that results in low oral bioavailability (Delmas et al. 2011; Francioso et al. 2014).

Various approaches have been investigated to improve the properties of resveratrol of which, nanotechnology was found to be the most promising one. Improvement in solubility, stability and bioavailability of resveratrol was reported by many scientists after encapsulating it into different nanocarriers (Amri et al. 2012; Santos et al. 2011). Ansari et al., prepared resveratrol nanosponges with the average particle size range of 400–500 nm by cross linking cyclodextrin for improved solubility, stability and permeation. The prepared nanosponges showed enhancement in dissolution due to improved solubility and wetting and increased cytotoxicity on HCPC-I cells due to improved solubility. These nanosponges also showed increased permeation through pig skin and higher accumulation (twofolds) in rabbit mucosa compared to the plain drug. Furthermore, the drug is in encapsulated in the inner cavity of the cyclodextrin molecule prevents photo degradation (Ansari et al. 2011). The authors concluded that, prepared resveratrol nanosponge has good potential for use in buccal and topical delivery. In another study, resveratrol was formulated into nano complex with soy protein isolate. The resultant nano complex showed twofolds improvement in solubility compared to plain drug and also retained the activity of resveratrol *in-vitro* (Pujara et al. 2017). Singh and Pai formulated polymeric nanoparticles of resveratrol with size less than 100 nm and encapsulation efficiencies reaching 90%. These polymeric nanoparticles can be easily internalized into cells thereby improving the permeability of encapsulated resveratrol. Moreover, prepared nanoparticles enhanced the stability of drug by protecting it from degradation in the external environment (Singh and Pai 2014). Lindner et al., evaluated the protective effect of resveratrol loaded poly-lactic acid nanoparticles in Parkinson's disease. The administered nanocarriers showed greater protective effects compared to pure drug due to improved targeting and stability (da Rocha Lindner et al. 2015).

Apart from these formulations, other formulations were also explored for resveratrol. Kobierski et al., studied effect of various surfactants (Tween and Poloxamer) and their concentrations (1% and 2%) on the stability of resveratrol (5%) loaded nanosuspension for dermal application. The obtained nanosuspensions have a particle size in the range of 150–220 nm. Nanosuspensions with 1% stabilizer concentrations had better short term (30 days) stability compared with 2% stabilizers. The authors also reported the stability was due to stearic effect as they observed less zeta potential in case of suspensions (Kobierski et al. 2009). In another study by Zhang et al., nano solid dispersions of resveratrol were prepared by spray drying method. The nano solid dispersion showed high stability in sunlight than the raw resveratrol (Zhang et al. 2013c). Gokce Eh et al., compared both SLN and NLC for anti-oxidative efficiency and permeation of resveratrol. Both the formulations showed similar activity but, nanostructured lipid carriers showed better skin penetration owing to their low particle size, reduced electrical charge and flexible fluidic structure (Gokce et al. 2012). In another study by Loureiro et al., resveratrol solid lipid nanoparticles were administered to evaluate its anti-Alzheimer's activity. Three different solid lipid nanoparticles were prepared such as drug loaded solid lipid nanoparticles, solid lipid nanoparticles coated with unspecific antibody and solid lipid nanoparticles coated with specific OX26 antibody. Functionalized SLNs with OX26 showed highest cellular uptake and improved brain internalization of the drug as it was internalized using active targeting by OX26 antibody. Furthermore, the solid lipid nanoparticles prevented Aβ fibrillation thus proving efficient anti-Alzheimer's effect (Loureiro et al. 2017).

Liposomes are bi-layered lipid vesicles capable of encapsulating or entrapping both lipophilic and hydrophilic molecules. These improve the delivery of drugs by fusing with the cell membrane and thereby unloading the active cargo into the target site. Various liposomal formulations of resveratrol were formulated to improve the delivery. It was observed that the resveratrol entrapment was high in cationic liposomes when compared with the neutral liposomes as reported by Bonechi et al., (Bonechi et al. 2012). The drug loaded liposomes have improved skin penetration of the drug. Since the encapsulated drug can be protected from the external environment, encapsulation of resveratrol in liposomes has protected drug from UV radiation, metabolism, and biological degradation (Caddeo et al. 2008). Resveratrol liposomes were also used in treating diabetic retinopathy through its anti-oxidative effect through which free radical scavenging is done. Other nano formulations that were evaluated for effective delivery of resveratrol are nanoemulsions, niosomes, ethosomes, and transferosomes (Pangeni et al. 2014).

Though resveratrol was regarded as a potential blockbuster candidate and garnered interest among the scientific community, seldom is the translation to the clinical stage. Most of the clinical studies are related to protective effect and also included plain drug. Currently, there is no clinical trial involving resveratrol nanocarriers. Still, numerous clinical studies need to be done to establish the beneficial effects of resveratrol.

3.5.2 Artemisinin

Artemisinin is an anti-malarial sesquiterpene lactone isolated from Chinese medicinal herb *Artemisia annua*. It was first discovered by Youyou Tu in 1972 (Mabberley 2008; You-you et al. 1982). It has proven efficacy in treating multidrug-resistant malaria. World Health Organization (WHO) has included artemisinin as one of the essential medicine in tackling global malaria. The drug indicated for drug-resistant *Plasmodium falciparum* infections. Apart from malaria the drug and its derivatives have activity against other infections caused by *Leishmania* species, *Trypanosoma* species, cytomegalovirus, and Hepatitis B and C viruses (Krishna et al. 2008). Recent studies also indicated it also has anti-cancer activity. Despite its multipolar activity, the drug has several drawbacks that limit its capacity as clinical candidate. The drug has poor aqueous solubility which is a limiting factor for its bioavailability. Shorter half-life and rapid metabolism also contribute to the low bioavailability of the drug. Anti-infective treatments require high plasma concentrations of active compounds which translate to high bioavailability. Hence there is a need to improve the bioavailability of artemisinin (Efferth et al. 2016). Nanotechnology may serve as an effective platform for solving the problems associated with the molecule and enhancing its clinical potency.

Surfactant micelles offer improvement in solubility; these are one of the first products to be marketed based on nanotechnology. Artemisinin was formulated into micelles using sodium dodecyl sulfate in different molar ranges which showed significant improvement (25 to 50- folds) in aqueous solubility of artemisinin (Lapenna et al. 2009). Kakran et al. formulated artemisinin nanosuspension by evaporative precipitation using polyethylene glycol as stabiliser. The resultant composite formulation showed a linear increase in artemisinin solubility with an increase in stabiliser content. The dissolution rate also increased when compared with pure artemisinin powder (Kakran et al. 2010). In another study, conventional and PEGylated artemisinin liposomes were formulated and compared with the free drug for pharmacokinetic profiling. It was found that formulating the drug into liposomes increased the residence time of the drug in mice. The area under the curve increased by six times when administered as liposomes (Isacchi et al. 2011). Even artemisinin was formulated into solid lipid nanoparticles and nanostructured lipid carriers where the residence time of drug was drastically increased with increased anti-malarial efficacy in in-vivo (Dwivedi et al. 2014; Joshi et al. 2008). In one study, the drug was loaded into poly(lactic-co-glycolic acid) nanoparticles and administered into the BALB/c mice model of visceral *Leishmania*. The treatment reduced the splenomegaly and overall parasite load in the liver (Want et al. 2015). Albumin nanoparticles of artemisinin were prepared and administered to humanized mice infected with P. *falciparum*. Results showed the superiority of nanoparticles in reducing microbial load and improved safety (Ibrahim et al. 2015). Other formulations that are evaluated include nanotubes (Zhang et al. 2015b), fullerenes (Zhang et al. 2015a), and dendrimers (Fröhlich et al. 2018). Though nano formulations of artemisinin have many advantages over the administration of pure drug or conventional delivery systems, clinical translation is required to make it an effective therapeutic agent.

3.5.3 Thymoquinone

Thymoquinone is a benzoquinone derivative obtained from the essential oil of *Nigella sativa*. The compound was found to have multifarious pharmacological indications like anti-cancer, anti-oxidative, free radical scavenging, anti-inflammatory, anti-hyperlipidemic, anti-microbial and anti-psoriatic activity. It is also successfully used for hepatic and gastro-protective purposes (Gholamnezhad et al. 2016). Preclinical studies reported, a high safety profile of thymoquinone even at high doses of 90 mg/kg (Badary et al. 1998). The above reasons prove that thymoquinone can be a potent alternative for various ailments and diseases. Despite having several merits, thymoquinone has yet to find its clinical use. The major reasons for thymoquinone not being aggressively pursued by industry are its hydrophobicity, poor aqueous solubility, and poor permeability. Moreover, the drug is unstable in presence of water, light and high temperatures (Salmani et al. 2014). Formulating thymoquinone into nanocarriers can protect the drug from degradation and improve its properties to be useful for clinical application.

Thymoquinone was formulated into various nanocarriers for the improvement of its physicochemical properties and stability. Tubesha et al. prepared thymoquinone nanoemulsions by high-pressure homogenization method. The obtained nanoemulsion was stable up to 6 months with intact drug in place (Tubesha et al. 2013). Alam et al., prepared chitosan nanoparticles of thymoquinone and administered to rats intranasally to evaluate brain bioavailability. The concentration of thymoquinone was more in the brain when compared to plasma. Area under the curve has been improved 2 and 20-folds when compared with intranasal solution and intravenous solution respectively. Enhanced absorption was due to interaction between positively charged terminal amino groups of chitosan with negatively charged cell membrane which improved transit through tight junctions (Alam et al. 2012b). Singh et al. evaluated the stability of thymoquinone from solid lipid nanoparticles under accelerated conditions. The prepared solid lipid nanoparticles showed improved stability, enhanced area under the curve (approximately fivefolds) more hepatoprotective effect over the free drug (Singh et al. 2013).

Nallamuthu et al. prepared thymoquinone poly (lactic-co-glycolic acid) nanoparticles and evaluated for anti-oxidative and anti-microbial effects. The prepared nanoparticles showed improved 2,2-diphenyl-1-picrylhydrazyl free radical scavenging, improved anti-microbial effect against *Escherichia coli, Salmonella typhi* and *Staphylococcus aureus* (Nallamuthu et al. 2013). Apart from oral and parenteral delivery, the efficiency of thymoquinone nanocarriers was also explored in topical application. Kausar et al., prepared thymoquinone loaded ethosomes for treating skin acne. It was observed that, ethosomal formulation showed higher skin permeability, anti-inflammatory and anti-acne efficiency compared to the free drug and similar activity like that of marketed formulations (Kausar et al. 2019). Jain et al., formulated stable thymoquinone lipospheres and evaluated for anti-psoriatic efficacy in Imiquimod induced psoriasis mice model. Animals treated with lipospheres showed reduced psoriatic lesions, improved skin physiology and reduced inflammatory mediator's level by 1.4 to 2-folds thus proving as an efficient alternative in psoriasis management (Jain et al. 2017).

Rushmi et al., prepared thymoquinone liposomes using egg lecithin and cholesterol. The obtained liposomes had high entrapment efficiency. When administered to mice the liposomal formulation showed improved analgesic activity compared with free drug (Rushmi et al. 2017). Though it has multiple advantages still the compound didn't see clinical approval. The shortcomings include not translating the preclinical success to the clinical stage for both conventional and nanoformulations. More emphasis on clinical testing can improve the position of the drug in the scientific community.

3.5.4 Rapamycin

Rapamycin is a macrolide compound isolated from bacteria *Streptomyces hygroscopicus*. It is widely used for immunosuppression to prevent organ rejection after transplantation, immune disorders and as a coating on stents to prevent hyperplasia (Li et al. 2014). Rapamycin was approved for use in organ transplant rejection by the United States Food and Drug Administration in 1999. The drug was also approved for coronary artery disease as a drug-coated coronary stent (Khan et al. 2013; Virmani et al. 2004). The drug has disadvantages like poor aqueous solubility and low bioavailability (Simamora et al. 2001; Zimmerman et al. 1999). Moreover, side effects may occur if the drug is distributed to non-target organs. Hence there is a need for developing a localized, site specific delivery system that may reduce these problems.

In another study by Haddadi et al., rapamycin loaded polylactide-co-glycolide nanoparticles were evaluated for enhanced immunosuppressive effect in dendritic cells. It was observed that the nanoparticle formulation showed better inhibition of dendritic cell maturation than that of free drug due to improved cell uptake. The authors concluded that, enhanced rapamycin activity was due to improved permeability of drug from nanoformulations compared to pure drug (Haddadi et al. 2008). Chitosan and poly-lactic acid nanoparticles of rapamycin were used to facilitate corneal transplantation in rabbit model by Yaun et al. It was found that, the nanocarrier group has higher graft survivability than that of plain drug suspension thus improving the outcome (Yuan et al. 2008). Localized delivery was also evaluated for rapamycin where rapamycin was loaded into bioadhesive carbopol/ polylactide-co-glycolide nanoparticles and evaluated for bio adhesion. The results showed improved drug transfer in *ex-vivo* models due to enhanced bioadhesion of nanocarrier systems (Zou et al. 2009b).

3.5.5 Naringenin

Naringenin, a flavonoid obtained from various citrus fruits is a part of traditional Chinese medicine and is widely used for multiple indications including, anti-inflammatory, anti-cancer, anti-atherogenic, anti-fibrogenic, and antioxidant actions (Erlund 2004). Despite multiple indications, poor water solubility and poor bioavailability hampered its development into the clinical stage (Felgines et al. 2000; Zhang et al. 2013b). Efforts have been put to increase the water solubility through various conventional approaches but with little use. Nanotechnology has been proved to be the best alternative to improve the properties of hydrophobic drugs like naringenin. These nanoformulations were also able to control the release of drugs, can improve stability and targeting.

Kumar and Abraham prepared poly vinyl pyrrolidone coated naringenin nanoparticles to evaluate its protective effect on the liver, kidney, and heart of rats. The nanoparticle treated group showed protective effects compared to the control group which is evident from reduced inflammatory markers and positive histopathological findings. Consequently, the authors suggested that, naringenin loaded nanocarriers can reduce the overall dose required for action and thereby improving the therapeutic potential of the drug (Kumar and Abraham 2016). Yen et al., developed naringenin loaded Eudragit® E nanoparticles by the nanoprecipitation method. The obtained nanoparticles had a particle size of 66.2 ± 0.38 nm and PDI 0.29 ± 0.04. Nanoparticles showed enhanced drug release properties owing to improved solubility of the loaded drug molecule. These nanoparticles were also tested for their hepatoprotective effect in carbon tetrachloride induced acute liver failure in rats. Results showed, the orally administered nanoparticle treated group has better liver function index and lipid peroxidation protection which is evident by inhibition of activation of caspase signalling pathway responsible for liver damage (Yen et al. 2009). Khan et al., prepared self-nanoemulsifying drug delivery system of naringenin to improve the bioavailability. Emulsion droplets with particle sizes as low as 38 nm were obtained that showed improved release (100% in 2 h compared with 15% in case of plain drug). Pharmacokinetic data showed 2.82-folds increase in AUC of naringenin from the self-nanoemulsifying drug delivery system compared to that of a plain drug (Khan et al. 2015).

Ji et al., prepared naringenin loaded solid lipid nanoparticles for pulmonary delivery. The resultant formulation was stable for a longer period. Pulmonary administration of solid lipid nanoparticles improved the drug bioavailability by 2.53-folds compared with that of plain drug administered in the form of suspension (Ji et al. 2016). Apart from protective effects, the naringenin nanoformulations were also explored for their anti-diabetic effects. Maity et al., administered naringenin loaded core-shell chitosan nanoparticles coated with alginate orally to diabetic rats. Results showed higher hypoglycemic effect in nanoformulation treated group compared to the group received pure drug. Blood parameters and histopathological observations concluded that the formulation is non-toxic thereby, can be used as a better alternative for diabetes treatment (Maity et al. 2017). Topical application of nanoformulations also investigated for naringenin which showed significant improvement in permeation and efficacy. Tsai et al. developed stable naringenin nanoemulsions for topical application and tested for permeation and safety. The nanoemulsion formulation was found to be more effective as it deposited more amount of compound in the skin than plain drug solution. Skin toxicity studies of nanoemulsions showed low skin irritation proving that nanoformulations are safer alternatives compared to conventional formulations (Tsai et al. 2015b). The same authors have evaluated the effect of elastic liposomes on naringenin topical delivery. The elastic liposomes were topically administered and evaluated for skin permeation. The drug deposition in the skin was improved by 7.3~11 fold from elastic liposomes compared with drug solution thus improving the topical delivery (Tsai et al. 2015a). Thus nanoformulations improve the outcome of naringenin treatment. With much emphasis on the clinical studies, it can be possible that the compound may get regulatory approval for various applications.

3.5.6 Silymarin and Silibinin

Silymarin and silibinin are natural flavonoids isolated from milk thistle *Silybum marianum*. Silymarin exists as a mixture of three isomers one of which is silibinin. Silibinin was found to be most potent of all silymarin isomers. These compounds are widely used for hepatoprotection, where there is functional impairment of liver. It is also used as antioxidant, anti-viral, anti-inflammatory and as an antidote for mushroom poisoning (*Amanita phalloides*). The compound was also found to be hepatoprotective when intoxicated with phalloidin, galactosamine, and thioacetamide (Valenzuela and Garrido 1994). Consider having high therapeutic potential also, the compound has not found its clinical use as it suffers from low water solubility and low bioavailability. Nanoformulations of silibinin and silymarin were evaluated to improve protective effects.

Lutsenko et al., formulated silibinin liposomes and evaluated them for hepatoprotective effect in carbon tetrachloride induced hepatotoxicity in mice. Systemic alanine aminotransferase and aspartate aminotransferase levels were calculated after treatment with silibinin liposomes. It was found that the liposomal formulation reduced alanine aminotransferase and aspartate aminotransferase activity by 2.2 and 1.8-folds respectively compared with free drug. The intravenous liposomal formulation thus showed a higher hepatoprotective effect compared to oral administration of the same formulation (Lutsenko et al. 2018). Zhang et al., also prepared stable solid lipid nanoparticles (SLN) where the controlled release of silibinin was observed from SLN (Zhang et al. 2007). In an attempt to improve the anti-fibrotic activity of silibinin, Ying et al., formulated silibinin SLN and administered both orally and intravenously to rats with liver fibrosis. It was observed that the SLN have markedly increased the anti-fibrotic activity of silibinin when administered by both the routes with intravenous route exerting the highest activity compared with the oral suspension group (Yingchao et al. 2007). Song et al., prepared stable nanoemulsions of silibinin and evaluated for pharmacokinetics in rabbits. Intramuscularly injected nanoemulsion showed controlled release and improved bioavailability over intravenous injection of drug solution (Song et al. 2005). Marchiori et al., prepared silibinin and pomegranate oil nanoemulsions loaded hydrogel and evaluated for anti-inflammatory activity on ultraviolet B radiation-induced skin damage in mice. The positive effect of formulation lasted for 48 h with a reduction in ear edema, leukocyte infiltration and increased the anti-edematogenic effect. The nano formulation showed a similar therapeutic benefit as available standard synthetic treatment of 1% silver sulfadiazine (Marchiori et al. 2017). Shetty et al., loaded silibinin and epigallocatechin-3-gallate into peptide dendrimers and evaluated for enhanced skin permeation and deposition. *In-vitro* skin permeation studies showed higher cumulative drug deposition in the skin from peptide dendrimers with improved protective effect on the skin indicating the usefulness of nanocarriers in enhancing the therapeutic efficacy of silibinin. Apart from the protective effect, silibinin was also explored for anti-viral activity, especially against Hepatitis C infections (Shetty et al. 2017). Ripoli et al. prepared silibinin phytoliposomes for improved activity against the Hepatitis C virus. These liposomes were tested against human hepatocellular carcinoma cells transfected with Hepatitis C virus DNA. It was observed that, silibinin absorption into the cells was increased by 2.4-folds with 300-folds improvement in pharmacological activity thus showing the effectiveness of nanocarrier system as a suitable delivery platform of silibinin in Hepatitis C treatment (Ripoli et al. 2016). Extensive clinical studies need to be done to evaluate further the effects of silibinin nanocarriers in humans thus, providing a platform for clinical approval.

3.5.7 Cyclosporin

Cyclosporin is an undecapeptide isolated from fungus *Tolypocladium inflatum* and approved for medical use in 1983. It acts by decreasing lymphocyte function thereby exerting immunosuppressant activity. The substance is widely used to prevent graft vs. host disease during organ transplantation, rheumatoid arthritis, psoriasis, Crohn's disease, and keratoconjunctivitis sicca (Borel et al. 1994). Despite great therapeutic benefit, the drugs poor aqueous solubility, poor permeability and resulting lower bioavailability hindered its clinical effectiveness (Ismailos et al. 1991; Wu et al. 1995). The drug also has severe toxic effects on sudden exposure like, compromising the immune system, high blood pressure, kidney problems, and preterm birth. The marketed formulation Sandimmune® is an oily solution and other product Sandimmune® Neoral is a microemulsion formulation. The oily solution shows high inter-subject variability in bioavailability. Though microemulsion formulation reduced the variation it produced nephrotoxicity due to the presence of Cremophor EL® (Portman et al. 2000). To overcome these difficulties novel formulations are being considered for cyclosporin.

Muller et al., compared oral bioavailability profile of cyclosporin from solid lipid nanoparticles (SLN), nanocrystals and marketed microemulsion formulation Sandimmune® Neoral. Nanocrystals showed less bioavailability compared to microemulsion whereas, SLN showed approximately similar pharmacokinetic profile as that of Sandimmune® Neoral formulation with less variability and reduced toxicity. The slower release of encapsulated drug from SLNs that prevented sudden increase in plasma concentration as observed from microemulsion was the reason for reduced side effects. They concluded that SLN may improve uniformity between subjects and prevent unwanted effects due to immediate exposure of the drug (Müller et al. 2006). Dai et al., have prepared pH-sensitive cyclosporin loaded poly (methacrylic acid-co-methyl methacrylate) copolymer nanoparticles by quasi-emulsion solvent diffusion technique. Various polymer grades were used such as Eudragit® E100, L100, S100 and L100–55. All formulations showed pH-dependent drug release whereas, the marketed product Sandimmune® Neoral showed pH independent drug release. *In-vivo* pharmacokinetic studies revealed that all the formulations except that of E100 have increased bioavailability compared with marketed microemulsion formulation (Dai et al. 2004).

In another study by Ankola et al., cyclosporin loaded polylactide-co-glycolide nanoparticles were administered to rats over 30 days and compared pharmacokinetic profile with marketed microemulsion formulation. When administered as nanoparticles cyclosporin relative bioavailability was improved from 31 to 89%. The nanoparticle formulation has also showed less nephrotoxicity compared to marketed formulation proving that, it is an effective and safer alternative (Ankola et al. 2011). Cyclosporin nanoformulations were also evaluated for ocular delivery. Gokce et al., prepared cyclosporin solid lipid nanoparticles using Compritol 888 ATO, Poloxamer 188 and Tween 80 by high shear homogenization followed by ultrasonication. *In-vitro* cellular uptake was studied in rabbit corneal epithelial cells and *ex-vivo* uptake was studied in pig cornea. Results showed, improvement in penetration of cyclosporine from SLN in cornea thus improving the ocular delivery (Gokce et al. 2008). Battaglia et al., prepared cyclosporin solid lipid nanoparticles with different surface charges and evaluated for ocular toxicity and distribution. No toxicity was observed with any of the prepared formulations. Similarly, positively charged SLN showed higher corneal penetration compared with neutral and negatively charged particles. It was assumed that the interaction between positively charged SLN surface and negatively charged corneal surface increases the residence of carrier system on the cornel surface thus improving drug release (Battaglia et al. 2012). Basaran et al., also showed enhanced residence time and penetration of drug from SLN in sheep eyes compared to pure drug solution thus improving ocular delivery (Başaran et al. 2010).

Vadiei et al., formulated cyclosporin liposomes and administered intravenously to rats and compared the safety with the marketed formulation. The liposomal group showed low nephrotoxicity compared with marketed formulation thus improving the safety of the drug (Vadiei et al. 1989). Waranuch et al., prepared cyclosporin loaded neutral liposomes and studied for skin distribution. Results showed, improved skin permeability through pilosebaceous pathway (Waranuch et al. 1998). Egbaria et al., prepared four different liposomal formulations of cyclosporin and compared their skin penetration with emulsion based formulation in both rodent and human skin. It was observed that, all the liposomal formulations were superior in improving skin penetration than simple emulsion formulation (Egbaria et al. 1990). Puigdemont et al., prepared chitosan nanocapsules and topically administered in dogs to treat atopic dermatitis. It was shown that, the nanoformulation has reduced dermatitis in 87.5% subjects compared to the placebo (28.6%) indicating the role of nanocarriers in improving the efficacy of loaded cyclosporine (Puigdemont et al. 2013).

Walunj et al., prepared cyclosporin cationic liposomes and tested their anti-psoriatic efficiency in BALB/c mice. It was proved from the study that, the liposomes reduced psoriatic symptoms and also reduced levels of various inflammatory mediators in mice (Walunj et al. 2019). Musa et al., prepared cyclosporin nanoemulsion and tested on healthy human volunteers to check the hydration potential. The nanoemulsion formulation improved the skin hydration thus facilitating improved skin penetration. This can be useful in dry conditions such as psoriasis (Musa et al. 2017). Kumar et al., investigated the anti-psoriatic effect of cyclosporin liposomal gel on human volunteers with chronic plaque psoriasis. It was observed that, the liposomal formulation reduced the dermatological sum score which is a proportional indicator of dermatological damage severity in comparison to plain drug, where there is no significant improvement compared with the control group (Kumar et al. 2016).

3.5.8 Berberine

Berberine is a quaternary benzylisoquinoline alkaloid obtained from various medicinal plants especially from root and bark of Barberry (*Berberis vulgaris*). Over the years the compound was found to have many therapeutic benefits such as anti-inflammatory, antioxidant, anti-microbial, anti-hyperlipidemic and anxiolytic activities. Moreover, the drug was found to exert a protective effect on the heart, liver, kidneys and nervous system. Various studies have also indicated potential anti-Alzheimer, anti-Parkinson, and anti-diabetic activity (Amritpal et al. 2010; Imanshahidi and Hosseinzadeh 2008). This made berberine a compound of interest to the research community. Though it has potential, the road to clinical approval has been barricaded by poor solubility and poor bioavailability. This was further complicated, as it is a substrate for P-gp efflux pump. New approaches need to be explored to improve the delivery of berberine for which nanotechnology can prove to be a better alternative.

One of the approaches to improve the berberine properties was particle size reduction to nanometre range. Sahibzada et al., prepared nanosuspension of berberine and evaluated for solubility and *in-vitro* drug release. The resultant nanosuspension showed twofolds increase in solubility and more than fourfold increase in dissolution (Sahibzada et al. 2018). Xie et al. evaluated the renal protective effect of berberine nanoparticles where, they found that formulating berberine into nanoparticles prevented oxidative stress and apoptosis in rat renal injury (Xie et al. 2017). Zou et al., prepared nanohydroxyapatite/chitosan nanoparticles loaded with berberine to inhibit the microorganism, *Staphylococcus aureus*. Results showed that the nanoformulation could effectively inhibit microbial growth (Zou et al. 2009a). In another study, Lin et al., prepared fucose, chitosan/heparin nanoparticle loaded with berberine to treat *Helicobacter pylori* infection. These nanoparticles released berberine in a controlled manner and allowed accumulation of excess drug concentrations at gastric epithelium thus inhibiting microbial infection (Lin et al. 2015). Zhou et al., evaluated berberine loaded chitosan nanoparticles in osteoarthritis. Results proved that the retention time of berberine as nanoformulation was high in synovial fluid than that of free berberine. This increased residence time directly reduced the apoptosis thus improving treatment outcome (Zhou et al. 2015).

Mehra et al., prepared berberine loaded sodium alginate nanoparticles to investigate its anti-microbial effect. The nanoformulation was found to have more anti-microbial effect than plain drug solution (Mehra et al. 2016). In another study, Xu et al. entrapped berberine nanosuspension in hydrogel-grafted fabrics. This grafted fabric effectively reduced the infection at the infected wound site (Xu et al. 2014). Xue et al., loaded berberine into solid lipid nanoparticles and evaluated them for pharmacokinetics and pharmacodynamics. It was found that, berberine oral bioavailability increased when administered as SLN. Data analysis showed significant brain localization also paving the way for brain targeting. Similarly, the SLN formulation was able to produce a higher anti-diabetic effect than that of free drug (Xue et al. 2013). Gupta et al. conjugated berberine with poly(amidoamine) G4 dendrimer and investigated pharmacokinetic parameters in comparison with free berberine. It was observed that the conjugated dendrimers maintained concentrations of the drug over a long period compared with free berberine (Gupta et al. 2017). Despite the huge advantage with nanotechnology, special focus is needed on the toxicity assessment so that the formulation may reach the market with lesser difficulty.

3.5.9 Quercetin

Quercetin is a natural flavonoid having diverse distribution in various plants. It was discovered as a part of the French paradox. The compound has multiple applications like antioxidant, anti-allergic, antiplatelet, anti-inflammatory, anti-microbial, neuro-protective and hepatoprotective effects (Formica and Regelson 1995). Though, the compound has numerous beneficial effects, the compound suffers from poor solubility and permeability. Hence nanotechnology can be used as a platform to improve the properties of quercetin.

Most of the studies targeted at improving bioavailability while few focussed on dissolution improvement. Kakran et al., prepared nano sized quercetin suspensions, solid dispersions and inclusion complexes for improved dissolution. All the nano-formulations showed improved dissolution compared to the raw drug (Kakran et al. 2011). The same authors prepared quercetin nanoparticles by anti-solvent precipitation method which improved the drug dissolution compared to the plain drug (Kakran et al. 2012). Sahoo et al., prepared quercetin nanocrystals by high-pressure homogenization and evaluated for dissolution and antioxidant activity. It was observed that, the nanocrystals showed a marked increase in dissolution rate and antioxidant activity than the free drug (Sahoo et al. 2011). Wu et al., prepared quercetin nanoparticles using different carriers like Eudragit® E100 and polyvinyl alcohol and characterized for drug release and antioxidant efficiency. Results proved the ability of nanoparticles to improve drug release as the release of drug from the nanoparticle was 74 times higher than the free drug. Also, the nanoparticles showed increased free radical scavenging activity in comparison to plain drug (Wu et al. 2008). Antcnio et al., prepared bovine serum albumin nanoparticles loaded with quercetin. The nanoparticles have a mean particle size of 130 nm and retained the free radical scavenging activity even after 96 h compared with the plain drug (Antçnio et al. 2016).

Nanoformulations of quercetin were also prepared by synthetic polymers. For example, Kumari et al., prepared quercetin loaded poly-lactic acid nanoparticles by solvent evaporation method to improve solubility and stability. The nanoparticles showed biphasic drug release with initial burst release followed by controlled release for a longer period. It also showed improved anti-oxidant activity compared with plain drug (Kumari et al. 2010). Li et al., prepared quercetin solid lipid nanoparticles (SLN) with an average diameter of 155.3 nm. The formulations were orally dosed to rats at a dose of 50 mg/kg. Pharmacokinetic data showed that, the relative bioavailability of SLN to plain drug was 571.4% thus proving nanoformulations as effective carriers for quercetin delivery (Li et al. 2009).

Tan et al. prepared lecithin-chitosan nanoparticles of quercetin for topical delivery. The *in-vivo* topical delivery study revealed that, the nanoparticles showed significant enhancement in penetration of quercetin by disrupting the stratum corneum layer (Tan et al. 2011). Quercetin nanoparticles were also studied for treating Alzheimer's disease by improved brain delivery. Dhawan et al. prepared quercetin SLN for application in Alzheimer's disease. Particles obtained have less than

200 nm size and zeta potential of +21.5 mV. Behavioural studies in rats showed better memory retention in the group treated with SLN than the free drug suspension group indicating improved brain delivery of quercetin using nanocarriers (Dhawan et al. 2011). Sun et al., prepared poly(lactic-co-glycolic acid) functionalized quercetin nanoparticles and tested for anti-Alzheimer's activity. It was found that the quercetin nanoparticles have low cytotoxicity towards SH-SY5Y cells. Moreover, these nanoparticles markedly decreased the neurotoxic effects of the Zn^{2+}-Aβ42 system and increased neuronal cell viability. Behavioural tests in APP/PS1 mice showed improved cognition and memory compared with free drug. Thus it can be assumed that, the increase in activity of quercetin can be attributed to increased brain penetration by nanocarriers (Sun et al. 2016). Further clinical studies need to be done to establish the effectiveness of quercetin in treating various disorders thus making it a potential clinical candidate.

3.5.10 Epigallocatechin-3-gallate

Epigallocatechin-3-gallate is a principle green tea polyphenol with a wide variety of activities. It was found to have antioxidant, anti-inflammatory, anti-cancer, anti-neurodegenerative and protective effects. With these activities the compound is widely used in treating skin damage, Alzheimer's disease, wound healing, anti-aging, amyotrophic lateral sclerosis and reducing organ toxicity (Jung and Ellis 2001; Lu et al. 2003; Mandel et al. 2004). The compound has good water solubility but suffers from drawbacks like poor stability in gastric fluids, poor absorption, rapid metabolism and clearance which reduced its clinical potential (Takagaki and Nanjo 2009; Zhu et al. 1997). Nonetheless, new approaches were considered for improving the biopharmaceutical properties of Epigallocatechin-3-gallate of which, nanotechnology is widely considered to be enabling enhanced biopharmaceutical properties.

Initial studies were focussed on improving stability and gastric permeability. Zhang et al., prepared EGCG nanostructured lipid carriers and evaluated for stability improvement. It was found that, prepared nanostructured lipid carriers showed enhancement in stability of Epigallocatechin-3-gallate to both acidic and basic environment. Moreover, it was found that the Epigallocatechin-3-gallate deposition in THP-1-derived macrophages was high with nanostructured lipid carriers than the unencapsulated drug. Results showed that, the nanostructured lipid carriers formulation was able to reduce atherosclerotic lesions more precisely than the free drug (Zhang et al. 2013a). The same authors evaluated Epigallocatechin-3-gallate nanoparticles in the mice model for atherosclerosis. It was found that, the nanoformulation was able to produce more effectiveness in atherosclerotic mice than the free drug (Zhang et al. 2015c). Dube et al., prepared stable Epigallocatechin-3-gallate loaded chitosan nanoparticles for improved permeation. The ex-vivo jejunal perfusion studies proved that, the chitosan nanoparticles improved the drug transfer into the jejunum. Similarly, in another study by the same authors, the

Epigallocatechin-3-gallate nanoparticles were administered into mice to determine pharmacokinetic parameters. It was found that, the area under the curve improved by 2.3-folds when administered as nanoparticles thus improving oral delivery (Dube et al. 2010; Dube et al. 2011).

Avadhani et al., have prepared Epigallocatechin-3-gallate and hyaluronic acid transferosomes and evaluated for anti-aging and antioxidant effects on damaged skin induced by UV irradiation. It was found that, the nanoformulation was able to improve the viability and free radical scavenging in skin cells. Also, the nanoformulation enabled increased deposition of Epigallocatechin-3-gallate in the skin (Avadhani et al. 2017). Fang et al., prepared liposomes for various catechins and tested them for *in-vitro* and *in-vivo* permeation. It was observed that Epigallocatechin-3-gallate loaded liposomes showed more skin deposition than plain counterparts. Also, the formulations were found to be stable and had less skin disruptive activity thus reducing toxicity (Fang et al. 2006). Similarly, Shetty et al., have prepared peptide dendrimers of Epigallocatechin-3-gallate and found that the dendrimers formulation improved Epigallocatechin-3-gallate deposition into the skin compared with free drug (Shetty et al. 2017). Gharib et al. prepared Epigallocatechin-3-gallate liposome with three different surface charges (cationic, anionic and neutral). These were tested against skin wound infection by methicillin-resistant *Staphylococcus aureus*. It was found that cationic Epigallocatechin-3-gallate liposomes showed high killing efficiency compared with plain drug and other charged liposomes (Gharib et al. 2013).

Brain delivery of Epigallocatechin-3-gallate is also widely researched for the treatment of neurodegenerative diseases. Zhang et al., synthesized Epigallocatechin-3-gallate stabilized selenium nanoparticles coated with Tet-1 peptide to treat Alzheimer's disease. It was observed that nanoparticles inhibited Aβ-fibrillation compared to free drug. Also, the Tet-1 peptide has improved the cellular internalization of nanoparticles in PC-12 cells paving a way for Alzheimer's treatment (Zhang et al. 2014). Smith et al., prepared nanolipidic particles of Epigallocatechin-3-gallate and evaluated for improvement in oral bioavailability. The nanoparticles improved the oral bioavailability by twofolds compared with free drug. *In-vitro* cell line study on N2a cells showed that the nanoparticles improved α-secretase induction to 91% more than that induced by the free drug (Smith et al. 2010). Kaur et al., prepared Epigallocatechin-3-gallate solid lipid nanoparticles and evaluated for neuroprotective effect in cerebral ischemia-induced memory impairment. It was found that, the SLN treatment reduced memory impairment (Kaur et al. 2019). In another study by Italia et al., Epigallocatechin-3-gallate nanoparticles reduced cyclosporin-induced nephrotoxicity while the free drug didn't show any protective effect. The combination approach was also used to improve its anti-Alzheimer's activity (Italia et al. 2008). Cano et al., prepared Epigallocatechin-3-gallate /Ascorbic acid nanoparticles and evaluated its anti-Alzheimer's activity in APPswe/PS1dE9 Alzheimer's disease mice model. The nanoparticles reduced neuroinflammation and Aβ burden in mice. This protective effect improved spatial learning and memory in diseased mice thus proving to be efficient in treating Alzheimer's (Cano et al. 2019). The above examples prove that

Epigallocatechin-3-gallate is a potential clinical candidate for various conditions. However, extensive clinical investigations are required to establish uniform safety and efficacy of Epigallocatechin-3-gallate nanocarriers for therapeutic approval.

3.5.11 Natural Antibiotics

Antibiotics have a long history of use in treating various life-threatening infections. These hold a major stake of infectious disease management. In time newer antibiotics from new sources like marine organisms have been introduced. However, the majority of legacy antibiotics lost their effectiveness due to the development of microbial resistance (Davies and Davies 2010; Taylor and Webster 2011). Antibiotic resistance became a major global healthcare issue as the majority of the existing antibiotics were ineffective against these resistant microbial strains. Few microbes are extremely resistant even to potent antibiotics like methicillin where there are only a few alternatives left to tackle infections thus causing major concern for World Health Organisation. It was observed that microorganisms had developed various mechanisms by which they attained antibiotic resistance (Pelgrift and Friedman 2013). These include, decreasing uptake and increasing efflux of drug, inactivation by covalent modification, engineering false substrates, biofilm formation and swarming.

Developing new molecules is a tedious process to tackle antibiotic resistance. New approaches such as nanotechnology have to be considered so that the microbial resistance can be overcome (Seil and Webster 2012). Amphotericin B is an anti-fungal antibiotic isolated from *Streptomyces nodosus*. It was also approved to treat leishmaniasis, aspergillosis, blastomysosis, candidiasis, and cryptococcosis (Ellis 2002). However, the compound has poor solubility and severe side effects due to rapid exposure. Various clinical studies suggested that immediate exposure of the drug caused lethal side effects like high fever, chills, dyspnea, anorexia, hypotension, cardiac arrhythmias, and weakness. It is also associated with multiple organ damage especially kidney and liver damage (Cohen 1998; Laniado-Laborín and Cabrales-Vargas 2009). Owing to its poor solubility and permeability the drug has poor oral bioavailability. Though the salt formation into deoxycholate improved solubility the side effects were not reduced. Novel approaches have been investigated to improve pharmacokinetic parameters and to reduce toxic side effects. One of the first formulations widely considered was liposomes. Amphotericin B, when administered in the form of liposomes, showed improved safety profile and increased the circulating time. The formulation found regulatory approval in the late '90s as Ambisome®. Other products followed suit like Abelcet® (lipid complex) and Fungisome® (liposome) (Hiemenz and Walsh 1996) (Dupont 2002). Though nanoformulations are available in the market a considerable amount of research is being carried out regarding nanocarrier mediated delivery of Amphotericin B.

De Carvalho et al., prepared deoxycholate Amphotericin B nanoparticles using poly(lactic-co-glycolic acid). The new nanocarrier has been evaluated for its efficacy in cutaneous leishmaniasis. It was found that the nanoparticles have more anti-parasitic activity compared to the plain drug. Moreover, the dosing frequency was also reduced for nanoparticle group to attain therapeutic concentrations (de Carvalho et al. 2013). Mehrizi et al., prepared Amphotericin B loaded chitosan nanoparticles and evaluated for efficacy and safety. It was found that the formulation increased activity by 80% compared with untreated animals. Safety studies showed no sign of toxicity indicating that the formulation is a safer alternative. Moreover, the efficacy of the nanoparticle group was found to be similar to that of marketed liposomal formulation with improved safety profile thus making it a viable delivery platform (Mehrizi et al. 2018).

Polymyxin B is a non-ribosomal peptide isolated from *Paenibacillus polymyxa*. The natural drug is approved for use in treating various gram-negative bacterial infections (Storm et al. 1977). The drug was found to be less active against gram-positive bacteria (Hsu Chen and Feingold 1973). The drug has higher water solubility but no oral absorption as it degrades in gastric conditions. Similarly, the microbes also developed resistance to the drug (Fernández et al. 2010). These made it difficult to achieve therapeutic benefit. Chauhan et al. have prepared polymyxin B niosomes by thin film hydration method. The resultant niosomes had a particle size of less than 200 nm. These were tested for anti-fungal activity, pharmacokinetics and toxicity studies. The niosomal formulation showed improved anti-fungal activity. Moreover, the niosomes protected the drug from degradation in the gastric environment. Similarly, the drug showed enhanced oral absorption with complete absorption taking place within 90 min. Toxicity studies showed the formulation to be tolerable (Chauhan and Bhatt 2019). Park et al., produced nanocomplexes of polymyxin B with gold and cadmium telluride nanoparticles and investigated for antimicrobial activity. It was found that the nanoformulations have similar activity like a free drug but the cadmium telluride complexed polymyxin B nanoparticles showed improved toxicity towards *E. Coli* (Park et al. 2011). Severino et al., prepared polymyxin B solid lipid nanoparticles (SLN) crosslinked to sodium alginate and evaluated for improved anti-microbial activity. The SLN were tested against HaCat and NIH/3T3 cell lines and showed no toxicity on these cell lines. Similarly, antimicrobial evaluation against *Pseudomonas aeruginosa* showed enhanced minimum inhibitory concentration (MIC) than the free drug (Severino et al. 2015). Insua et al., prepared polyion complexes of polymyxin B with poly(styrene sulphonate) and tested for anti-microbial activity against *Pseudomonas aeruginosa*. Results showed improved activity in the magnitude of reducing colonies by 10,000 times more compared with previously developed formulations (Insua et al. 2017).

Vancomycin is a glycopeptide antibiotic obtained from bacterium *Amycolatopsis orientalis*. It was approved for use in severe infections caused by methicillin resistant *Staphylococcus aureus* strains. It is also used in the treatment of severe *Clostridium difficile* colitis in the mouth (Levine 2006). The drug has problems like poor oral bioavailability. Recent studies suggested that the microbes have acquired resistance towards vancomycin (Smith et al. 1999). Bacterium developed resistance to vancomycin by altering terminal cell surface peptides thus preventing the binding of vancomycin. Various approaches have been used to improve the outcome of vancomycin treatment. Milani et al., prepared vancomycin loaded poly(lactic-co-glycolic acid) nanoparticles and evaluated them for improving intestinal permeability by single-pass intestinal perfusion technique (Zakeri-Milani et al. 2013). It was found that the nanoparticles improved the perfusion of the drug through intestine compared with a solution. Cerchiara et al., prepared chitosan nanoparticles of vancomycin for colon targeting and improved anti-microbial activity. *In-vitro* drug release studies revealed colonic pH-selective release. Anti-microbial studies revealed that the inhibitive activity increased by 3 fold (Cerchiara et al. 2015). Gu et al., prepared vancomycin gold nanoparticles and tested for activity against vancomycin-resistant bacterial strains. The nanoparticles lowered the minimum inhibitory concentration of drug using polyvalent inhibition (Gu et al. 2003).

Tobramycin is a newer aminoglycoside antibiotic obtained from *Streptomyces tenebrarius*. It is approved for combating infections caused by *Pseudomonas* species. It is not absorbed from epithelial barriers thus doesn't have any bioavailability through extravascular or topical route. It is available as an intravenous injection (Neu 1976). Various approaches have been used to improve drug permeability through different membranes. Cavalla et al., have prepared solid lipid nanoparticles loaded with tobramycin to improve the corneal permeation. It was found that the tobramycin content in aqueous humor in solid lipid nanoparticles treated rabbits was high when compared with the simple solution (Cavalli et al. 2002). In a similar study by the same authors' tobramycin loaded solid lipid nanoparticles were administered duodenally to improve drug permeation. From the transmission electron microscopy of blood and lymph and pharmacokinetic evaluation, it was confirmed that the nanoparticles improved the oral permeation by translocation through lymphatic drainage (Cavalli et al. 2003). Another problem encountered by tobramycin is, resistance by the biofilms (Anwar et al. 1989) to the drug in conditions like pulmonary fibrosis where the crosslinked biofilm network prevents localization of lethal concentrations of the drug thus reducing effectiveness. To prevent these issues nanotechnology was adopted by scientists. In a study by Deacon et al., tobramycin loaded polymeric nanoparticles were prepared to improve the drug transfer through the mucous layer in cystic fibrosis. These nanoformulations improved transfer through the mucous layer thus improving efficacy. In another study by Messiaen et al., the tobramycin liposomes were tested on *Burkholderia cepacia* complex biofilms that have intrinsic resistance to antibiotics. Three different liposomal (cationic, anionic and neutral) formulations were tested against biofilm. It was found that the anionic liposomes loaded with tobramycin overcame resistance and improved its anti-biofilm property compared with free drug (Messiaen et al. 2013). Still, more relevant and extensive studies have to be done to overcome such resistance.

3.5.12 Plant Extracts

Long before isolated drugs were introduced for therapeutic applications, natural plant extracts were the only available remedies to treat various diseases. Sometimes where it is difficult to isolate the active ingredient, the plant extract is directly administered for treatment. Similarly, some components of the extract may potentiate the activity of the active component and hence they cannot be ruled out from the formulation. In such cases, the extract shall be formulated and administered rather than the active ingredient alone. However due to problems like poor solubility, poor permeability, low stability and rapid systemic clearance these extracts may not provide intended effects thus reducing treatment effectiveness (Liu and Feng 2015). In such cases nanotechnology may be useful to overcome these drawbacks.

One of the first approaches to improve the solubility and bioavailability is nano suspension. *Cuscuta chinensis* is a Chinese herb widely used for its anti-oxidative and hepatoprotective effect (Yen et al. 2007). Though rich in flavonoids and lignins, the extract has poor solubility and permeability thus limiting its use. Yen et al. prepared nanoparticles by nano suspension method. The administration of nanosized extract reduced the required hepatoprotective dose by 5 times (Yen et al. 2008). Similarly, *Radix salvia* which has great cardioprotective effects was formulated into nanoparticles by spray drying. This size reduction approach has improved bioavailability compared to unmodified crude extract (Su et al. 2008). *Ginkgo biloba,* a Chinese herb has many beneficial effects like anti-inflammatory, cardioprotective, anti-hypertensive and neuroprotective effects (Kleijnen and Knipschild 1992). However, the extract is poorly soluble in water and has less oral bioavailability which limits its wide applicability. Wang et al., prepared nanoparticles of ethanolic extract of the plant by emulsion solvent evaporation followed by freeze drying. Compared to raw extract, the nanosized extract showed high solubility and dissolution. Bioavailability studies showed improved area under the curve of terpenoids and flavonoids of the plant (Wang et al. 2016). Naik et al., prepared *Ginkgo biloba* phytosomes and evaluated for hepatoprotective effects in carbon tetrachloride induced liver injury. The phytosomes (25 mg/kg) showed hepatoprotective effects similar to a high dose of silymarin (200 mg/kg) thus proving as a better alternative (Naik and Panda 2007). Haghighi et al., prepared *Ginkgo biloba* solid lipid nanoparticles and evaluated for activity on skin infections. The nanoformulation was found to be having better activity against both gram-negative and gram-positive bacteria. The nanoformulation also didn't show any sign of toxicity on rabbit skin indicating high safety profile (Haghighi et al. 2018). Wang et al., prepared nanoparticles of the same plant extract by anti-solvent precipitation method. Bioavailability study showed improvement in area under the curve by more than twofolds compared with raw extract (Wang et al. 2019).

Liquorice (*Glycyrrhiza glabra*) is a medicinal herb that is being used as a traditional medicine over a long time. Liquorice extract was found to be having expectorant, emollient, anti-inflammatory, anti-microbial, anti-diabetic and neuroprotective effects (Shibata 2000). However, like most of the plant extracts, it also suffers from poor solubility and permeability. Various approaches have been explored by the researchers to improve the properties of the liquorice extract. Damle et al., prepared phytophospholipid complex of liquorice and citrus extracts. Skin distribution studies revealed that the skin retention of the polyphenols present in the extract was high from phytophospholipid complex compared with the crude extract (Damle and Mallya 2016). In another study by Rani et al., the active constituent of liquorice, Glycyrrhizin was formulated into nanoparticles and evaluated for anti-diabetic activity. It was found that the nanoformulation group showed improved anti-diabetic effect compared to group received crude extract (Rani et al. 2017). Esmaeli et al. prepared chitosan-alginate nanocapsules loaded with liquorice extract. The nano-encapsulated extract showed improved anti-oxidant and anti-microbial activity compared to crude extract (Esmaeili and Rafiee 2015). Roque et al., prepared poly-lactic acid (oral gel), (oral film) and alginate (toothpaste) mucoadhesive nanoparticles loaded with liquorice extract. All the nanoformulations showed improved activity against *Candida albicans*. All of them have enhanced mucoadhesive properties (Roque et al. 2018). Vishwanathan et al., formulated inhalable liposomes of liquorice extract and evaluated for effect against tuberculosis. The formulation showed enhanced lung deposition and a significant reduction in bacterial load (Viswanathan et al. 2019).

Green tea has wide applicability in treating various ailments (Zaveri 2006). However, few of its components are unstable at gastric conditions and have poor permeability (Ananingsih et al. 2013; Zhu et al. 1997). Nanoformulations are assumed to improve stability and permeability. Manea et al., prepared solid lipid nanoparticles of green tea extract and evaluated for antioxidant and anti-microbial effect. It was found that the nanoformulation has higher antioxidant and anti-microbial efficacy compared with the crude extract (Manea et al. 2014). Dag et al., encapsulated green tea extract into liposomes. The formulation showed high stability of actives and improved antioxidant activity (Dag and Oztop 2017). Lu et al. prepared green tea polyphenols into liposomes to improve the stability of polyphenols which are unstable to oxidation and light. The formulation showed first order release kinetics and found to be protecting the polyphenols from degradation (Lu et al. 2011). Though green tea is consumed on a daily basis more clinical research has to be done to establish its beneficial effects which will pave a way for its clinical approval.

Withania somnifera, commonly called Ashwagandha is a native Indian plant that was used since ancient days. The extracts of Ashwagandha was found to have various activities like anti-diabetic, immune-modulatory, anti-inflammatory and anti-oxidant (Alam et al. 2012a; Mishra et al. 2000). It is a principal part of the historic Ayurvedic medicinal system. However, the extract has poor permeability which hinders its efficacy. Various nanocarriers have been evaluated to improve the same. Chinembiri et al., prepared stable Ashwagandha extract solid lipid nanoparticles and niosomes for improved topical delivery. *In-vitro* diffusion studies found that the nanoformulation improved dermal deposition of the components thereby improving topical delivery (Chinembiri et al. 2017). Similarly, Gauttam et al., prepared phospholipid vesicle system of a polyherbal composition containing Ashwagandha as one of the components for anti-diabetic activity. The nanoformulation had improved the anti-diabetic efficacy compared to crude extracts (Gauttam and Kalia 2013). Neog et al. prepared mannosylated liposomes of Withaferin A, an active constituent of Ashwagandha to target RAW 264.7 macrophages. It was found that the liposomes improved the uptake into cells. It was found that the liposomal formulation down-regulated the release of inflammatory cytokines and also reduced the release of nitric oxide and reactive oxygen species. The study found normalization of mitochondrial function thus imparting the protective effects (Neog et al. 2018). Though the extract has various benefits, extensive characterization and clinical investigation have to be one so that the product reaches to the standards required to be approved as a pharmaceutical product.

3.6 Challenges

Though considerable advantages have been postulated for use of nanotechnology in delivery of natural medicine, considerable challenges arise which need to be addressed to improve clinical outcome of such products. These include:

- Natural products are not always safer. Few of the natural products are highly potent (E.g. Atropine). In such cases care shall be taken to evaluate safety of the formulation
- Development of cost effective therapy is always a challenge with nanotechnology based products
- Developing and scaling of nanocarriers for natural product especially plant extracts is a very complicated process as the processing has to ensure uniform distribution of active constituents
- Stability and assay standardization are always a challenge if multiple components are involved
- Considerable amount of clinical data has to be generated for regulatory approval
- Clinical investigations are also complicated as the investigator has to take into consideration the toxicological parameters of both the nanocarrier and active components.

3.7 Conclusion

Natural products are effective and safer alternatives than synthetic compounds with high tolerability. Moreover, natural products proved their use since ancient times which makes them ideal option to treat various ailments. However, these natural products have issues such as poor physicochemical properties which hindered their active deployment into medicine. With introduction of nanotechnology various natural products having poor properties were effectively delivered when loaded into nanocarriers. Moreover, use of nanocarriers improved the therapeutic outcome of various natural compounds. All the above case studies proved nanotechnology can improve the delivery of various complicated molecules. With the introduction of newer nanocarriers the investigator will have a wide variety of choice to select the nanocarrier for the molecule according to the therapeutic requirements. These factors again increased the focus of scientific community on the natural medicine. Introduction of newer natural compounds from sources which were never explored increased the interest into natural product development. Further definitive clinical investigations needs to be done to establish the advantages of such nanotechnology based natural products. Extensive efforts like this can bring fruitful results with any nanocarrier based natural product getting regulatory approvals in due time.

References

Alam N, Hossain M, Khalil MI, Moniruzzaman M, Sulaiman SA, Gan SH (2012a) Recent advances in elucidating the biological properties of Withania somnifera and its potential role in health benefits. Phytochem Rev 11(1):97–112. https://doi.org/10.1007/s11101-011-9221-5

Alam S, Khan ZI, Mustafa G, Kumar M, Islam F, Bhatnagar A, Ahmad FJ (2012b) Development and evaluation of thymoquinone-encapsulated chitosan nanoparticles for nose-to-brain targeting: a pharmacoscintigraphic study. Int J Nanomedicine 7:5705. https://doi.org/10.2147/IJN.S35329

Allen TM, Cullis PR (2013) Liposomal drug delivery systems: from concept to clinical applications. Adv Drug Deliv Rev 65(1):36–48. https://doi.org/10.1016/j.addr.2012.09.037

Amri A, Chaumeil J, Sfar S, Charrueau C (2012) Administration of resveratrol: what formulation solutions to bioavailability limitations? J Control Release 158(2):182–193. https://doi.org/10.1016/j.jconrel.2011.09.083

Amritpal S, Sanjiv D, Navpreet K, Jaswinder S (2010) Berberine: alkaloid with wide spectrum of pharmacological activities. J Nat Prod (India) 3:64–75

Ananingsih VK, Sharma A, Zhou W (2013) Green tea catechins during food processing and storage: a review on stability and detection. Food Res Int 50(2):469–479. https://doi.org/10.1016/j.foodres.2011.03.004

Ankola D, Wadsworth R, Ravi Kumar M (2011) Nanoparticulate delivery can improve peroral bioavailability of cyclosporine and match Neoral Cmax sparing the kidney from damage. J Biomed Nanotechnol 7(2):300–307. https://doi.org/10.1166/jbn.2011.1278

Ansari KA, Vavia PR, Trotta F, Cavalli R (2011) Cyclodextrin-based nanosponges for delivery of resveratrol: in vitro characterisation, stability, cytotoxicity and permeation study. AAPS PharmSciTech 12(1):279–286. https://doi.org/10.1208/s12249-011-9584-3

Antçnio E, Khalil NM, Mainardes RM (2016) Bovine serum albumin nanoparticles containing quercetin: characterization and antioxidant activity. J Nanosci Nanotechnol 16(2):1346–1353. https://doi.org/10.1166/jnn.2016.11672

Anwar H, Dasgupta M, Lam K, Costerton JW (1989) Tobramycin resistance of mucoid Pseudomonas aeruginosa biofilm grown under iron limitation. J Antimicrob Chemother 24(5):647–655. https://doi.org/10.1093/jac/24.5.647

Avadhani KS, Manikkath J, Tiwari M, Chandrasekhar M, Godavarthi A, Vidya SM, Hariharapura RC, Kalthur G, Udupa N, Mutalik S (2017) Skin delivery of epigallocatechin-3-gallate (EGCG) and hyaluronic acid loaded nano-transfersomes for antioxidant and anti-aging effects in UV radiation induced skin damage. Drug Deliv 24(1):61–74. https://doi.org/10.1080/10717544.2016.1228718

Badary OA, Al-Shabanah OA, Nagi MN, Al-Bekairi AM, Elmazar M (1998) Acute and subchronic toxicity of thymoquinone in mice. Drug Dev Res 44(2–3):56–61. https://doi.org/10.1002/(SICI)1098-2299(199806/07)44:2/3<56::AID-DDR2>3.0.CO;2-9
Bajwa S, Munawar A, Khan W (2017) Nanotechnology in medicine: innovation to market. Pharm Bioprocess 5(2):11–15
Barenholz YC (2012) Doxil®—the first FDA-approved nano-drug: lessons learned. J Control Release 160(2):117–134. https://doi.org/10.1016/j.jconrel.2012.03.020
Başaran E, Demirel M, Sırmagül B, Yazan Y (2010) Cyclosporine-A incorporated cationic solid lipid nanoparticles for ocular delivery. J Microencapsul 27(1):37–47. https://doi.org/10.3109/02652040902846883
Battaglia L, D'Addino I, Peira E, Trotta M, Gallarate M (2012) Solid lipid nanoparticles prepared by coacervation method as vehicles for ocular cyclosporine. J Drug Delivery Sci Technol 22(2):125–130. https://doi.org/10.1016/S1773-2247(12)50016-X
Baur JA, Sinclair DA (2006) Therapeutic potential of resveratrol: the in vivo evidence. Nat Rev Drug Discov 5(6):493. https://doi.org/10.1038/nrd2060
Bianco A, Kostarelos K, Prato M (2005) Applications of carbon nanotubes in drug delivery. Curr Opin Chem Biol 9(6):674–679. https://doi.org/10.1016/j.cbpa.2005.10.005
Bilia AR, Piazzini V, Guccione C, Risaliti L, Asprea M, Capecchi G, Bergonzi MC (2017) Improving on nature: the role of nanomedicine in the development of clinical natural drugs. Planta Med 83(05):366–381. https://doi.org/10.1055/s-0043-102949
Bittner B, Mountfield R (2002) Intravenous administration of poorly soluble new drug entities in early drug discovery: the potential impact of formulation on pharmacokinetic parameters. Curr Opin Drug Discov Devel 5(1):59–71
Blessy M, Patel RD, Prajapati PN, Agrawal Y (2014) Development of forced degradation and stability indicating studies of drugs—a review. J Pharm Anal 4(3):159–165. https://doi.org/10.1016/j.jpha.2013.09.003
Boisseau P, Loubaton B (2011) Nanomedicine, nanotechnology in medicine. Comptes Rendus Physique 12(7):620–636. https://doi.org/10.1016/j.crhy.2011.06.001
Bonechi C, Martini S, Ciani L, Lamponi S, Rebmann H, Rossi C, Ristori S (2012) Using liposomes as carriers for polyphenolic compounds: the case of trans-resveratrol. PLoS One 7(8):e41438. https://doi.org/10.1371/journal.pone.0041438
Borel JF, Feurer C, Gubler H, Stähelin H (1994) Biological effects of cyclosporin A: a new antilymphocytic agent. Agents Actions 43(3–4):179–186. https://doi.org/10.1007/BF01986686
Broach JR, Thorner J (1996) High-throughput screening for drug discovery. Nature 384(6604):14–16
Caddeo C, Teskač K, Sinico C, Kristl J (2008) Effect of resveratrol incorporated in liposomes on proliferation and UV-B protection of cells. Int J Pharm 363(1–2):183–191. https://doi.org/10.1016/j.ijpharm.2008.07.024
Cano A, Ettcheto M, Chang J-H, Barroso E, Espina M, Kühne BA, Barenys M, Auladell C, Folch J, Souto EB (2019) Dual-drug loaded nanoparticles of Epigallocatechin-3-gallate (EGCG)/Ascorbic acid enhance therapeutic efficacy of EGCG in a APPswe/PS1dE9 Alzheimer's disease mice model. J Control Release. https://doi.org/10.1016/j.jconrel.2019.03.010
Caster JM, Patel AN, Zhang T, Wang A (2017) Investigational nanomedicines in 2016: a review of nanotherapeutics currently undergoing clinical trials. Wiley Interdiscip Rev Nanomed Nanobiotechnol 9(1):e1416. https://doi.org/10.1002/wnan.1416
Cavalli R, Gasco MR, Chetoni P, Burgalassi S, Saettone MF (2002) Solid lipid nanoparticles (SLN) as ocular delivery system for tobramycin. Int J Pharm 238(1–2):241–245. https://doi.org/10.1016/S0378-5173(02)00080-7
Cavalli R, Bargoni A, Podio V, Muntoni E, Zara GP, Gasco MR (2003) Duodenal administration of solid lipid nanoparticles loaded with different percentages of tobramycin. J Pharm Sci 92(5):1085–1094. https://doi.org/10.1002/jps.10368
Cerchiara T, Abruzzo A, Di Cagno M, Bigucci F, Bauer-Brandl A, Parolin C, Vitali B, Gallucci M, Luppi B (2015) Chitosan based micro-and nanoparticles for colon-targeted delivery of

vancomycin prepared by alternative processing methods. Eur J Pharm Biopharm 92:112–119. https://doi.org/10.1016/j.ejpb.2015.03.004

Chan LM, Lowes S, Hirst BH (2004) The ABCs of drug transport in intestine and liver: efflux proteins limiting drug absorption and bioavailability. Eur J Pharm Sci 21(1):25–51. https://doi.org/10.1016/j.ejps.2003.07.003

Chauhan MK, Bhatt N (2019) Bioavailability enhancement of polymyxin B with novel drug delivery: development and optimization using quality-by-design approach. J Pharm Sci 108(4):1521–1528. https://doi.org/10.1016/j.xphs.2018.11.032

Chinembiri TN, Gerber M, Du Plessis LH, Du Preez JL, Hamman JH, Du Plessis J (2017) Topical delivery of Withania somnifera crude extracts in niosomes and solid lipid nanoparticles. Pharmacogn Mag 13(Suppl 3):S663. https://doi.org/10.4103/pm.pm_489_16

Cohen B (1998) Amphotericin B toxicity and lethality: a tale of two channels. Int J Pharm 162(1–2):95–106. https://doi.org/10.1016/S0378-5173(97)00417-1

Cragg GM (1998) Paclitaxel (Taxol®): a success story with valuable lessons for natural product drug discovery and development. Med Res Rev 18(5):315–331. https://doi.org/10.1002/(SICI)1098-1128(199809)18:5<315::AID-MED3>3.0.CO;2-W

Crowley P, Martini LG (2001) Drug-excipient interactions. Pharm Technol 4:7–12

da Rocha Lindner G, Bonfanti Santos D, Colle D, Gasnhar Moreira EL, Daniel Prediger R, Farina M, Khalil NM, Mara Mainardes R (2015) Improved neuroprotective effects of resveratrol-loaded polysorbate 80-coated poly (lactide) nanoparticles in MPTP-induced Parkinsonism. Nanomedicine 10(7):1127–1138. https://doi.org/10.2217/nnm.14.165

Dag D, Oztop MH (2017) Formation and characterization of green tea extract loaded liposomes. J Food Sci 82(2):463–470. https://doi.org/10.1111/1750-3841.13615

Dai J, Nagai T, Wang X, Zhang T, Meng M, Zhang Q (2004) pH-sensitive nanoparticles for improving the oral bioavailability of cyclosporine A. Int J Pharm 280(1–2):229–240. https://doi.org/10.1016/j.ijpharm.2004.05.006

Damle M, Mallya R (2016) Development and evaluation of a novel delivery system containing phytophospholipid complex for skin aging. AAPS PharmSciTech 17(3):607–617. https://doi.org/10.1208/s12249-015-0386-x

Davies J, Davies D (2010) Origins and evolution of antibiotic resistance. Microbiol Mol Biol Rev 74(3):417–433. https://doi.org/10.1128/MMBR.00016-10

de Carvalho RF, Ribeiro IF, Miranda-Vilela AL, de Souza Filho J, Martins OP, e Silva DOC, Tedesco AC, Lacava ZGM, Báo SN, Sampaio RNR (2013) Leishmanicidal activity of amphotericin B encapsulated in PLGA–DMSA nanoparticles to treat cutaneous leishmaniasis in C57BL/6 mice. Exp Parasitol 135(2):217–222. https://doi.org/10.1016/j.exppara.2013.07.008

Delmas D, Aires V, Limagne E, Dutartre P, Mazué F, Ghiringhelli F, Latruffe N (2011) Transport, stability, and biological activity of resveratrol. Ann N Y Acad Sci 1215(1):48–59. https://doi.org/10.1111/j.1749-6632.2010.05871.x

Devi VK, Jain N, Valli KS (2010) Importance of novel drug delivery systems in herbal medicines. Pharmacogn Rev 4(7):27. https://doi.org/10.4103/0973-7847.65322

Dhawan S, Kapil R, Singh B (2011) Formulation development and systematic optimization of solid lipid nanoparticles of quercetin for improved brain delivery. J Pharm Pharmacol 63(3):342–351. https://doi.org/10.1111/j.2042-7158.2010.01225.x

Dias DA, Urban S, Roessner U (2012) A historical overview of natural products in drug discovery. Meta 2(2):303–336. https://doi.org/10.3390/metabo2020303

Dube A, Nicolazzo JA, Larson I (2010) Chitosan nanoparticles enhance the intestinal absorption of the green tea catechins (+)-catechin and (−)-epigallocatechin gallate. Eur J Pharm Sci 41(2):219–225. https://doi.org/10.1016/j.ejps.2010.06.010

Dube A, Nicolazzo JA, Larson I (2011) Chitosan nanoparticles enhance the plasma exposure of (−)-epigallocatechin gallate in mice through an enhancement in intestinal stability. Eur J Pharm Sci 44(3):422–426. https://doi.org/10.1016/j.ejps.2011.09.004

Dupont B (2002) Overview of the lipid formulations of amphotericin B. J Antimicrob Chemother 49(1):31–36. https://doi.org/10.1093/jac/49.suppl_1.31

Dwivedi P, Khatik R, Khandelwal K, Taneja I, Raju KSR, Paliwal SK, Dwivedi AK, Mishra PR (2014) Pharmacokinetics study of arteether loaded solid lipid nanoparticles: an improved oral bioavailability in rats. Int J Pharm 466(1–2):321–327. https://doi.org/10.1016/j.ijpharm.2014.03.036

Efferth T, Bilia A, Osman A, Elsohly M, Wink M, Bauer R, Khan I, Bergonzi M, Marin J (2016) Expanding the therapeutic spectrum of artemisinin: activity against infectious diseases beyond malaria and novel pharmaceutical developments. World J Tradit Chin Med 2:1–23. https://doi.org/10.15806/j.issn.2311-8571.2016.0002

Egbaria K, Ramachandran C, Weiner N (1990) Topical delivery of ciclosporin: evaluation of various formulations using in vitro diffusion studies in hairless mouse skin. Skin Pharmacol Physiol 3(1):21–28. https://doi.org/10.1159/000210837

Ellis D (2002) Amphotericin B: spectrum and resistance. J Antimicrob Chemother 49(1):7–10. https://doi.org/10.1093/jac/49.suppl_1.7

Erlund I (2004) Review of the flavonoids quercetin, hesperetin, and naringenin. Dietary sources, bioactivities, bioavailability, and epidemiology. Nutr Res 24(10):851–874. https://doi.org/10.1016/j.nutres.2004.07.005

Esmaeili A, Rafiee R (2015) Preparation and biological activity of nanocapsulated Glycyrrhiza glabra L. var. glabra. Flavour Fragr J 30(1):113–119. https://doi.org/10.1002/ffj.3225

Fang J-Y, Hwang T-L, Huang Y-L, Fang C-L (2006) Enhancement of the transdermal delivery of catechins by liposomes incorporating anionic surfactants and ethanol. Int J Pharm 310(1–2):131–138. https://doi.org/10.1016/j.ijpharm.2005.12.004

Farjadian F, Ghasemi A, Gohari O, Roointan A, Karimi M, Hamblin MR (2018) Nanopharmaceuticals and nanomedicines currently on the market: challenges and opportunities. Nanomedicine 14(1):93–126. https://doi.org/10.2217/nnm-2018-0120

Felgines C, Texier O, Morand C, Manach C, Scalbert A, Régerat F, Rémésy C (2000) Bioavailability of the flavanone naringenin and its glycosides in rats. Am J Physiol Gastrointest Liver Physiol 279(6):G1148–G1154. https://doi.org/10.1152/ajpgi.2000.279.6.G1148

Fernández L, Gooderham WJ, Bains M, McPhee JB, Wiegand I, Hancock RE (2010) Adaptive resistance to the "last hope" antibiotics polymyxin B and colistin in Pseudomonas aeruginosa is mediated by the novel two-component regulatory system ParR-ParS. Antimicrob Agents Chemother 54(8):3372–3382. https://doi.org/10.1128/AAC.00242-10

Formica J, Regelson W (1995) Review of the biology of quercetin and related bioflavonoids. Food Chem Toxicol 33(12):1061–1080. https://doi.org/10.1016/0278-6915(95)00077-1

Francioso A, Mastromarino P, Masci A, d'Erme M, Mosca L (2014) Chemistry, stability and bioavailability of resveratrol. Med Chem 10(3):237–245

Freitas RA Jr (2005) What is nanomedicine? Nanomedicine 1(1):2–9. https://doi.org/10.1016/j.nano.2004.11.003

Frémont L (2000) Biological effects of resveratrol. Life Sci 66(8):663–673. https://doi.org/10.1016/S0024-3205(99)00410-5

Fröhlich T, Hahn F, Belmudes L, Leidenberger M, Friedrich O, Kappes B, Couté Y, Marschall M, Tsogoeva SB (2018) Synthesis of artemisinin-derived dimers, trimers and dendrimers: investigation of their antimalarial and antiviral activities including putative mechanisms of action. Chem Eur J 24(32):8103–8113. https://doi.org/10.1002/chem.201800729

Gauttam VK, Kalia AN (2013) Development of polyherbal antidiabetic formulation encapsulated in the phospholipids vesicle system. J Adv Pharm Technol Res 4(2):108. https://doi.org/10.4103/2231-4040.111527

Ghalioungui P (1987) The Ebers papyrus: a new English translation, commentaries and glossaries. Academy of Scientific Research and Technology, Cairo

Gharib A, Faezizadeh Z, Godarzee M (2013) Therapeutic efficacy of epigallocatechin gallate-loaded nanoliposomes against burn wound infection by methicillin-resistant Staphylococcus aureus. Skin Pharmacol Physiol 26(2):68–75. https://doi.org/10.1159/000345761

Gholamnezhad Z, Havakhah S, Boskabady MH (2016) Preclinical and clinical effects of Nigella sativa and its constituent, thymoquinone: a review. J Ethnopharmacol 190:372–386. https://doi.org/10.1016/j.jep.2016.06.061

Gillies ER, Frechet JM (2005) Dendrimers and dendritic polymers in drug delivery. Drug Discov Today 10(1):35–43. https://doi.org/10.1016/S1359-6446(04)03276-3

Gokce EH, Sandri G, Bonferoni MC, Rossi S, Ferrari F, Güneri T, Caramella C (2008) Cyclosporine A loaded SLNs: evaluation of cellular uptake and corneal cytotoxicity. Int J Pharm 364(1):76–86. https://doi.org/10.1016/j.ijpharm.2008.07.028

Gokce EH, Korkmaz E, Dellera E, Sandri G, Bonferoni MC, Ozer O (2012) Resveratrol-loaded solid lipid nanoparticles versus nanostructured lipid carriers: evaluation of antioxidant potential for dermal applications. Int J Nanomedicine 7:1841. https://doi.org/10.2147/IJN.S29710

Gopi S, Amalraj A, Haponiuk JT, Thomas S (2016) Introduction of nanotechnology in herbal drugs and nutraceutical: a review. J Nanomedicine Biotherapeutic Discov 6(2):1–8. https://doi.org/10.4172/2155-983X.1000143

Group IEW ICH guideline Q1A (R2) stability testing of new drug substances and products. In: International Conference on Harmonization, 2003. vol 24. sn

Gu H, Ho P, Tong E, Wang L, Xu B (2003) Presenting vancomycin on nanoparticles to enhance antimicrobial activities. Nano Lett 3(9):1261–1263. https://doi.org/10.1021/nl034396z

Gupta L, Sharma AK, Gothwal A, Khan MS, Khinchi MP, Qayum A, Singh SK, Gupta U (2017) Dendrimer encapsulated and conjugated delivery of berberine: a novel approach mitigating toxicity and improving in vivo pharmacokinetics. Int J Pharm 528(1–2):88–99. https://doi.org/10.1016/j.ijpharm.2017.04.073

Haddadi A, Elamanchili P, Lavasanifar A, Das S, Shapiro J, Samuel J (2008) Delivery of rapamycin by PLGA nanoparticles enhances its suppressive activity on dendritic cells. J Biomed Mater Res A 84(4):885–898. https://doi.org/10.1002/jbm.a.31373

Haghighi P, Ghaffari S, Arbabi Bidgoli S, Qomi M, Haghighat S (2018) Preparation, characterization and evaluation of Ginkgo biloba solid lipid nanoparticles. Nanomedicine Res J 3(2):71–78. https://doi.org/10.22034/nmrj.2018.02.003

Hansch C, Clayton JM (1973) Lipophilic character and biological activity of drugs II: the parabolic case. J Pharm Sci 62(1):1–21. https://doi.org/10.1002/jps.2600620102

Harvey AL (2008) Natural products in drug discovery. Drug Discov Today 13(19–20):894–901. https://doi.org/10.1016/j.drudis.2008.07.004

Hiemenz JW, Walsh TJ (1996) Lipid formulations of amphotericin B: recent progress and future directions. Clin Infect Dis 22(Supplement_2):S133–S144. https://doi.org/10.1093/clinids/22.Supplement_2.S133

Hillaireau H, Couvreur P (2009) Nanocarriers' entry into the cell: relevance to drug delivery. Cell Mol Life Sci 66(17):2873–2896. https://doi.org/10.1007/s00018-009-0053-z

Hsu Chen CC, Feingold DS (1973) Mechanism of polymyxin B action and selectivity toward biologic membranes. Biochemistry 12(11):2105–2111. https://doi.org/10.1021/bi00735a014

Huyck TK, Gradishar W, Manuguid F, Kirkpatrick P (2011) Eribulin mesylate. Nature Publishing Group

Ibrahim N, Ibrahim H, Sabater AM, Mazier D, Valentin A, Nepveu F (2015) Artemisinin nanoformulation suitable for intravenous injection: preparation, characterization and antimalarial activities. Int J Pharm 495(2):671–679. https://doi.org/10.1016/j.ijpharm.2015.09.020

Imanshahidi M, Hosseinzadeh H (2008) Pharmacological and therapeutic effects of Berberis vulgaris and its active constituent, berberine. Phytother Res 22(8):999–1012. https://doi.org/10.1002/ptr.2399

Insua I, Zizmare L, Peacock AF, Krachler AM, Fernandez-Trillo F (2017) Polymyxin B containing polyion complex (PIC) nanoparticles: improving the antimicrobial activity by tailoring the degree of polymerisation of the inert component. Sci Rep 7(1):9396. https://doi.org/10.1038/s41598-017-09667-3

Isacchi B, Arrigucci S, Gl M, Bergonzi MC, Vannucchi MG, Novelli A, Bilia AR (2011) Conventional and long-circulating liposomes of artemisinin: preparation, characterization, and pharmacokinetic profile in mice. J Liposome Res 21(3):237–244. https://doi.org/10.3109/08982104.2010.539185

Ismailos G, Reppas C, Dressman JB, Macheras P (1991) Unusual solubility behaviour of cyclosporin A in aqueous media. J Pharm Pharmacol 43(4):287–289. https://doi.org/10.1111/j.2042-7158.1991.tb06688.x

Italia J, Datta P, Ankola D, Kumar M (2008) Nanoparticles enhance per oral bioavailability of poorly available molecules: epigallocatechin gallate nanoparticles ameliorates cyclosporine induced nephrotoxicity in rats at three times lower dose than oral solution. J Biomed Nanotechnol 4(3):304–312. https://doi.org/10.1166/jbn.2008.341

Iyer AK, Khaled G, Fang J, Maeda H (2006) Exploiting the enhanced permeability and retention effect for tumor targeting. Drug Discov Today 11(17–18):812–818. https://doi.org/10.1016/j.drudis.2006.07.005

Jain A, Pooladanda V, Bulbake U, Doppalapudi S, Rafeeqi TA, Godugu C, Khan W (2017) Liposphere mediated topical delivery of thymoquinone in the treatment of psoriasis. Nanomedicine 13(7):2251–2262. https://doi.org/10.1016/j.nano.2017.06.009

Ji P, Yu T, Liu Y, Jiang J, Xu J, Zhao Y, Hao Y, Qiu Y, Zhao W, Wu C (2016) Naringenin-loaded solid lipid nanoparticles: preparation, controlled delivery, cellular uptake, and pulmonary pharmacokinetics. Drug Des Devel Ther 10:911. https://doi.org/10.2147/DDDT.S97738

Joshi M, Pathak S, Sharma S, Patravale V (2008) Design and in vivo pharmacodynamic evaluation of nanostructured lipid carriers for parenteral delivery of artemether: nanoject. Int J Pharm 364(1):119–126. https://doi.org/10.1016/j.ijpharm.2008.07.032

Jung YD, Ellis LM (2001) Inhibition of tumour invasion and angiogenesis by epigallocatechin gallate (EGCG), a major component of green tea. Int J Exp Pathol 82(6):309–316. https://doi.org/10.1046/j.1365-2613.2001.00205.x

Kakran M, Sahoo NG, Li L, Judeh Z (2010) Dissolution of artemisinin/polymer composite nanoparticles fabricated by evaporative precipitation of nanosuspension. J Pharm Pharmacol 62(4):413–421. https://doi.org/10.1211/jpp.62.04.0002

Kakran M, Sahoo N, Li L (2011) Dissolution enhancement of quercetin through nanofabrication, complexation, and solid dispersion. Colloids Surf B: Biointerfaces 88(1):121–130. https://doi.org/10.1016/j.colsurfb.2011.06.020

Kakran M, Sahoo NG, Li L, Judeh Z (2012) Fabrication of quercetin nanoparticles by anti-solvent precipitation method for enhanced dissolution. Powder Technol 223:59–64. https://doi.org/10.1016/j.powtec.2011.08.021

Kaur H, Kumar B, Chakrabarti A, Medhi B, Modi M, Radotra BD, Aggarwal R, Sinha VR (2019) A new therapeutic approach for brain delivery of epigallocatechin gallate: development and characterization studies. Curr Drug Deliv 16(1):59–65. https://doi.org/10.2174/1567201815666180926121104

Kausar H, Mujeeb M, Ahad A, Moolakkadath T, Aqil M, Ahmad A, Akhter MH (2019) Optimization of ethosomes for topical thymoquinone delivery for the treatment of skin acne. J Drug Delivery Sci Technol 49:177–187. https://doi.org/10.1016/j.jddst.2018.11.016

Keck CM, Müller RH (2006) Drug nanocrystals of poorly soluble drugs produced by high pressure homogenisation. Eur J Pharm Biopharm 62(1):3–16. https://doi.org/10.1016/j.ejpb.2005.05.009

Khan W, Farah S, Nyska A, Domb AJ (2013) Carrier free rapamycin loaded drug eluting stent: in vitro and in vivo evaluation. J Control Release 168(1):70–76. https://doi.org/10.1016/j.jconrel.2013.02.012

Khan AW, Kotta S, Ansari SH, Sharma RK, Ali J (2015) Self-nanoemulsifying drug delivery system (SNEDDS) of the poorly water-soluble grapefruit flavonoid Naringenin: design, characterization, in vitro and in vivo evaluation. Drug Deliv 22(4):552–561. https://doi.org/10.3109/10717544.2013.878003

Kleijnen J, Knipschild P (1992) Ginkgo biloba. Lancet 340(8828):1136–1139. https://doi.org/10.1016/0140-6736(92)93158-J

Klockgether-Radke A (2002) FW Sertürner and the discovery of morphine. 200 years of pain therapy with opioids. Anästhesiol Intensivmed Notfallmed Schmerzther 37(5):244–249

Kobierski S, Ofori-Kwakye K, Müller R, Keck C (2009) Resveratrol nanosuspensions for dermal application–production, characterization, and physical stability. Die Pharmazie Int J Pharm Sci 64(11):741–747. https://doi.org/10.1691/ph.2009.9097
Krishna S, Bustamante L, Haynes RK, Staines HM (2008) Artemisinins: their growing importance in medicine. Trends Pharmacol Sci 29(10):520–527. https://doi.org/10.1016/j.tips.2008.07.004
Kumar RP, Abraham A (2016) PVP-coated naringenin nanoparticles for biomedical applications–in vivo toxicological evaluations. Chem Biol Interact 257:110–118. https://doi.org/10.1016/j.cbi.2016.07.012
Kumar R, Dogra S, Amarji B, Singh B, Kumar S, Vinay K, Mahajan R, Katare O (2016) Efficacy of novel topical liposomal formulation of cyclosporine in mild to moderate stable plaque psoriasis: a randomized clinical trial. JAMA Dermatol 152(7):807–815. https://doi.org/10.1001/jamadermatol.2016.0859
Kumari A, Yadav SK, Pakade YB, Singh B, Yadav SC (2010) Development of biodegradable nanoparticles for delivery of quercetin. Colloids Surf B: Biointerfaces 80(2):184–192. https://doi.org/10.1016/j.colsurfb.2010.06.002
Laniado-Laborín R, Cabrales-Vargas MN (2009) Amphotericin B: side effects and toxicity. Rev Iberoam Micol 26(4):223–227. https://doi.org/10.1016/j.riam.2009.06.003
Lapenna S, Bilia AR, Morris GA, Nilsson M (2009) Novel artemisinin and curcumin micellar formulations: drug solubility studies by NMR spectroscopy. J Pharm Sci 98(10):3666–3675. https://doi.org/10.1002/jps.21685
Levine DP (2006) Vancomycin: a history. Clin Infect Dis 42(Supplement_1):S5–S12. https://doi.org/10.1086/491709
Li H, Zhao X, Ma Y, Zhai G, Li L, Lou H (2009) Enhancement of gastrointestinal absorption of quercetin by solid lipid nanoparticles. J Control Release 133(3):238–244. https://doi.org/10.1016/j.jconrel.2008.10.002
Li J, Kim SG, Blenis J (2014) Rapamycin: one drug, many effects. Cell Metab 19(3):373–379. https://doi.org/10.1016/j.cmet.2014.01.001
Lin Y-H, Lin J-H, Chou S-C, Chang S-J, Chung C-C, Chen Y-S, Chang C-H (2015) Berberine-loaded targeted nanoparticles as specific Helicobacter pylori eradication therapy: in vitro and in vivo study. Nanomedicine 10(1):57–71. https://doi.org/10.2217/nnm.14.76
Liu Y, Feng N (2015) Nanocarriers for the delivery of active ingredients and fractions extracted from natural products used in traditional Chinese medicine (TCM). Adv Colloid Interf Sci 221:60–76. https://doi.org/10.1016/j.cis.2015.04.006
Logothetidis S (2006) Nanotechnology in medicine: the medicine of tomorrow and nanomedicine. Hippokratia 10(1):7–21
Loureiro J, Andrade S, Duarte A, Neves A, Queiroz J, Nunes C, Sevin E, Fenart L, Gosselet F, Coelho M (2017) Resveratrol and grape extract-loaded solid lipid nanoparticles for the treatment of Alzheimer's disease. Molecules 22(2):277. https://doi.org/10.3390/molecules22020277
Lu H, Meng X, Li C, Sang S, Patten C, Sheng S, Hong J, Bai N, Winnik B, Ho C-T (2003) Glucuronides of tea catechins: enzymology of biosynthesis and biological activities. Drug Metab Dispos 31(4):452–461. https://doi.org/10.1124/dmd.31.4.452
Lu Q, Li D-C, Jiang J-G (2011) Preparation of a tea polyphenol nanoliposome system and its physicochemical properties. J Agric Food Chem 59(24):13004–13011. https://doi.org/10.1021/jf203194w
Lutsenko SV, Gromovykh TI, Krasnyuk II, Vasilenko IA, Feldman NB (2018) Antihepatotoxic activity of liposomal silibinin. BioNanoScience 8(2):581–586. https://doi.org/10.1007/s12668-018-0512-9
Maas J, Kamm W, Hauck G (2007) An integrated early formulation strategy–from hit evaluation to preclinical candidate profiling. Eur J Pharm Biopharm 66(1):1–10. https://doi.org/10.1016/j.ejpb.2006.09.011
Mabberley D (2008) Mabberlev's plant-book. Cambridge University Press, Cambridge

Maity S, Mukhopadhyay P, Kundu PP, Chakraborti AS (2017) Alginate coated chitosan core-shell nanoparticles for efficient oral delivery of naringenin in diabetic animals—an in vitro and in vivo approach. Carbohydr Polym 170:124–132. https://doi.org/10.1016/j.carbpol.2017.04.066

Mandel S, Weinreb O, Amit T, Youdim MB (2004) Cell signaling pathways in the neuroprotective actions of the green tea polyphenol (−)-epigallocatechin-3-gallate: implications for neurodegenerative diseases. J Neurochem 88(6):1555–1569. https://doi.org/10.1046/j.1471-4159.2003.02291.x

Manea A-M, Andronescu C, Meghea A (2014) Green tea extract loaded into solid lipid nanoparticles. UPB Sci Bull B Chem Mater Sci 76(2):125–136

Marchiori MCL, Rigon C, Camponogara C, Oliveira SM, Cruz L (2017) Hydrogel containing silibinin-loaded pomegranate oil based nanocapsules exhibits anti-inflammatory effects on skin damage UVB radiation-induced in mice. J Photochem Photobiol B Biol 170:25–32. https://doi.org/10.1016/j.jphotobiol.2017.03.015

Masimirembwa CM, Bredberg U, Andersson TB (2003) Metabolic stability for drug discovery and development. Clin Pharmacokinet 42(6):515–528. https://doi.org/10.2165/00003088-200342060-00002

McBain SC, Yiu HH, Dobson J (2008) Magnetic nanoparticles for gene and drug delivery. Int J Nanomedicine 3(2):169

Mehra M, Sheorain J, Kumari S (2016) Synthesis of berberine loaded polymeric nanoparticles by central composite design. In: AIP Conference Proceedings, vol 1724. AIP Publishing, p 020060

Mehrizi TZ, Ardestani MS, Molla Hoseini MH, Khamesipour A, Mosaffa N, Ramezani A (2018) Novel nano-sized chitosan amphotericin B formulation with considerable improvement against Leishmania major. Nanomedicine 13(24):3129–3147. https://doi.org/10.2217/nnm-2018-0063

Merisko-Liversidge EM, Liversidge GG (2008) Drug nanoparticles: formulating poorly water-soluble compounds. Toxicol Pathol 36(1):43–48. https://doi.org/10.1177/0192623307310946

Messiaen A-S, Forier K, Nelis H, Braeckmans K, Coenye T (2013) Transport of nanoparticles and tobramycin-loaded liposomes in Burkholderia cepacia complex biofilms. PLoS One 8(11):e79220. https://doi.org/10.1371/journal.pone.0079220

Mishra L-C, Singh BB, Dagenais S (2000) Scientific basis for the therapeutic use of Withania somnifera (ashwagandha): a review. Altern Med Rev 5(4):334–346

Mout R, Moyano DF, Rana S, Rotello VM (2012) Surface functionalization of nanoparticles for nanomedicine. Chem Soc Rev 41(7):2539–2544. https://doi.org/10.1039/C2CS15294K

Müller R, Runge S, Ravelli V, Mehnert W, Thünemann AF, Souto E (2006) Oral bioavailability of cyclosporine: solid lipid nanoparticles (SLN®) versus drug nanocrystals. Int J Pharm 317(1):82–89. https://doi.org/10.1016/j.ijpharm.2006.02.045

Musa SH, Basri M, Masoumi HRF, Shamsudin N, Salim N (2017) Enhancement of physicochemical properties of nanocolloidal carrier loaded with cyclosporine for topical treatment of psoriasis: in vitro diffusion and in vivo hydrating action. Int J Nanomedicine 12:2427. https://doi.org/10.2147/IJN.S125302

Naik SR, Panda VS (2007) Antioxidant and hepatoprotective effects of Ginkgo biloba phytosomes in carbon tetrachloride-induced liver injury in rodents. Liver Int 27(3):393–399. https://doi.org/10.1111/j.1478-3231.2007.01463.x

Nallamuthu I, Parthasarathi A, Khanum F (2013) Thymoquinone-loaded PLGA nanoparticles: antioxidant and anti-microbial properties. Int Curr Pharm J 2(12):202–207. https://doi.org/10.3329/icpj.v2i12.17017

Neog MK, Sultana F, Rasool M (2018) Targeting RAW 264.7 macrophages (M1 type) with Withaferin-A decorated mannosylated liposomes induces repolarization via downregulation of NF-κB and controlled elevation of STAT-3. Int Immunopharmacol 61:64–73. https://doi.org/10.1016/j.intimp.2018.05.019

Neu HC (1976) Tobramycin: an overview. J Infect Dis 134:S3–S19

Pangeni R, Sahni JK, Ali J, Sharma S, Baboota S (2014) Resveratrol: review on therapeutic potential and recent advances in drug delivery. Expert Opin Drug Deliv 11(8):1285–1298. https://doi.org/10.1517/17425247.2014.919253

Pardeike J, Hommoss A, Müller RH (2009) Lipid nanoparticles (SLN, NLC) in cosmetic and pharmaceutical dermal products. Int J Pharm 366(1–2):170–184. https://doi.org/10.1016/j.ijpharm.2008.10.003

Park S, Chibli H, Wong J, Nadeau JL (2011) Antimicrobial activity and cellular toxicity of nanoparticle–polymyxin B conjugates. Nanotechnology 22(18):185101. https://doi.org/10.1088/0957-4484/22/18/185101

Patwardhan B, Warude D, Pushpangadan P, Bhatt N (2005) Ayurveda and traditional Chinese medicine: a comparative overview. Evid Based Complement Alternat Med 2(4):465–473. https://doi.org/10.1093/ecam/neh140

Pelgrift RY, Friedman AJ (2013) Nanotechnology as a therapeutic tool to combat microbial resistance. Adv Drug Deliv Rev 65(13–14):1803–1815. https://doi.org/10.1016/j.addr.2013.07.011

Portman R, Meier-Kriesche H, Swinford R, Brannan P, Kahan B (2000) Reduced variability of Neoral pharmacokinetic studies in pediatric renal transplantation. Pediatr Nephrol 15(1–2):2–6. https://doi.org/10.1007/s004670000435

Puigdemont A, Brazís P, Ordeix L, Dalmau A, Fuertes E, Olivar A, Pérez C, Ravera I (2013) Efficacy of a new topical cyclosporine A formulation in the treatment of atopic dermatitis in dogs. Vet J 197(2):280–285. https://doi.org/10.1016/j.tvjl.2013.02.018

Pujara N, Jambhrunkar S, Wong KY, McGuckin M, Popat A (2017) Enhanced colloidal stability, solubility and rapid dissolution of resveratrol by nanocomplexation with soy protein isolate. J Colloid Interface Sci 488:303–308. https://doi.org/10.1016/j.jcis.2016.11.015

Qi L, Gao X (2008) Emerging application of quantum dots for drug delivery and therapy. Expert Opin Drug Deliv 5(3):263–267. https://doi.org/10.1517/17425247.5.3.263

Rani R, Dahiya S, Dhingra D, Dilbaghi N, Kim K-H, Kumar S (2017) Evaluation of anti-diabetic activity of glycyrrhizin-loaded nanoparticles in nicotinamide-streptozotocin-induced diabetic rats. Eur J Pharm Sci 106:220–230. https://doi.org/10.1016/j.ejps.2017.05.068

Ripoli M, Angelico R, Sacco P, Ceglie A, Mangia A (2016) Phytoliposome-based silibinin delivery system as a promising strategy to prevent hepatitis C virus infection. J Biomed Nanotechnol 12(4):770–780. https://doi.org/10.1166/jbn.2016.2161

Roque L, Duarte N, Bronze MR, Garcia C, Alopaeus J, Molpeceres J, Hagesaether E, Tho I, Rijo P, Reis C (2018) Development of a bioadhesive nanoformulation with Glycyrrhiza glabra L. extract against Candida albicans. Biofouling 34(8):880–892. https://doi.org/10.1080/08927014.2018.1514391

Rushmi ZT, Akter N, Mow RJ, Afroz M, Kazi M, de Matas M, Rahman M, Shariare MH (2017) The impact of formulation attributes and process parameters on black seed oil loaded liposomes and their performance in animal models of analgesia. Saudi Pharm J 25(3):404–412. https://doi.org/10.1016/j.jsps.2016.09.011

Sahibzada MUK, Sadiq A, Faidah HS, Khurram M, Amin MU, Haseeb A, Kakar M (2018) Berberine nanoparticles with enhanced in vitro bioavailability: characterization and antimicrobial activity. Drug Des Devel Ther 12:303. https://doi.org/10.2147/DDDT.S156123

Sahoo N, Kakran M, Shaal L, Li L, Müller R, Pal M, Tan L (2011) Preparation and characterization of quercetin nanocrystals. J Pharm Sci 100(6):2379–2390. https://doi.org/10.1002/jps.22446

Salmani J, Asghar S, Lv H, Zhou J (2014) Aqueous solubility and degradation kinetics of the phytochemical anticancer thymoquinone; probing the effects of solvents, pH and light. Molecules 19(5):5925–5939. https://doi.org/10.3390/molecules19055925

Santos AC, Veiga F, Ribeiro AJ (2011) New delivery systems to improve the bioavailability of resveratrol. Expert Opin Drug Deliv 8(8):973–990. https://doi.org/10.1517/17425247.2011.581655

Seil JT, Webster TJ (2012) Antimicrobial applications of nanotechnology: methods and literature. Int J Nanomedicine 7:2767. https://doi.org/10.2147/IJN.S24805

Severino P, Chaud MV, Shimojo A, Antonini D, Lancelloti M, Santana MHA, Souto EB (2015) Sodium alginate-cross-linked polymyxin B sulphate-loaded solid lipid nanoparticles: antibiotic resistance tests and HaCat and NIH/3T3 cell viability studies. Colloids Surf B: Biointerfaces 129:191–197. https://doi.org/10.1016/j.colsurfb.2015.03.049

Sharma P, Garg S (2010) Pure drug and polymer based nanotechnologies for the improved solubility, stability, bioavailability and targeting of anti-HIV drugs. Adv Drug Deliv Rev 62(4–5):491–502. https://doi.org/10.1016/j.addr.2009.11.019

Shetty PK, Manikkath J, Tupally K, Kokil G, Hegde AR, Raut SY, Parekh HS, Mutalik S (2017) Skin delivery of EGCG and silibinin: potential of peptide dendrimers for enhanced skin permeation and deposition. AAPS PharmSciTech 18(6):2346–2357. https://doi.org/10.1208/s12249-017-0718-0

Shibata S (2000) A drug over the millennia: pharmacognosy, chemistry, and pharmacology of licorice. Yakugaku Zasshi 120(10):849–862. https://doi.org/10.1248/yakushi1947.120.10_849

Shoeb M (2006) Anti-cancer agents from medicinal plants. Bangladesh J Pharmacol 1(2):35–41. https://doi.org/10.3329/bjp.v1i2.486

Siddiqui AA, Iram F, Siddiqui S, Sahu K (2014) Role of natural products in drug discovery process. Int J Drug Dev Res 6(2):172–204

Simamora P, Alvarez JM, Yalkowsky SH (2001) Solubilization of rapamycin. Int J Pharm 213(1–2):25–29. https://doi.org/10.1016/S0378-5173(00)00617-7

Singh G, Pai RS (2014) Optimized PLGA nanoparticle platform for orally dosed trans-resveratrol with enhanced bioavailability potential. Expert Opin Drug Deliv 11(5):647–659. https://doi.org/10.1517/17425247.2014.890588

Singh A, Ahmad I, Akhter S, Jain GK, Iqbal Z, Talegaonkar S, Ahmad FJ (2013) Nanocarrier based formulation of Thymoquinone improves oral delivery: stability assessment, in vitro and in vivo studies. Colloids Surf B: Biointerfaces 102:822–832. https://doi.org/10.1016/j.colsurfb.2012.08.038

Slowing II, Vivero-Escoto JL, Wu C-W, Lin VS-Y (2008) Mesoporous silica nanoparticles as controlled release drug delivery and gene transfection carriers. Adv Drug Deliv Rev 60(11):1278–1288. https://doi.org/10.1016/j.addr.2008.03.012

Smith TL, Pearson ML, Wilcox KR, Cruz C, Lancaster MV, Robinson-Dunn B, Tenover FC, Zervos MJ, Band JD, White E (1999) Emergence of vancomycin resistance in Staphylococcus aureus. N Engl J Med 340(7):493–501. https://doi.org/10.1056/NEJM199902183400701

Smith A, Giunta B, Bickford PC, Fountain M, Tan J, Shytle RD (2010) Nanolipidic particles improve the bioavailability and α-secretase inducing ability of epigallocatechin-3-gallate (EGCG) for the treatment of Alzheimer's disease. Int J Pharm 389(1–2):207–212. https://doi.org/10.1016/j.ijpharm.2010.01.012

Song Y, Ping Q, Wu Z (2005) Preparation of silybin nanoemulsion and its pharmacokinetics in rabbits. J China Pharm Univ 36(5):427

Soppimath KS, Aminabhavi TM, Kulkarni AR, Rudzinski WE (2001) Biodegradable polymeric nanoparticles as drug delivery devices. J Control Release 70(1–2):1–20. https://doi.org/10.1016/S0168-3659(00)00339-4

Storm DR, Rosenthal KS, Swanson PE (1977) Polymyxin and related peptide antibiotics. Annu Rev Biochem 46(1):723–763. https://doi.org/10.1146/annurev.bi.46.070177.003451

Su Y, Fu Z, Zhang J, Wang W, Wang H, Wang Y, Zhang Q (2008) Microencapsulation of Radix salvia miltiorrhiza nanoparticles by spray-drying. Powder Technol 184(1):114–121. https://doi.org/10.1016/j.powtec.2007.08.014

Sun D, Li N, Zhang W, Zhao Z, Mou Z, Huang D, Liu J, Wang W (2016) Design of PLGA-functionalized quercetin nanoparticles for potential use in Alzheimer's disease. Colloids Surf B: Biointerfaces 148:116–129. https://doi.org/10.1016/j.colsurfb.2016.08.052

Surendiran A, Sandhiya S, Pradhan S, Adithan C (2009) Novel applications of nanotechnology in medicine. Indian J Med Res 130(6):689–701

Sutton D, Nasongkla N, Blanco E, Gao J (2007) Functionalized micellar systems for cancer targeted drug delivery. Pharm Res 24(6):1029–1046. https://doi.org/10.1007/s11095-006-9223-y

Takagaki A, Nanjo F (2009) Metabolism of (−)-epigallocatechin gallate by rat intestinal flora. J Agric Food Chem 58(2):1313–1321. https://doi.org/10.1021/jf903375s

Tan Q, Liu W, Guo C, Zhai G (2011) Preparation and evaluation of quercetin-loaded lecithin-chitosan nanoparticles for topical delivery. Int J Nanomedicine 6:1621. https://doi.org/10.2147/IJN.S22411

Taylor E, Webster TJ (2011) Reducing infections through nanotechnology and nanoparticles. Int J Nanomedicine 6:1463. https://doi.org/10.2147/IJN.S22021

Tsai M-J, Huang Y-B, Fang J-W, Fu Y-S, Wu P-C (2015a) Preparation and characterization of naringenin-loaded elastic liposomes for topical application. PLoS One 10(7):e0131026. https://doi.org/10.1371/journal.pone.0131026

Tsai M-J, Huang Y-B, Fang J-W, Fu Y-S, Wu P-C (2015b) Preparation and evaluation of submicron-carriers for naringenin topical application. Int J Pharm 481(1–2):84–90. https://doi.org/10.1016/j.ijpharm.2015.01.034

Tubesha Z, Bakar ZA, Ismail M (2013) Characterization and stability evaluation of thymoquinone nanoemulsions prepared by high-pressure homogenization. J Nanomater 2013:126. https://doi.org/10.1155/2013/453290

Uchegbu IF, Vyas SP (1998) Non-ionic surfactant based vesicles (niosomes) in drug delivery. Int J Pharm 172(1–2):33–70. https://doi.org/10.1016/S0378-5173(98)00169-0

Vadiei K, Perez-Soler R, Lopez-Berestein G, Luke DR (1989) Pharmacokinetic and pharmacodynamic evaluation of liposomal cyclosporine. Int J Pharm 57(2):125–131. https://doi.org/10.1016/0378-5173(89)90300-1

Valenzuela A, Garrido A (1994) Biochemical bases of the pharmacological action of the flavonoid silymarin and of its structural isomer silibinin. Biol Res 27:105–105

Virmani R, Farb A, Guagliumi G, Kolodgie FD (2004) Drug-eluting stents: caution and concerns for long-term outcome. Coron Artery Dis 15(6):313–318

Viswanathan V, Pharande R, Bannalikar A, Gupta P, Gupta U, Mukne A (2019) Inhalable liposomes of Glycyrrhiza glabra extract for use in tuberculosis: formulation, in vitro characterization, in vivo lung deposition, and in vivo pharmacodynamic studies. Drug Dev Ind Pharm 45(1):11–20. https://doi.org/10.1080/03639045.2018.1513025

Von Nussbaum F, Brands M, Hinzen B, Weigand S, Häbich D (2006) Antibacterial natural products in medicinal chemistry—exodus or revival? Angew Chem Int Ed 45(31):5072–5129. https://doi.org/10.1002/anie.200600350

Walunj M, Doppalapudi S, Bulbake U, Khan W (2019) Preparation, characterization and in vivo evaluation of cyclosporine cationic liposomes for the treatment of psoriasis. J Liposome Res (Just-Accepted):1–25. https://doi.org/10.1080/08982104.2019.1593449

Wang L, Zhao X, Zu Y, Wu W, Li Y, Zu C, Zhang Y (2016) Enhanced dissolution rate and oral bioavailability of ginkgo biloba extract by preparing nanoparticles via emulsion solvent evaporation combined with freeze drying (ESE-FR). RSC Adv 6(81):77346–77357. https://doi.org/10.1039/C6RA14771B

Wang L, Zhao X, Yang F, Wu W, Liu Y, Wang L, Wang L, Wang Z (2019) Enhanced bioaccessibility in vitro and bioavailability of Ginkgo biloba extract nanoparticles prepared by liquid anti-solvent precipitation. Int J Food Sci Technol. https://doi.org/10.1111/ijfs.14141

Want MY, Islamuddin M, Chouhan G, Ozbak HA, Hemeg HA, Dasgupta AK, Chattopadhyay AP, Afrin F (2015) Therapeutic efficacy of artemisinin-loaded nanoparticles in experimental visceral leishmaniasis. Colloids Surf B: Biointerfaces 130:215–221. https://doi.org/10.1016/j.colsurfb.2015.04.013

Waranuch N, Ramachandran C, Weiner ND (1998) Controlled topical delivery of cyclosporin-A from nonionic liposomal formulations: mechanistic aspects. J Liposome Res 8(2):225–238. https://doi.org/10.3109/08982109809035528

Watkins R, Wu L, Zhang C, Davis RM, Xu B (2015) Natural product-based nanomedicine: recent advances and issues. Int J Nanomedicine 10:6055. https://doi.org/10.2147/IJN.S92162

Wu CY, Benet LZ, Hebert MF, Gupta SK, Rowland M, Gomez DY, Wacher VJ (1995) Differentiation of absorption and first-pass gut and hepatic metabolism in humans: studies with cyclosporine. Clin Pharmacol Ther 58(5):492–497. https://doi.org/10.1016/0009-9236(95)90168-X

Wu T-H, Yen F-L, Lin L-T, Tsai T-R, Lin C-C, Cham T-M (2008) Preparation, physicochemical characterization, and antioxidant effects of quercetin nanoparticles. Int J Pharm 346(1–2):160–168. https://doi.org/10.1016/j.ijpharm.2007.06.036
Xie D, Xu Y, Jing W, Juxiang Z, Hailun L, Yu H, Zheng D-H, Lin Y-T (2017) Berberine nanoparticles protects tubular epithelial cells from renal ischemia-reperfusion injury. Oncotarget 8(15):24154. https://doi.org/10.18632/oncotarget.16530
Xu H, Yuan X-D, Shen B-D, Han J, Lv Q-Y, Dai L, Lin M-G, Yu C, Bai J-X, Yuan H-L (2014) Development of poly (N-isopropylacrylamide)/alginate copolymer hydrogel-grafted fabrics embedding of berberine nanosuspension for the infected wound treatment. J Biomater Appl 28(9):1376–1385. https://doi.org/10.1177/0885328213509503
Xue M, Yang M-X, Zhang W, Li X-M, Gao D-H, Ou Z-M, Li Z-P, Liu S-h, Li X-j, Yang S-Y (2013) Characterization, pharmacokinetics, and hypoglycemic effect of berberine loaded solid lipid nanoparticles. Int J Nanomedicine 8:4677. https://doi.org/10.2147/IJN.S51262
Yalkowsky SH (1999) Solubility and solubilization in aqueous media, vol 3. American Chemical Society, Washington, DC
Yen F-L, Wu T-H, Lin L-T, Lin C-C (2007) Hepatoprotective and antioxidant effects of Cuscuta chinensis against acetaminophen-induced hepatotoxicity in rats. J Ethnopharmacol 111(1):123–128. https://doi.org/10.1016/j.jep.2006.11.003
Yen F-L, Wu T-H, Lin L-T, Cham T-M, Lin C-C (2008) Nanoparticles formulation of Cuscuta chinensis prevents acetaminophen-induced hepatotoxicity in rats. Food Chem Toxicol 46(5):1771–1777. https://doi.org/10.1016/j.fct.2008.01.021
Yen F-L, Wu T-H, Lin L-T, Cham T-M, Lin C-C (2009) Naringenin-loaded nanoparticles improve the physicochemical properties and the hepatoprotective effects of naringenin in orally-administered rats with CCl 4-induced acute liver failure. Pharm Res 26(4):893–902. https://doi.org/10.1007/s11095-008-9791-0
Yingchao L, Lei D, Ai J, Xinming C, Hui X (2007) Preparation and anti-fibrotic effects of solid lipid nanoparticles loaded with silibinin [J]. J Xi'an Jiaotong Univ (Med Sci) 5
You-you T, Mu-Yun N, Yu-Rong Z, Lan-Na L, Shu-Lian C, Mu-Qun Z, Xiu-Zhen W, Zheng J, Xiao-Tian L (1982) Studies on the constituents of Artemisia annua Part II. Planta Med 44(03):143–145. https://doi.org/10.1055/s-2007-971424
Yuan X-B, Yuan Y-B, Jiang W, Liu J, Tian E-J, Shun H-M, Huang D-H, Yuan X-Y, Li H, Sheng J (2008) Preparation of rapamycin-loaded chitosan/PLA nanoparticles for immunosuppression in corneal transplantation. Int J Pharm 349(1–2):241–248. https://doi.org/10.1016/j.ijpharm.2007.07.045
Yuan H, Ma Q, Ye L, Piao G (2016) The traditional medicine and modern medicine from natural products. Molecules 21(5):559. https://doi.org/10.3390/molecules21050559
Zakeri-Milani P, Loveymi BD, Jelvehgari M, Valizadeh H (2013) The characteristics and improved intestinal permeability of vancomycin PLGA-nanoparticles as colloidal drug delivery system. Colloids Surf B: Biointerfaces 103:174–181. https://doi.org/10.1016/j.colsurfb.2012.10.021
Zaveri NT (2006) Green tea and its polyphenolic catechins: medicinal uses in cancer and noncancer applications. Life Sci 78(18):2073–2080. https://doi.org/10.1016/j.lfs.2005.12.006
Zhang J, Liu J, Li X, Jasti B (2007) Preparation and characterization of solid lipid nanoparticles containing silibinin. Drug Deliv 14(6):381–387. https://doi.org/10.1080/10717540701203034
Zhang J, Nie S, Wang S (2013a) Nanoencapsulation enhances epigallocatechin-3-gallate stability and its antiatherogenic bioactivities in macrophages. J Agric Food Chem 61(38):9200–9209. https://doi.org/10.1021/jf4023004
Zhang P, Lin R, Yang G, Zhang J, Zhou L, Liu T (2013b) Solubility of naringenin in ethanol and water mixtures. J Chem Eng Data 58(9):2402–2404. https://doi.org/10.1021/je4000718
Zhang X-P, Le Y, Wang J-X, Zhao H, Chen J-F (2013c) Resveratrol nanodispersion with high stability and dissolution rate. LWT-Food Sci Technol 50(2):622–628. https://doi.org/10.1016/j.lwt.2012.07.041
Zhang J, Zhou X, Yu Q, Yang L, Sun D, Zhou Y, Liu J (2014) Epigallocatechin-3-gallate (EGCG)-stabilized selenium nanoparticles coated with Tet-1 peptide to reduce amyloid-β aggrega-

tion and cytotoxicity. ACS Appl Mater Interfaces 6(11):8475–8487. https://doi.org/10.1021/am501341u
Zhang H, Hou L, Jiao X, Ji Y, Zhu X, Zhang Z (2015a) Transferrin-mediated fullerenes nanoparticles as Fe2+–dependent drug vehicles for synergistic anti-tumor efficacy. Biomaterials 37:353–366. https://doi.org/10.1016/j.biomaterials.2014.10.031
Zhang H, Ji Y, Chen Q, Jiao X, Hou L, Zhu X, Zhang Z (2015b) Enhancement of cytotoxicity of artemisinin toward cancer cells by transferrin-mediated carbon nanotubes nanoparticles. J Drug Target 23(6):552–567. https://doi.org/10.3109/1061186X.2015.1016437
Zhang J, Nie S, Hossen MN, Sun M, Martinez-Zaguilan R, Sennoune S, Wang S (2015c) Anti-atherogenic effects of lesion-targeted epigallocatechin gallate (EGCG)-loaded nanoparticles. FASEB J 29(1_supplement):271.3
Zhou Y, S-q L, Peng H, Yu L, He B, Zhao Q (2015) In vivo anti-apoptosis activity of novel berberine-loaded chitosan nanoparticles effectively ameliorates osteoarthritis. Int Immunopharmacol 28(1):34–43. https://doi.org/10.1016/j.intimp.2015.05.014
Zhu QY, Zhang A, Tsang D, Huang Y, Chen Z-Y (1997) Stability of green tea catechins. J Agric Food Chem 45(12):4624–4628. https://doi.org/10.1021/jf9706080
Zimmerman JJ, Ferron GM, Lim HK, Parker V (1999) The effect of a high-fat meal on the oral bioavailability of the immunosuppressant sirolimus (rapamycin). J Clin Pharmacol 39(11):1155–1161. https://doi.org/10.1177/00912700990390l107
Zou Q, Li Y, Zhang L, Zuo Y, Li J, Li J (2009a) Antibiotic delivery system using nano-hydroxyapatite/chitosan bone cement consisting of berberine. J Biomed Mater Res A 89(4):1108–1117. https://doi.org/10.1002/jbm.a.32199
Zou W, Cao G, Xi Y, Zhang N (2009b) New approach for local delivery of rapamycin by bioadhesive PLGA-carbopol nanoparticles. Drug Deliv 16(1):15–23. https://doi.org/10.1080/10717540802481307

Chapter 4
Nanoemulsions as Optimized Vehicles for Essential Oils

Thaís Nogueira Barradas and Kattya Gyselle de Holanda e Silva

Abstract During the last few years, a great interested has been given in for nanotechnology applications in pharmaceutical technology. At the same time, there is a growing research effort in the search for healthier and safer products. In this context, the use of naturally originate raw materials, essential oils for instance, rises as an interesting approach for replacing several synthetic active pharmaceutical ingredients. However, these materials often feature low bioavailability, uncontrolled volatility or low long-term stability, which require novel encapsulation techniques. Therefore, research efforts have been focusing on nanoemulsion technology that is particularly suited to produce novel products. Nanoemulsions constitute one interesting vehicle for enhancing solubility, stability and delivering natural oils, by encapsulating them into nanosized micelles with sizes ranging from 20–200 nm. They gather some unique characteristics as small size, increased surface area and stability which can increase efficiency and biological effects of pharmaceutical dosage forms.

The wide application of nanoemulsions to the encapsulation of bioactive molecules still require much development in order to achieve the optimization of the obtention methods for large-scale production. This chapter includes an overview about nanoemulsion stability characteristics and the different approaches for obtaining nanoemulsions, including high energy methods like high-pressure homogenization, microfluidizers and ultrasonic homogenization, which are the most used approaches, and low energy methods such as phase inversion composition and phase inversion temperature, simple but still less reproduceable methods. In addition, we present the main aspects of nanoemulsion formulations, type of surfactants and oil phase, and the techniques for characterizing nanoemulsions, for example,

T. N. Barradas (✉)
Instituto Federal de Educação, Ciência e Tecnologia do Rio de Janeiro, Rio de Janeiro, Brazil

Instituto de Macromoléculas Professora Eloisa Mano, Universidade Federal do Rio de Janeiro (IMA/UFRJ), Laboratório de Macromoléculas e Colóides aplicados à Indústria do Petróleo, Rio de Janeiro, Brazil
e-mail: thais.barradas@ima.ufrj.br

K. G. de Holanda e Silva
Faculdade de Farmácia, Universidade Federal do Rio de Janeiro, Rio de Janeiro, Brazil

A. Saneja et al. (eds.), *Sustainable Agriculture Reviews 44*, Sustainable Agriculture Reviews 44, https://doi.org/10.1007/978-3-030-41842-7_4

dynamic light scattering, zeta potential, microscopy and X-ray diffraction. Moreover, applications of nanoemulsions to pharmaceutical technological approaches for essential oils encapsulation, such as the improvement of bioactive oils bioavailability and solubilization, masking unpleasant aspects of oils and enhancement of essential oils pharmacological activity are also discussed.

Keywords Nanoemulsions · Nanoencapsulation · Surfactants · Emulsification methods · Essential oils · Drug delivery

4.1 Introduction

Recently, the use of naturally originate bioactive molecules, especially natural antioxidants and anti-microbial agents, has been rising. The search for natural products has been triggering more and more research in the development of suitable and stable formulations. Essential oils are bioactive molecules, being considered safe and biocompatible, offering a great range of benefits due to a complex composition of fatty acids, terpenes, triterpenes and many other lipophilic components, that can offer protection against dehydration, solar radiation, inflammation, insect attack, microorganisms, and viruses (Donsì and Ferrari 2016; Andreu et al. 2015; Bonferoni et al. 2017).

Many essential oils present proved antioxidant potential, preventing oxidation reaction and reducing the formation of free radicals, which are risk factors for cellular damage and for chronicle diseases as cancer. However, one of the main challenges to formulate essential oils into a pharmaceutical product lies on low water solubility of most of oils, which impairs bioavailability of lipophilic components and stability in hydrophilic formulations. Moreover, many essential oils are highly volatile and labile, being unstable under many environmental conditions, as light exposure and oxidation. Those restrictions must be overcome during the development of pharmaceutical formulations (Badgujar et al. 2014; Herman and Herman 2015; Moghimi et al. 2016).

Nanoemulsions are nanosized colloidal dispersions with droplets within the nanometric scale, more often between 20 and 200 nm. Nanoemulsions are thermodynamically stable isotropic systems in which two immiscible liquids are mixed to form a single phase by means of an emulsifying agent, i.e., a surfactant (Helgeson 2016; Solans et al. 2005). Nanoemulsions have been considered of a great potential of industrial application in pharmaceutical, food, agrochemical, cosmetics and personal care products (McClements and Jafari 2018). Nanoemulsions are widely applied to encapsulate, protect and promote controlled release of lipophilic bioactive molecules. Besides, nanoemulsions can also act as templates to produce other types of nanoparticles, such as solid lipid nanoparticles (SLN), nanostructured lipid carriers (NLC), multiple emulsions and hydrogel-thickened nanoemulsions (Sutradhar and Amin 2013; McClements and Jafari 2018; Barradas et al. 2017).

Because of the reduced droplet size, nanoemulsions promote interesting characteristics in comparison to conventional emulsions, i.e. macroemulsions. For instance, nanoemulsions present much larger surface area of dispersed phase in relation to the total volume of the dispersion than that observed in macroemulsions. Thus, droplet deformation-derived phenomena are typically higher for nanoemulsions than for conventional emulsions (Helgeson 2016). Besides, macroemulsions exhibit multiple light scattering, which provide them a white opaque appearance. In contrast, in nanoemulsions, droplets are much smaller than the wavelength of visible light. Hence, most of the nanoemulsions are optically transparent systems when droplet size is small enough, usually below 100 nm (McClements and Rao 2011).

The development of nanoemulsions as nanostructured carriers for pharmaceutical application requires an accurate selection of the components such as, surfactants, and oils as well as the preparation method selected for nanoemulsion production. Special attention must be given to oily phase used as inner phase in oil-in-water nanoemulsions. In order to obtain nanoemulsions, a minimum energy input is necessary. Energy input can be achieved by applying work to the system, usually called "high energy" methods, of by modulating composition or temperature of the system, referred as "low energy" methods. Both methods require the same amount of energy for providing nanosized droplets, regardless if the energy is provided mechanically or thermodynamically (Helgeson 2016; McClements and Jafari 2018).

Nanoemulsions and polymeric micelles can be considered promising systems for encapsulation of essential oils since because of various advantages such as: sustained and controlled release of bioactive molecules; solubilization of lipophilic substances as oils; they are suitable for administration under different routes; protection from chemical, environmental and enzymatic degradation of labile molecules; controlled volatilization of essential oils, reduction of side effects and dose.

In this chapter, we provide an overview focused on formulation of oil-in-water nanoemulsions, as they can provide encapsulation of lipophilic components, and diverse approaches for nanoemulsions production. Additionally, we summarize the instability phenomena related to nanoemulsions and the recent applications of nanoemulsions for the encapsulation of natural oils.

4.2 Nanoemulsions: General Aspects

Emulsions are heterogeneous colloidal systems, consisting of an immiscible liquid dispersed as droplets, i.e. disperse or inner phase in another immiscible liquid, i.e., continuous or outer phase, stabilized by a surfactant. When the continuous phase is an aqueous solution and the dispersed phase is an oil, the emulsion is called oil-in-water emulsion (Fig. 4.1a). On the other hand, when the dispersed phase is an aqueous solution and the continuous phase is constituted by an oil, the emulsion is called

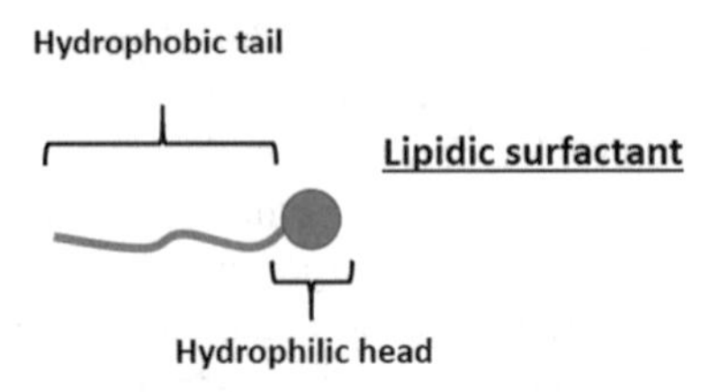

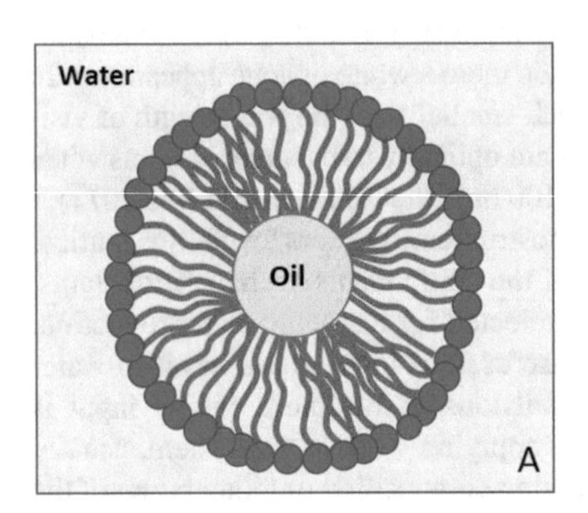

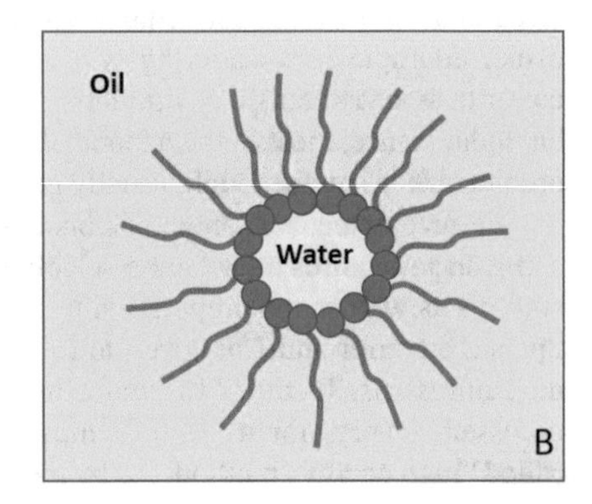

Fig. 4.1 Oil-in-water nanoemulsion (**a**) and water-in-oil nanoemulsion (**b**), composed by oil and aqueous phases, stabilized by surfactants. In oil-in-water nanoemulsion aqueous phase constitutes the outer phase and oil phase is the inner phase, and the surfactant molecule self-assemble into a micelle structure, with hydrophilic portion oriented towards the aqueous phase. In water-in-oil nanoemulsions oil phase and water phase constitute the outer phase and inner phase, respectively, and surfactants hydrophobic portion is in contact with the outer phase

water-in-oil emulsion (Fig. 4.1b). In all cases, an emulsifier agent or surfactant is needed to reduce interfacial tension between dispersed and continuous phases and provide stable systems. The nature of the surfactant and the stabilizing film provided by it, greatly influence general aspects of nanoemulsions, as long-term stability, droplet size and charge, for example. In general, emulsified systems can be classified according to some physicochemical properties such as droplet size and preparation methods into macroemulsions, microemulsions and nanoemulsions.

The most important difference between nanoemulsions and macroemulsions lies in the droplet size, since both are thermodynamically unstable systems. Nanoemulsions usually show mean droplet size smaller than 200 nm, although there is no agreement regarding the size range that comprise nanoemulsion and distinguish them from micro-sized and conventional-sized macroemulsions. According to the literature, droplet size can vary from 20 nm to either 500, 300, 200, or yet 100 nm, being until now a controversial issue regarding nanoemulsions classification (Solans and Solé 2012; McClements and Rao 2011).

In general, macroemulsions are optically opaque, due to the large droplet size, which produces multiple light scattering, while nanoemulsions are often transparent or translucent, since they present small droplet size scatter light weakly (McClements and Jafari 2018; McClements 2011). Nanoemulsions are kinetically stable systems and do not require the addition of a cosurfactant, unlike microemulsions. Kinetic stability is attributed to the Brownian motion of the droplets which overcomes the gravity force and prevents droplets sedimentation and coalescence. In addition, the use of nonionic or polymeric surfactants may cause steric repulsion between the droplets and contribute to the stability of the system (Tadros et al. 2004; Solans et al. 2005).

Besides storage stability, there are other interesting physical properties that differentiate nanoemulsions from macroemulsions. For example, nanoemulsions have dispersed phase surface area relative to the total volume of the dispersion much larger than that observed in macroemulsions. Due to the high relative surface area, all phenomena related to droplet deformation are typically more relevant to nanoemulsions than to macroemulsions (Mason et al. 2006).

It is noteworthy the common misconception between nano and microemulsions. There has been much effort to clarify the difference between these two colloidal systems in this scientific field (Sonneville-Aubrun et al. 2004; Anton and Vandamme 2011). Both microemulsions and nanoemulsions contain droplets of diameters ranging from 20 to 200 nm, differing in certain aspects, especially regarding the preparation method (Abismail et al. 1999; McClements 2011). Although both systems can feature similar characteristics, as nanosized droplets as inner phase dispersed in a continuous phase, they differ in therms of stability and in physicochemical concepts (Anton and Vandamme 2011; McClements 2012). Microemulsions is a term usually used to refer to thermodynamically stable isotropic oil/surfactant/water systems, while nanoemulsions are kinetically stable conventional emulsions constituted by nanosized micelles (McClements 2012). Microemulsions can feature a wide range of structures with one, two, three ore more phases in equilibrium, depending on the concentrations between all the components and upon certain temperatures. These different structures can be water-continuous, oil-continuous or bicontinuous and nanometric swollen micelles, which gives them a bluish and translucent nanoemulsion-like aspect. However, micelles can form different geometries such as, worm-like, bicontinuous sponge-like, liquid crystalline, hexagonal, lamellar, and spherical swollen micelles, which is the most often confused with nanoemulsions (Anton and Vandamme 2011; McClements 2012).

Moreover, nanosized micelles from microemulsion are deeply afected in morphology and size upon temperature and dilution, while nano-micelles from nanoemulsions remain unaltered when submited to the same conditions, which is an important aspect to be considered when microemulsions are intended to be applied as parenteral prooducts. Parenteral route can pose some challenges as infinite dillution conditions and temperature, pH and osmolarity variations. Under biological conditions, only nanoemulsions remain stable (Lefebvre et al. 2017; Hörmann and Zimmer 2016; Anton and Vandamme 2011).

These misinterpretations are often found in literature, where one can find inadequate methods for obtaining and characterizing micro and nanoemulsions. For example, ternary phase diagrams to produce nanoemulsions constitutes a very frequent but inapropriate methodology that can be more suitable to produce microemulsions. The simplicity of low-energy emulsifying methods is the main aspect that can induce the confusion between nano and microemulsions. Besides, in order produce nanoemulsions by low-energy methods, surfactant must be mixed in oil phase before adding water, not being possible inverting the order of components incorporation. Microemulsions can be formed no matter the order in which components are added (Anton and Vandamme 2011). This difference between both systems is an easy and simple preliminary test for classification of a system in whether nano or microemulsions.

The large surface area provided by the nanostructured droplets increases drug absorption in biological membranes such as the intestinal epithelium, the cornea and skin for example (Singh et al. 2017). Moreover, because they are transparent formulations, nanoemulsions present an interesting aesthetic characteristic, even containing significant quantities of oil in the composition (Sonneville-Aubrun et al. 2004; Tadros et al. 2004).

4.3 Nanoemulsions: Composition

In general, a classical nanoemulsion formulation comprises an aqueous and an oil phase stabilized by a surfactant. The physical-chemical properties of the components will greatly impact the formation and stability of the nanoemulsions obtained from them. Typically, nanoemulsions require relatively high amounts of surfactants, usually from 10 to 15wt.%, in order to stabilize the high surface area of nanosized micelles (Azeem et al. 2009; Mason et al. 2006). The composition of nanoemulsions can be tuned by careful selection of the ingredients used to obtain them.

4.3.1 Surfactants

Surfactants are molecules whose structures comprise two parts of opposing affinities, one having a hydrophilic character and the other with a hydrophobic character. Both hydrophilic and lipophilic moieties should be in equilibrium in the surfactant molecule, so that it constitutes a good emulsifier. This hydrophilic-lipophilic equilibrium of a both moieties in the surfactant molecular structure is called hydrophilic-lipophilic balance (Abismail et al. 1999; Porras et al. 2004). Hydrophilic-lipophilic balance is a classifying system for all surfactants and provides a good indication for the right the choice of emulsifiers suitable for obtaining stable emulsions. During the production of nanoemulsions, surfactants molecules should be able to quickly adsorb on newly formed droplet surface, while maintaining droplet integrity once two droplets collide, avoiding droplet deformation or coalescence (Tadros 1994).

The formation of a stable nanoemulsion involves selecting an appropriate composition, controlling the order of addition of the components and applying a minimum energy input capable of promoting the deformation of droplets. This minimum energy input is related to the Young-Laplace Theory, which shows that the difference between the external and internal pressures of a drop is a direct function of the droplet radius (Eq. 4.1). Therefore, to break a drop into droplets of smaller size, the difference between the internal and external pressure to the drop should be considerable (Tadros et al. 2004).

$$Pi - Pe = \frac{\gamma}{2r} \tag{4.1}$$

where,

Pi = Internal pressure
Pe = External pressure
γ = Interfacial tension
r = Radius of droplet curvature

According to Eq. 4.1, large amounts of surfactants are required to reduce surface tension and stabilize the great interfacial surface produced. The production of nanoemulsions, therefore, is a thermodynamic unfavorable and non-spontaneous process. Besides, to obtain droplets of nanometric size it is necessary to supply energy to the system, usually provided by mechanical devices or even by chemical potential of the constituents. Nanoemulsions often present spherical droplets because both the high interfacial tension and small droplet size combined provide high Laplace pressures, which triggers the minimization of the oil–water interfacial area into a spherical shape (McClements 2011).

The success of developing a stable nanoemulsion depends on the right selection of the surfactant used, which shall act on the interface between dispersed and continuous phases. The choice of a surfactant for pharmaceutical purposes is one of the most critical steps and involves consideration of the toxicity of substances that can be applied in large quantities. (Tadros 2009; Wiedmann 2003). The selection of a surfactant can be considered the most important step to produce nanoemulsions, since emulsifiers can influence both fabrication method and performance of nanoemulsions.

Due to the amphiphilic molecular constitution, surfactants tend to migrate to the interfaces between two immiscible fluids, in a way that polar groups are oriented to the aqueous phase and the non-polar, to the organic phase. The polar group is of great importance because it defines some properties, such as water solubility and the classification of the surfactants which can be divided into ionic, which can be anionic and cationic, nonionic and amphoteric, or zwitterionic.

At low concentrations, the surfactant molecules are solubilized within the solution as free soluble surfactant molecules (Fig. 4.2a). As the surfactant concentration increases, a decrease in surface tension of the solution occurs, representing the surfactant adsorption at air-water surface (Fig. 4.2b). When reaching a certain concentration, it is observed that the variation of the surface tension is minimal in relation to the increase of the concentration, which means that, the saturation of the water-air interface is reached. At this stage, the adsorption of the surfactant on the surface is no longer observed. The surfactant concentration that causes this phenomenon is

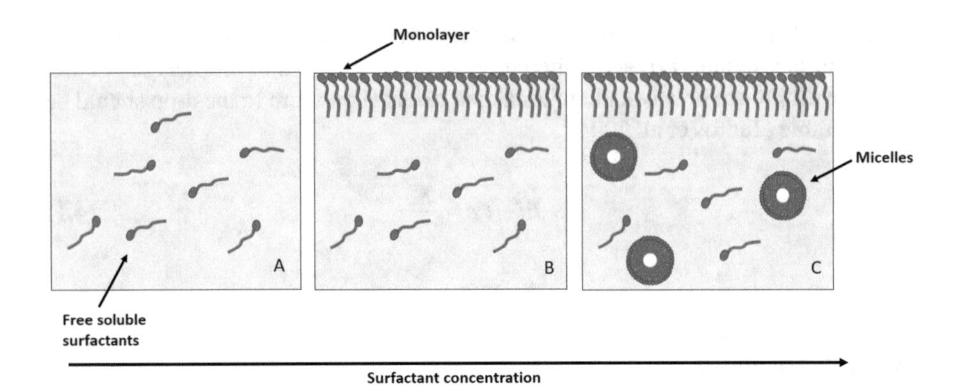

Fig. 4.2 Micellization steps: surfactants as free soluble molecules, below critical micelle concentration (**a**). As surfactant concentration raises, surfactants start to self-assemble as monolayer in water-air interface (**b**). Above critical micelle concentration, water-air interface is saturated with surfactant molecules and interfacial tension no longer reduces. Then surfactants start to self-assemble as micelles, in a thermodynamically favorable phenomenon (**c**)

called the critical micelle concentration, where it can be observed the formation of molecular aggregates, known as micelles (Fig. 4.2c).

The moieties that are not soluble in the surrounding solvent attract strongly and produce a stable compact vesicular morphology, as to expel the solvent from this environment. On the other hand, groups that are very soluble, tend to be externally exposed producing a soluble particle, i.e., a micelle (Myers 2005b). The size of the micelles and the number of surfactant molecules per micelle depends on the type of surfactant and the physicochemical environment. When the surfactant is adsorbed at the oil-water interface, interfacial tension decreases causing a steric hindrance effect or electrostatic repulsion, both able to prevent coalescence or aggregation phenomena. These barriers not only prevent emulsion droplets from coming into direct contact, but also serve to stabilize the liquid film between two adjacent droplets (Myers 2005a).

The interfacial tension is related to the amount of surfactant adsorbed at the interface and the nature of the interfacial layer. The interfacial tension decreases with increasing surface charge, which is directly related to the concentration and size of the surfactant. However, depending on the type of surfactant, many other effects are important (Tadros et al. 2004; Myers 2005a). It can be said, then, that surfactants play a very important role in the stabilization or destabilization of nanoemulsions, increasing or reducing the electrostatic or steric repulsions of the interface, which are dependent on structural aspects, such as double electric layer, branching, aromaticity. Other factors like the presence and type of electrolytes, pH, temperature and presence of additives (Myers 2005a). The most studied for the formation of nanoemulsions are non-ionic surfactants.

Both liquids and solids feature surface tension due to the cohesive energy between their molecules. Surface tension is resultant from unbalanced forces of attraction or non-equilibrated in interfacial regions, in which there is a sudden variation of the density. Thus, the resultant force on a molecule close to the surface liquid/vapor is different from that on a molecule that is in a completely homogeneous region, in which the resulting force is null. The surface tension can also be defined as the excess of surface energy (Tsujii 1998; Shaw 1992).

Liquids tend to self-assemble into forms that minimize surface area, that is, a greater number of molecules are inside the liquid and, therefore, remain surrounded by other liquid molecules. Liquid droplets therefore tend to be spherical because a sphere is the shape with the smallest surface/volume ratio. Surface effects can be expressed according to the Helmholtz and Gibbs energies (Eq. 4.2), in which the bond between these amounts of energy and the surface area is the work (force) needed to increase the surface area (σ) of a liquid in a given quantity (Hunter 2001)

$$\mathrm{dW} = \gamma \cdot d\theta \tag{4.2}$$

where, W means work or applied force; γ ($\mathrm{N.m^{-1}}$) means superficial tension and θ superficial tension.

The effect of surface tension is to minimize surface area, resulting in a curved surface, as in a vesicle. The intensity of this tension depends on the liquid, solvent purity and temperature in which the surfactant lies (Hunter 2001). The same considerations apply to the interface between two immiscible liquids. Again, intermolecular forces in non-equilibrium are presented, however with decreased intensity. The interfacial tensions usually lie between the individual surface tension of the two liquids above mentioned (Shaw 1992).

4.3.2 *Types of Surfactants*

There are several surfactants that have been used in the production of nanoemulsions, since small-molecule synthetic surfactants, which may be non-ionic and ionic, and natural surfactants such as phospholipids, proteins and polysaccharides. In general, the surfactants may be: (i) non-ionic, (ii) amphoteric (iii) cationic, or (iv) anionic.

The class of the non-ionic surfactants comprise small molecules such as sorbitan esters, like Span® and Tween® series and polymeric surfactants, such as poly(oxyethylene) ethers, for example, Brij® or block copolymers as poly(ethylene oxide)-poly(propylene oxide). Small molecules can form small-sized droplets because of the facility on easily adsorb on the surface of droplets that they feature (Jafari et al. 2017).

Natural surfactants exhibit excellent surface activity, although they present molecular structure being very voluminous compared to synthetic surfactants. Since natural surfactants are naturally produced substances, they show advantages like biodegradability, are easily produced from renewable resources, present high specificity and tend to less toxic (Salem and Ezzat 2018). Table 4.1 summarizes the most common types of natural and synthetic surfactants.

Table 4.1 Main types of natural and synthetic surfactants and their general aspects

Type	Examples	Characteristics/ charge	General aspects	References
Natural	Phospholipids for example: lecithin	Zwitterionic	Mixture of natural polar phospholipids obtained from eggs or soybean. Amphoteric behavior	Klang and Valenta (2011), Hoeller et al. (2009), Khan and Krishnaraj (2014)
Natural	Polysaccharides for example: gum arabic	Negatively or positively charged	Provide good stability either by electrostatic or steric effects. Not suitable for low-energy emulsification methods.	Bai et al. (2016), Jafari et al. (2017), Dickinson et al. (1991)
Natural	Amphiphilic proteins, as caseinate or whey protein isolate	Charge depends on pH relative to isoelectric point	Provide good stability mainly by electrostatic repulsion and also steric effects under neutral pH. Not suitable for low-energy emulsification methods	M. Sharma et al. (2017), Dickinson et al. (1998), Kuhn and Cunha (2012), David Julian McClements (2011)
Ionic surfactant	Sodium lauryl sulfate (SLS) or sodium dodecyl sulfate (SDS)	Negatively charged	Used in both low and high energy methods High irritation and acute toxicity potential	Tian et al. (2016), Lémery et al. (2015), Rosety et al. (2001), Bondi et al. (2015)
Non-ionic surfactant	Sucrose monopalmitate Sorbitan monooleate	Sugar esters	Non-toxic, biodegradable and hydrophilic. Used to stabilize oil-in-water nanoemulsions by high pressure homogenization with relatively low surfactant-to-oil ratios.	Strickley (2004), Cerqueira-Coutinho et al. (2015), Rao and McClements (2011)
Non-ionic surfactant	Brij	Polyoxyethylene alkyl esters	Relatively low phase inversion temperature. Used in low-energy methods	David Julian McClements (2011)
Non-ionic surfactant	Tweens and spans	Ethoxylated sorbian esters	Used in high-energy methods. Differences aliphatic chain length varies, providing several types of Tween and Span. Ex: Tween 20, 40, and 60.	Pathania et al. (2018), Dias et al. (2014), Salvia-Trujillo et al. (2015), Speranza et al. (2013)

The use of natural surfactants, due to the innocuous character, provides some advantages over synthetic surfactants, including biocompatibility and biodegradability, multifunctional characteristics, stable activity under different environmental conditions, such as high or low temperatures, pH, high pressure and osmolarity. Thus, natural surfactants may be more effective in stabilizing nanoemulsions for the most diverse purposes (Dickinson 2003; Qian and McClements 2011). In general, most of natural surfactants are large molecules, which can delay the ability to adsorb on the droplets surface and provide larger droplets. However, they can provide good stability by both steric and electrostatic stabilization (Qian and McClements 2011).

One of the most common natural surfactants used in the development of nanoemulsions are phospholipids. Phospholipids can be classified as natural amphoteric surfactants and have excellent biocompatibility because they are similar in composition to cell membranes. The term lecithin is commonly used to designate a mixture of phospholipids obtained from both animal and vegetable sources, as eggs and soybean, respectively (Shchipunov 2015). The composition of lecithin may vary upon several environmental and processing conditions, however, the phospholipids most commonly found in lecithin are: phosphatidylcholine, phosphatidylethanolamine, phosphatidylinositol, sphingomyelin, glycerol phospholipids of complex fatty acid composition (Klang and Valenta 2011; Shchipunov 2015).

The polar portion of lecithin is comprised by positively charged phosphate groups, which are bounded to lipophilic chains by ester groups. Moreover, nitrogen-containing moieties bounded to phosphate groups can assume negative charges, providing phospholipids such as lecithin an amphoteric character (Shchipunov 2015).

Biopolymers such as polysaccharides have been attracting attention as natural surfactants for producing stable nanoemulsions (Dickinson 2003; Bai et al. 2016). Polysaccharides feature a complex structure, water-soluble, with good emulsifying properties and thickening properties, are widely used in food, pharmaceutical and paper industries. Polysaccharides and gums such as gum arabic are often used as thickening or gelling agents in pharmaceutical industry (Prajapati et al. 2013). The polysaccharides most commonly found are hydrocolloids such as xanthan, modified starch, galactomannans and pectin (Dickinson 2003; Prabaharan 2011; Sweedman et al. 2013; Prajapati et al. 2013). Recently, they have been considered promising components to be used as natural surfactants as there is a growing demand for natural and sustainable ingredients. The emulsification properties of polysaccharides are related to the presence of non-polar groups and hydrophilic groups, providing the interaction with both oil and water. The stabilization method related to polysaccharides can occur by both electrostatic repulsion and steric effect, preventing droplet flocculation and coalescence (Jafari et al. 2017).

Gum arabic is one of the most used polysaccharides for producing stable emulsions, due to the formation of a stable layer around the droplets, providing steric repulsion among droplets by forming steric layer around them (Yadav et al. 2007; McNamee et al. 1998). Other naturally occurring polysaccharides used in producing nanoemulsion comprise pectin, a heterogenous anionic polysaccharide already used as gelling, thickening and stabilizing properties (Bai et al. 2016) and maltodextrin (Sonu et al. 2018).

Some proteins, especially those obtained from bovine milk, as casein and whey proteins, can be used as emulsifying and surfactants agents, since they feature both hydrophilic and lipophilic residues and surface activity, providing cohesive and strong films around droplets (Yerramilli and Ghosh 2017; Mayer et al. 2013; Kuhn and Cunha 2012; Adjonu et al. 2014). Proteins can stabilize droplets mainly by electrostatical repulsion because of the presence of negatively charged groups, although steric effect can also be attributed to them (Adjonu et al. 2014; Dickinson et al. 1998). As proteins are very large molecules, the adsorption on droplet surface takes longer. On the other hand, proteins feature the ability of forming a viscoelastic coating that prevent droplet deformation and, thus aggregation, providing higher stability under several conditions (McClements 2011). One of the most studied proteins used as natural surfactants is casein.

Casein comprises almost 80% of milk protein and it is a heterogeneous protein composed by four main proteins fractions: α_{s1} (~44 wt.%), α_{s2} (~11 wt.%), β (~32 wt.%) and k (~11 wt.%) with a strong aggregation behavior, which is fundamental to stabilize casein-based micelles (Wang and Zhang 2017; Sharma et al. 2017). Casein provides both electrostatic and steric stabilization against coalescence under neutral pH values (Dickinson et al. 1998). At acidic pH values, the negative charge on caseinate structure is reduced and aggregation between casein-based micelles occurs (Sharma et al. 2017).

Several works have reported the combination of proteins and polysaccharides-based emulsifiers based on the ability to form polyelectrolyte complexes between these molecules and the possibility of Maillard reaction, which contribute to increase stability of the layer around nanosized micelles end prevent coalescence (Sonu et al. 2018; Farshi et al. 2019).

There is a growing demand on replacing synthetic surfactants for natural molecules. However, the application of natural surfactants still is limited by several issues. For example, Phospholipids and protein-based surfactants, are unstable under environmental conditions such as pH, osmotic pressure and heating. Polysaccharides, on the other hand, provide stability under environmental stressful conditions (Zhang et al. 2015). Besides, polysaccharides can stabilize nanoemulsion by increasing the viscosity oh aqueous phase by stablishing hydrogen bonds water molecules (Dickinson 2009). Moreover, according to Qian and McClements 2011, small-molecule surfactants can provide smaller droplets than proteins, which can be related to the ability to rapidly adsorb to the droplet surfaces during homogenization and providing a more stable interface.

The solubility of synthetic or ionic surfactants is related to the interactions between the ionic group, which is the polar part of the surfactant, and water. The ionic surfactants most used in the production of nanoemulsion are summarized in Table 4.1. In general, ionic surfactants are resistant to changes in temperature and quite soluble in water. The applications of ionic surfactants are originated from these properties. Also, pH has great influence on the solubility of the ionic surfactants due to the neutralization reactions that occur with this type of molecules.

In the case of anionic products, a reduction in pH causes a decrease in the degree of dissociation and in the solubility of the surfactant due to the formation of the corresponding undissociated acid, which has lower solubility. The electrical charge on droplet surface plays an important role in stabilizing nanosized droplets as electrostatic repulsion, aggregation stability etc. Moreover, surface charge can be explored to provide nanoemulsions novel application by triggering the interaction of charged surfactants with other components in biological media (Tian et al. 2016). Quaternary ammonium salts belong to the class of cationic surfactants, the best-known being hexadecyl trimethyl ammonium bromide and dodecyl ammonium bromide (Bouchemal et al. 2004; Lawrence and Rees 2000).

In the case of ionic surfactants, the stability of nanoemulsions is mainly governed by electrostatic interaction between the droplets. Surface charge and, consequently, zeta potential can be modulated by mixing non-ionic and ionic surfactants and varying the proportion of both of them, providing extremely long-term stable nanoemulsion (Tian et al. 2016; Babchin and Schramm 2012). Sodium lauryl sulfate (SLS) or sodium dodecyl sulfate is the most common anionic surfactant used to produce emulsions. However, they can disorganize several cell membrane organization and affect both protein and lipid structures, being highly irritant for skin application and toxic for systemic administration (Elmahjoubi et al. 2009; Bondi et al. 2015).

Non-ionic surfactants are widely used for pharmaceutical purposes because they feature low toxicity and greater stability against changes in ionic strength and pH of the common biological media. There are different types of non-ionic surfactants, the most used to provide pharmaceutical nanoemulsions are polyglycerol alkyl ethers, glucosyl dialkyl ethers, crownethers, ester-linked surfactants, polyoxyethylene alkyl ethers, ethoxylated hydrogenated castor oil, Brij, Spans, or sorbitan esters and Tweens, or Polysorbates.

Synthetic non-ionic surfactants show no charges on the polar part of the molecular structure, which is often composed of ethylene oxide groups. For this reason, solubility of synthetic non-ionic surfactants in aqueous solution differs from most solutes: with increasing temperature, these surfactants show phase separation, and this temperature is known as cloud point temperature. This behavior occurs because of the increasing the size of the molecular aggregates, known as micelles and the inter-micellar attraction. This can be explained by the dehydration of the outer layer of the micelle of the nonionic surfactant with the increase in temperature, which is promoted by the breaking of the hydrogen bonds between the ethylene oxide group and the water (Prud'homme et al. 1996; Alexandridis et al. 1994).

Compared with low molecular weight surfactants, non-ionic polymeric surfactants produce more stable micelles, have lower critical micelle concentration values, with a slower dissociation rate, which allows controlling the release of the molecules contained in the nucleus of nano-sized droplets and also achieve greater accumulation of the drug at a specific site. In addition, these copolymers are capable of forming micelles in a wide variety of solvents (Xing and Mattice 1997; Kataoka et al. 2001).

Like Mao et al. studied the effect of different surfactants on interfacial tension, droplet size, zeta potential and morphology of β-carotene-loaded nanoemulsions produced by high pressure homogenization. High molecular weight surfactants, such as Tween 20®, decaglycerol monolaurate, modified starch, whey protein isolate were evaluated. The results showed that Tween 20® and decaglycerol monolaurate provided smaller droplet size, although with poorer stability compared with modified starch and whey protein isolate, presenting droplets aggregates. On the other hand, modified starch and whey protein isolate provided larger droplets with higher stability against droplet aggregation due to the formation of stronger interfacial layers. Moreover, the combination of Tween 20 and whey protein isolate provided nanoemulsions with higher stability regarding β-carotene content under storage conditions of 55 °C for prolonged time (Like Mao et al. 2009).

4.3.3 Oil Phase

Oil phase composes the core of oil-in-water nanoemulsions and it can influence greatly both formation and stability of nanoemulsions. Several aspects as polarity, water miscibility, interfacial tension and viscosity can change nanoemulsions characteristics such as droplet size and stability (McClements 2011). Different types of oil phases can be used for the formation of nanoemulsions: synthetic, mineral, vegetable or essential oils. In this chapter the origin and the chemical and physicochemical properties of essential oils will be the focus. Triacylglycerols, free fatty acids, fixed/vegetable, essential and mineral oils, waxes alone or in associations can be used as oil phase upon nanoemulsion development (Jafari 2017). The most common the essential oils often found are clove, peppermint, sweet fennel and bergamot (Al-Subaie et al. 2015; Saberi et al. 2014; Pengon et al. 2018; Strickley 2004; Wan et al. 2019; Barradas et al. 2015). They can act as drug carriers or solvents or be used as the only constitute of oil phase due to pharmacological properties commonly attributed to essential oils.

Very often, oil phase of nanoemulsions are liquid, as natural oils. However, they can also be solid as natural waxes and solid lipids (in the case of SLN and NLC) (Sutradhar and Amin 2013; McClements and Jafari 2018). Physical state of oil droplets can influence creaming stability, optical properties and release pattern of the encapsulates bioactive molecules. The right selection of liquid and solid lipophilic components can provide the control over physical state of oil droplets (McClements 2015).

In both oil-in-water and water-in-oil, oil phase is one of the most important issues, since it can influence several physicochemical aspects, for example: viscosity, refractive index, transparency, interfacial tension stability, sensorial aspects and droplet size. As a consequence, a range of functional properties that are consequence of physicochemical parameters are affected, as bioavailability, miscibility and digestibility (McClements and Jafari 2018). Moreover, molecular weight of the components, interfacial tension, density and viscosity of the oil phase also influence the properties of nanoemulsions and can lead the choice of a certain type of emulsification method (McClements and Rao 2011).

Droplet stability is dependent on oil solubility, which is influenced by the amount of water soluble and lipophilic oil components. Great amounts of water soluble components in oil phase can lead to Ostwald ripening, the major source of instability in nanoemulsions (Chebil et al. 2013). There are some strategies to retard droplet growth due to Ostwald ripening. Ripening inhibitors can be added to oil phase of oil-in-water nanoemulsions prepared with highly soluble oils as essential oils (Rao and McClements 2012; McClements and Jafari 2018). Ripening inhibitors are usually highly hydrophobic molecules almost water insoluble able to increase hydrophobicity inside the droplets even upon diffusion of water-soluble molecules. As a consequence, there is an increase in the concentration of the ripening inhibitor inside droplets, promoting a concentration gradient in the system that triggers the diffusion of water-soluble components from large back to small droplets, as the opposite of what occurs in Ostwald ripening (McClements and Jafari 2018).

Oils and lipophilic components can be encapsulated inside nanomicelles with the aim of protecting labile and degradable bioactive molecules from environmental conditions, controlling release, increasing bioavailability and providing an easier manipulation or incorporation of oils into aqueous formulations. Moreover, encapsulation techniques can be applied to the modification of physicochemical properties of oils, which makes it possible to convert solutions into fine powders of to mask undesirable tastes when administrated by oral route.

4.3.4 Essential Oils

Essential oils are aromatic substances extracted from aromatic plants or plants parts, including leaves, fruits, barks and seeds. They considered raw materials of great importance for the cosmetic, pharmaceutical, fragrance and food industries. These pure and extremely potent organic substances are considered the main biochemical components of the therapeutic action of medicinal and aromatic plants (Pathania et al. 2018; Pérez-Recalde et al. 2018; Donsì and Ferrari 2016).

These are volatile substances extracted from aromatic plants, being extremely potent organic substances and considered the main biochemical components of the therapeutic action of medicinal and aromatic plants. The methods of extraction vary according to the location of the volatile oil in the plant and the proposed use of it. The most common methods are: enfleurage, steam distillation, or azeotropic distillation; extraction with organic solvent in a continuous and discontinuous way; pressing or supercritical CO_2 extraction (Asbahani et al. 2015).

The designation of oil is given thanks to some physical-chemical characteristics, for example, that they are often oily-appearing liquids at room temperature. The main characteristic of essential oils is the volatility, which differs them from the fixed/vegetable oils, which are mixtures of lipid substances normally obtained from seeds. Another major characteristic is given thanks to the pleasant and intense aroma of most of the volatile oils, being therefore also called essences. They are still soluble in apolar organic solvents, as ether, thus receiving the name of ethereal oils or, in Latin, *aetheroleum*. Although they have a limited solubility in water, they have been used to increase water solubility of poorly soluble drug by constituting the inner phase of oil-in-water nanoemulsions (Barradas et al. 2017).

Essential oils constituents range from terpene hydrocarbons, simple and terpene alcohols, aldehydes, ketones, phenols, esters, ethers, oxides, peroxides, furans, organic acids, lactones, coumarins, to sulfur compounds. In the mixture, these compounds are present in different concentrations, and usually one of them is the majority compound, others are in lower levels and some in very low quantities (Donsì and Ferrari 2016)

Essential oils are well-known as flavors, antioxidant and antimicrobial products and have been widely used as functional ingredients in food, pharmaceutical, and cosmetic formulations (Jin et al. 2016). In recent years, the application of essential oils in food products has generated great interest because essential oils are generally recognized as safe regulatory status, multiple functionalities, and wide acceptance by consumers. However, the poor solubility in aqueous solutions and high volatility during processing are two major obstacles of utilizing EOs in the industrial processes.

A superficial search in the main patent databases and indexed journal articles reveals the growing interest in the use of nanoemulsions in the last decade as well as the use of these systems in the pharmaceutical industry (Figs. 4.3 and 4.4). It is possible to notice a considerable increase in the number of publications describing essential oil-loaded nanoemulsions. The poor solubility in aqueous solutions and high volatility during processing are two major obstacles of utilizing essential oils as sanitizing agents or as preservatives in food matrices. Oil-in-water nanoemulsions carefully prepared using appropriate surfactants and emulsification processes

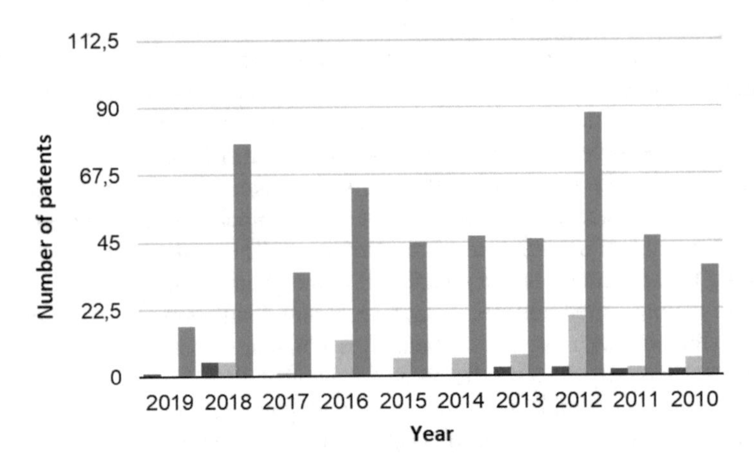

Fig. 4.3 Patents related to nanoemulsions formulations from 2010 to 2019 using Boolean operators AND (Espacenet search∗ 6/10/2019). Blue: nanoemulsions; Green: nanoemulsions AND drugs; Purple: Nanoemulsions AND essential oil. ∗https://worldwide.espacenet.com

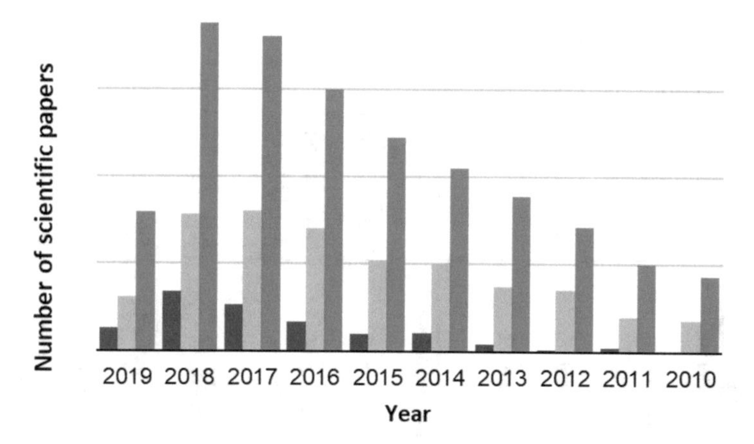

Fig. 4.4 Publications related to nanoemulsions formulations from 2010 to 2019 using Boolean operators AND (Web of Science search∗∗ 6/10/2019). Blue: nanoemulsions; Green: nanoemulsions AND drugs; Purple: nanoemulsions AND essential oil. ∗∗https://webofknowlegde.com

are common choices to deliver essential oils in aqueous systems, for several administration routes and distinct applications. Nevertheless, there are still few studies that deal with the particularities in the development, characterization and evaluation of the pharmaceutical properties of essential oils-loaded nanoemulsions, especially regarding the evaluation of the industrial scale production.

Essential oils are generally obtained from the peeling of fruits, flowers or leaves. Several essential oils are being used in the development of nanoemulsions, in Table 4.2 are listed some nanosystems with proven activity. For example, clove oils are the essential oils obtained by distillation of the flower buds, stems, and leaves of the clove tree (*Syzygium aromaticum*) (Goñi et al. 2016). Among clove oils, clove bud oil (CBO) is widely used and well-known as a potent antioxidant and for antibacterial, antifungal, and antiviral activities (Anwer et al. 2014; Chaieb et al. 2007). Eugenol, 4-allyl-2-methoxyphenol, is the primary constituent, responsible for more than 80% of CBO and is the major contributor of the above biological functions of CBO (Jirovetz et al. 2006; Chaieb et al. 2007).

Another example, *Rosmarinus officinalis L.* essential oil is usually isolated by hydrodistillation, steam distillation, or extraction with organic solvents. The Rosemary oil is mainly located in leaves and the flowers; it is known for antioxidant, antimicrobial, anti-inflammatory properties and studies suggest that the one chemical compounds more frequently reported molecules was 1,8-cineole, α-pinene, and camphor (Angioni et al. 2004; Hernández et al. 2016).

Basil and thyme are aromatic herbs that are also used extensively to add a distinctive aroma and flavor. The essential oils are extracted from fresh leaves and flowers are very recognized can be used as aroma additives in food, pharmaceuticals, and cosmetics (Q. X. Li and Chang 2016; Mandal and DebManda 2016). Major compounds found in volatile extracts of basil and thyme exhibited varying amounts of anti-oxidative activity in particular, eugenol, thymol, carvacrol and

Table 4.2 Essential oils used in nanoemulsions with proven pharmaceutical or biological activities

Scientific plant name	Common plant name	Surfactants	Production method	Energy supply	Activity	References
Allium sativum	Garlic	Polysorbate 80	Magnetic stirrer and ultrasonic water bath	Low energy	Antibacterial	Hasssanzadeh et al. (2018)
Anethum graveolens	Dill	Polysorbate 20 and ethanol	Spontaneous method without mechanical force	Low energy	Larvicidal	Osanloo et al. (2018b)
Baccharis reticularia DC	–	Polysorbate 80 and sorbitan monooleate	Magnetic stirring	Low energy	Larvicidal	Botas et al. (2017)
Backhousia citriodora/Syzygium anisatum	Lemon myrtle/ anise myrtle	Span 80 and Tween 80	Ultra-sonication	High energy	Antibacterial	Nirmal et al. (2018)
Calendula officinalis	Calendula	Tween 80 and Span 80	Ultrasonication	High energy	Antiinflametory	Kiaei et al. (2018)
Citrus maxima	Pomelo	Tween 20	High-speed emulsion homogenizer	High energy	Antibacterial	Lou et al. (2017)
Citrus medica L. var. sarcodactylis	Fingered citron	Different mixtures with Cremophor EL, 1, 2-propanediol, glycerol, Tween 80, ethanol, PEG-400	Self-emulsification	Low energy	Antioxidant/ antibacterial	Lou et al. (2017), Z. Li et al. (2018)
Cymbopogon densiflorus	Lemongrass	Sorbitan monooleate and PEG 40 Hydrogenated Castor Oil	Phase inversion emulsification	Low energy	Antioxidant	Seibert et al. (2019)
Eucaliptus	Eucaliptus	Polysorbate 80 and Labrasol	Self-nanoemulsification	Low energy	Wound healing	Alam et al. (2018)
Ocimum Basilicum	Basil	Sorbitan monooleate and Polysorbate 80	High-speed emulsion homogenizer	High energy	Antimicrobial	Hussain et al. (2008)
Ocimum Basilicum	Basil	Polysorbate 80	Magnetic stirring	Low energy	Larvicidal	Sundararajan et al. (2018)
Origanum vulgate L	Oregano	Pluronic F127	Self-nanoemulsification	Low energy	Antiacne	Taleb et al. (2018)

Pimpinella anisum L.	Anise	Polysorbate 80	Spontaneous method emulsification	Low energy	Insecticide	Hashem et al. (2018)
Pterodon emarginatus	Sucupira branca	Sorbitan monooleate/ polysorbate 80	Self-nanoemulsification	Low energy	Larvicidal	A. E. M. F. M. Oliveira et al. (2017)
Rosmarinus officinalis	Rosemary	Polysorbate 20	Magnetic stirring	Low energy	Larvicidal	Duarte et al. (2015)
Rosmarinus officinalis	Rosemary	Polysorbate 20	Self-nanoemulsification	Low energy	Antiinflammatory/ Antialgic	Borges et al. (2018)
Satureja khuzestanica	Satureja	Polysorbate 20 or 80 and sorbitan monolaurate or sorbitan monooleate	Ultra-sonication	High energy	Antibacterial	Mazarei and Rafati (2019)
Syzygium aromaticum	Clove	Polysorbate 80 and sorbitan monooleate	Ultra-sonication	High energy	Antibacterial	Shahavi et al. (2016)
Syzygium aromaticum/ Cinnamomum verum	Clove/ cinnamon	Polysorbate 80 and ethanol	Self-nanoemulsification	Low energy	Antimicrobial	Osanloo et al. (2018a, b)
Syzygium aromaticum/ Cymbopogon densiflorus	Clove/ lemongrass	Polysorbate-20 and Castor Oil Ethoxylate-40	Self-nanoemulsification	Low energy	Antifungal	S. Zhang et al. (2017)
Thymuns daenensis	Thyme	Polysorbate 80 and lecithin	Ultra-sonication	High energy	Antibacterial	A. Sharma et al. (2018)
Thymus capitatus	Conehead thyme	Sodium dodecyl sulfate and Polysorbate 20	High- pressure homogenizer	High energy	Antibacterial	Moghimi et al. (2016)
Zataria multiflora	Zataria	Polysorbate 80	Ultra-sonication	High energy	Antibacterial	Hashemi Gahruie et al. (2017)

4-allylphenol (Marotti et al. 1996; Hudaib et al. 2002; Politeo et al. 2007; Hussain et al. 2008).

Recognized in traditional medicine, lemongrass essential oil is a potent antimicrobial and antioxidant natural bioproduct. Lemongrass essential oil composition consists mainly of 70–85% geranial (citral), neral, geraniol, nerol, citronellol, 1,8-cineole (eucalyptol), α-terpineol, linalool, geranyl acetate. Lemongrass oil is collected by steam distillation of the herbage. Lemongrass essential oil is a viscous liquid, yellow to dark yellow or dark amber in color turning red on prolonged storage (Sharma et al. 2018).

Also, among the essential oils, there are citrus fruits, which also have countless therapeutic applications. Citrus fruits are well known and appreciated for centuries as they feature pleasant aroma and appetizing flavor. The essential oils from these fruits are generally obtained from the juice of the peels of fruits, but it can also be obtained from flowers or leaves (Asbahani et al. 2015). Among the best-known essential oils are those of orange, lemon, mint, eucalyptus, mint, citronella, clove, among others, as can be seen in Table 4.2.

The methods of extraction vary according to the location of the volatile oil in the plant and the proposed use of it. The most common methods are: enfleurage, steam distillation; extraction with organic solvent in a continuous and discontinuous way; pressing or supercritical CO2 extraction (Asbahani et al. 2015).

4.4 Preparation of Nanoemulsions

Typically, the emulsification process is based on the dispersion of an immiscible liquid into another immiscible liquid as small droplets which are surrounded by a thin interfacial layer of a surfactant that acts as an emulsifier, nanoemulsification methods are no different. In general, the methods to produce nanoemulsions can be classified into high or low-energy methods. High-energy methods have been widely used to produce nanoemulsions in large industrial scale (McClements and Jafari 2018). These methods use mechanical devices to produce nano-sized droplets. Low-energy methods depend on internal energy of the components to spontaneous produce nanoemulsions upon changes in the compositions or environmental conditions. In both type of methods, surfactants play an important role in reducing the interfacial tension of the system and thus they contribute to the reduction of the shear energy required to reduce the radius of curvature of the formed droplets (T. Tadros et al. 2004). The choice of the method to obtain nanoemulsion should be made based on the properties of the oil phase and the surfactant, physicochemical properties and functional aspects requires for the final application (McClements and Jafari 2018).

A minimum energy input required is the same no matter the nanoemulsification method. However, the preparation method may influence the properties of the produced nanoemulsions, such as: droplet size and stability. Without affecting the final nature of the dispersion phases, though (Gutiérrez et al. 2008). On the other hand, the choice of the technique adopted to produce nanoemulsions greatly affect droplet size and stability (Salem and Ezzat 2018). In this section, we have a brief overview on the most commonly used high-energy and low-energy approaches for nanoemulsion formation.

4.4.1 High-Energy Methods

High-energy methods are also known as work-based methods, since they are dependent of high amounts of intense energy, which are supplied by mechanical devices, like high pressure homogenizers, microfluidizers and ultrasonicators, which generate shear forces able to disrupt oil and water interface, providing nanosized droplets (Solans et al. 2005; Leong et al. 2009). The size and stability of nanoemulsions produced can be influenced by the type and processing conditions of the high-energy method, oil properties such as lipophilicity, viscosity and interfacial tension, surfactant type and concentration.

High energy methods are carried out by the input of mechanical energy using mechanical or ultrasonic equipment that generate high shear stress or pressure difference, disrupting and breaking the droplets into smaller sizes (Abismail et al. 1999; Sonneville-Aubrun et al. 2004; Tadros et al. 2004). In general, the preparation of nanoemulsions through high-energy methods can be divided into two distinct phases. In a first step the oil and aqueous phases are emulsified with an homogenizer as Turrax® or Politron® and the coarse emulsion obtained presents a submicron droplet size, between 500 and 1000 nm, depending on the equipment used and the operating conditions (McClements and Rao 2011; McClements 2011). Then, the droplet diameter is progressively reduced to its minimal value, ranging from 20 to 200 nm by means of high pressure homogenizers or microfluidizers, depending on the energy intensity of the homogenizer used, processing time and sample composition (T. Tadros et al. 2004; Salem and Ezzat 2018). It is worth noting that the minimum droplet size achievable may not be stabilized if the surfactant is insufficient to cover the newly created interface, as observed by Barradas et al. (2015). Final droplet size results from a balance between two phenomena that happen at the same time during homogenization: Breaking of drops into fine droplets and droplet coalescence after processing (Jafari et al. 2008). Higher surfactant concentration, increasing shear intensity or duration can contribute to reducing droplet size (Gupta et al. 2016b).

The obtention of nanoemulsions through high pressure homogenizers, ultrasonic cavitation and microfluidizers has been well described in the literature (Tadros et al. 2004; Helgeson 2016; Dias et al. 2014), and is considered as a safe method for keeping nanoemulsions from instability phenomena without the addition of stabilizers, thickeners, cosolvents or cosurfactants (Anton et al. 2008). Figure 4.5 summarizes the high-energy methods reported in this chapter.

Ultrasonication techniques for the preparation of nanoemulsions has also been described by several authors (Shahavi et al. 2015; Gupta et al. 2016a; Leong et al.

2009). The device comprises a sonication probe constituted by piezoelectric crystals able to expand and contract as a consequence to an altering electrical voltage (Fig. 4.5a). In consequence to sonicator probe vibration, ultrasonic high-frequency waves, higher than 20 KHz, are produced, which are able to wield a cavitation effect and causing mechanical vibration and the formation of micro-sized bubbles that collapse, causing the disruption of oil-water interface (McClements 2011). As a result, fine nanosized droplets are obtained after enough time of processing to ensure homogeneous size distribution and polydispersity index (Schwarz et al. 2012; Shahavi et al. 2015; Abismail et al. 1999; Sivakumar et al. 2014). The main parameters involved in this emulsification procedure are the interfacial properties of the emulsion, which can be controlled by the nature and concentration of surfactants and the oil properties, such as surface tension and viscosity (Gupta et al. 2016a). Moreover, droplet size is dependent on processing time, sonication intensity, type and concentration of surfactant (Dias et al. 2014).

High pressure homogenization features the advantage of being applied in the production of industrial scale nanoemulsions and greater control in droplet size reduction (Abismail et al. 1999). However, it requires large amounts of energy input, being more expensive to perform. The high-pressure valve homogenizer consists basically of a pump, which injects the liquid to be homogenized under very high pressure in a restrictive homogenizing valve. Many aspects may influence the physicochemical characteristics of the obtained nanoemulsions as temperature, viscosity and oil phase concentration of the emulsions, which affects the choice of the operational parameters that must be adjusted for each formulation (Lee and Norton 2013). In high pressure homogenizer, the sample is forced through small channels under a pressure ranging from 500 to 15,000 psi (Gupta et al. 2016a) (Fig. 4.5b).

The sample flows under high pressure through microchannels resulting in a very fine emulsion, which causes the disruption of the dispersed droplets. Therefore, the radius of the generated droplets decreases gradually according to the increase of the shear rate. However, due to the lack of homogeneity of the flow, it is often necessary to process this fluid through the device through various cycles, until adequate droplets of size and polydispersity index are obtained. The pressure and number of processing cycles can be adjusted to produce nanoemulsions with tunable droplet size (Ouzineb et al. 2006; Constantinides et al. 2008).

The use of high-pressure homogenizers and ultrasonicators can lead to nanoemulsions of equivalent physical-chemical properties upon the optimization of the operational conditions and the qualitative and quantitative composition of the formulations. However, some authors have reported some disadvantages regarding the use of ultrasonication, such as excessive heating of the sample, larger droplet distribution and low reproducibility in relation to the droplet diameter and polydispersity index, in addition to problems related to the difficulties of scale up (Tadros et al. 2004; Gutiérrez et al. 2008). Moreover, as they require high energy input, they are often considered cost-inefficient (Solans and Solé 2012).

Microfluidizers are composed of an interaction chamber where the fluid is injected and homogenized by cutting, impact and cavitation, in a design that resembles the high pressure homogenizers (Fig. 4.5c) (Lee and Norton 2013; Tadros et al.

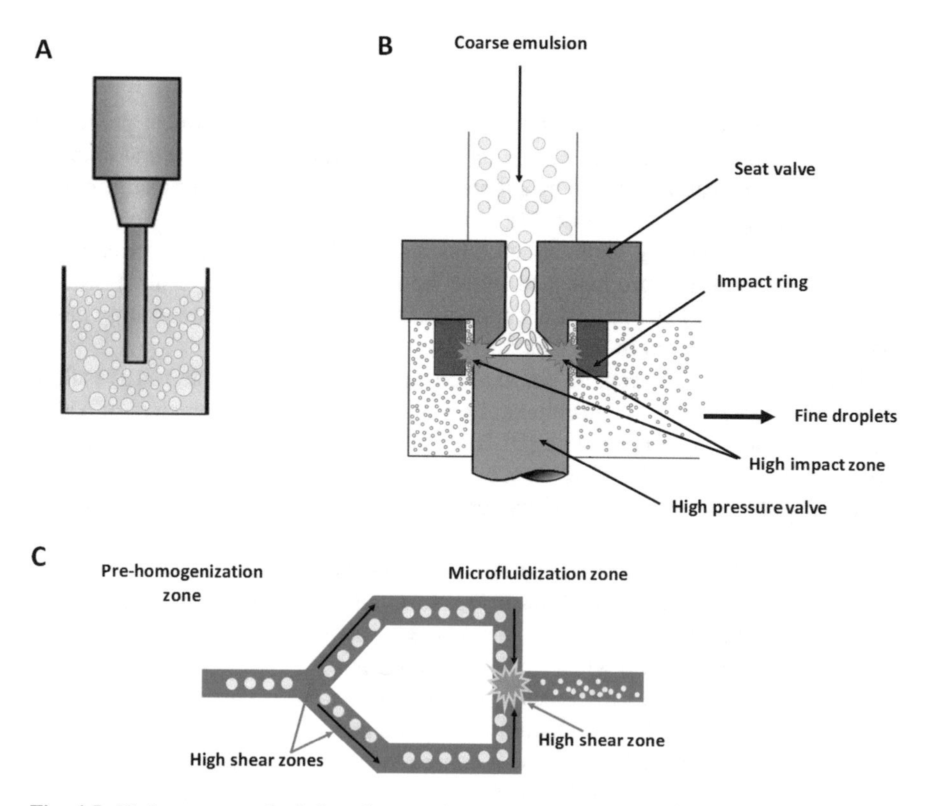

Fig. 4.5 High-energy methods based on mechanical devices, such as ultrasonicators (**a**), composed by a probe able to produce a cavitation effect and nanosized droplets are obtained after enough time of processing to ensure homogeneous size distribution and polydispersity index: High pressure homogenizators (**b**) consist basically of a pump, which injects the liquid to be homogenized under very high pressure in a restrictive homogenizing valve, which causes the disruption of the dispersed droplets. Microfluidizers (**c**) are composed of an interaction chamber where the fluid is injected and homogenized by cutting, impact and cavitation. Emulsification can occur inside the channels, since streams are conducted to an interaction chamber under high pressures, where they are submitted to disruptive conditions, which provide the formation of fine droplets. (Modified after Jafari 2017; Rao and McClements 2011)

2004; Salem and Ezzat 2018). In this case, emulsification can occur inside the channels, since both dispersed and continuous phases flow inside fine channels individually. Streams are conducted to an interaction chamber under high pressures, where they are submitted to disruptive conditions, which provide the formation of fine droplets (Lee and Norton 2013; McClements 2011).

In this method, several parameters can influence the obtention of small droplets: the viscosity of both dispersed and continuous phases and the type of the surfactant used (McClements 2011; Salem and Ezzat 2018). Droplet is shown to decrease as homogenization pressure, surfactant concentration and number of processing passes increase and viscosity decreases (McClements 2011).

4.4.2 Low-Energy Methods

All low-energy emulsification methods are based on physicochemical properties and uses the internal chemical energy of the components. In phase inversion methods nanoemulsification achieve by spontaneous inversion of the surfactant's curvature providing small size globules (Jin et al. 2016; Solans and Solé 2012). Low-energy emulsifying methods are advantageous because they are effective in providing small-sized droplets and allow nanoemulsification by simple stirring (Solans and Solé 2012). Figure 4.6 summarizes the most common low-energy methods, which are reviewed in this chapter.

Nanoemulsification occur by phase inversion in an coarse emulsion as a result of dramatic changes in the environmental conditions, in which parameters affecting the hydrophile-lipophilic balance (HLB) of the surfactant, as temperature and / or concentration are modified (Tadros et al. 2004).

The spontaneous formation of nanoemulsions is achievable by various methods based on diffusion of solutes between two phases, interfacial turbulence, surface tension gradient and dispersion or condensation mechanisms (Salem and Ezzat 2018). The spontaneous emulsification (Fig. 4.6a), also referred as self-emulsification method is based on the diffusion of a water-miscible component from the organic phase into aqueous phase when both phases are put into contact, and one of the phases contains a component miscible in both phases. Surfactant, cosurfactant or a polar organic solvent such as ethanol or acetone can be examples of dual-soluble components (McClements 2011). As a consequence, some of the components partially miscible in both phases diffuse from the original phase towards the other one in a rapid diffusion movement, without no phase transition or change in the surfactant spontaneous curvature (Solans and Solé 2012). The sudden diffusion of the components provides an increased oil-water interfacial area, which trigger other phenomena such as interfacial turbulence and thus, the spontaneous droplets assemble (Anton et al. 2008; Salem and Ezzat 2018).

Spontaneous nanoemulsification and droplet size can be highly influenced by the composition of the formulation, some physical chemical properties and the mixing conditions (Bouchemal et al. 2004; Rao and McClements 2012). The obtention of nanosized micelles can be performed with the addition of high concentrations of water-miscible component into the oil phase and very high solvent/oil ratio (McClements 2011; Solans and Solé 2012).

Spontaneous emulsification is often related to a special phenomenon called The Ouzo effect with beverages based on sweet fennel oil, rich in trans-anethole, that allow surfactant-free self-emulsification (Carteau et al. 2008). When water is added to the alcoholic oil solution, some of the ethanol molecules diffuse from the organic phase into aqueous phase, which reduce sweet fennel oil solubility and small oil droplets spontaneously assemble. Spontaneous emulsification has been recently explored to provide sweet fennel oil-based nanoemulsions with very reduced amounts of surfactants (Barradas et al. 2017).

Moreover, nanoemulsions can also be obtained from dilution of surfactants aggregates, as liquid crystalline particles and bicontinuous microemulsions. Solè et al. produced 20 nm oil-in-water nanoemulsions upon dilution of oil-in-water and water-in-oil microemulsions by self-emulsification method. The effect of dilution procedure, i.e., stepwise or at once, and cosurfactant nature on nanoemulsion formation was studied. Oil-in-water microemulsion provided small-sized 20 nm nanoemulsion regardless the dilution method and microemulsion composition. On the other hand, water-in-oil microemulsion resulted in nanoemulsion when water dilution was performed stepwise. Regarding cosurfactant nature, droplet size decreased as cosurfactant alkyl chain size increased, which enhanced nanoemulsion stability (Solè et al. 2012).

Self-emulsification is being applied by pharmaceutical industry to obtain oil-in-water nanoemulsions for being a low-cost technique. Moreover, spontaneous emulsification is being highly explored to produce self-emulsifying drug delivery systems (SEDDS) and self-nanoemulsifying drug delivery systems (SNEDDS). The main application of self-emulsifying drug delivery systems comprises the very self-emulsification obtention method increase in drug bioavailability and stability of the micelles (Wei et al. 2012; Shahba et al. 2012; Anton and Vandamme 2011). However, one of the major limitation of this method is the very high concentrations of surfactants needed to trigger self-emulsification, which can cause toxicity (McClements 2011; Azeem et al. 2009).

Phase inversion-based methods are low-energy techniques that involve the inversion of the surfactant curvature, passing through a transition phase in which surfactant curvature achieves zero curvature, such as bicontinuous microemulsions or lamellar liquid crystalline phases (Porras et al. 2008; Tadros et al. 2004; Fernandez et al. 2004; Sutradhar and Amin 2013; Mayer et al. 2013; Solans and Solé 2012). They are based on the chemical energy released from phase transitions phenomena during emulsification process. Phase inversion methods occur when some dramatic change in the environmental conditions take place, i.e., temperature or composition. These methods require an extensive control in terms of selecting the right surfactant, knowledge of surfactants phase behavior and the most adequate thermodynamic method. Droplet size provided depends on the selected surfactant, the properties of the intermediary phases formed and the interfacial properties between both fluids (Helgeson 2016; Maestro et al. 2008).

In phase inversion composition method (Fig. 4.6b), phase inversion occurs by a major change in composition of the system. There is an increase of the volumetric fraction of the dispersed phase that is added to a microemulsion and then, the curvature of the surfactant is altered. Maestro et al. showed that changes in salt concentration changed the spontaneous curvature of ionic surfactants and prepared water-in-oil nanoemulsions from oil-in-water emulsions by means of phase inversion composition. Phase inversion was achieved by the ability of salt ions to screen the charges on surfactant groups (Maestro et al. 2008).

Changing pH is also a useful approach to produce nanoemulsions by phase inversion composition. It can provide changes in electrical charge a stability of surfactants. Solè et al. reported the production of nanoemulsions by phase inversion

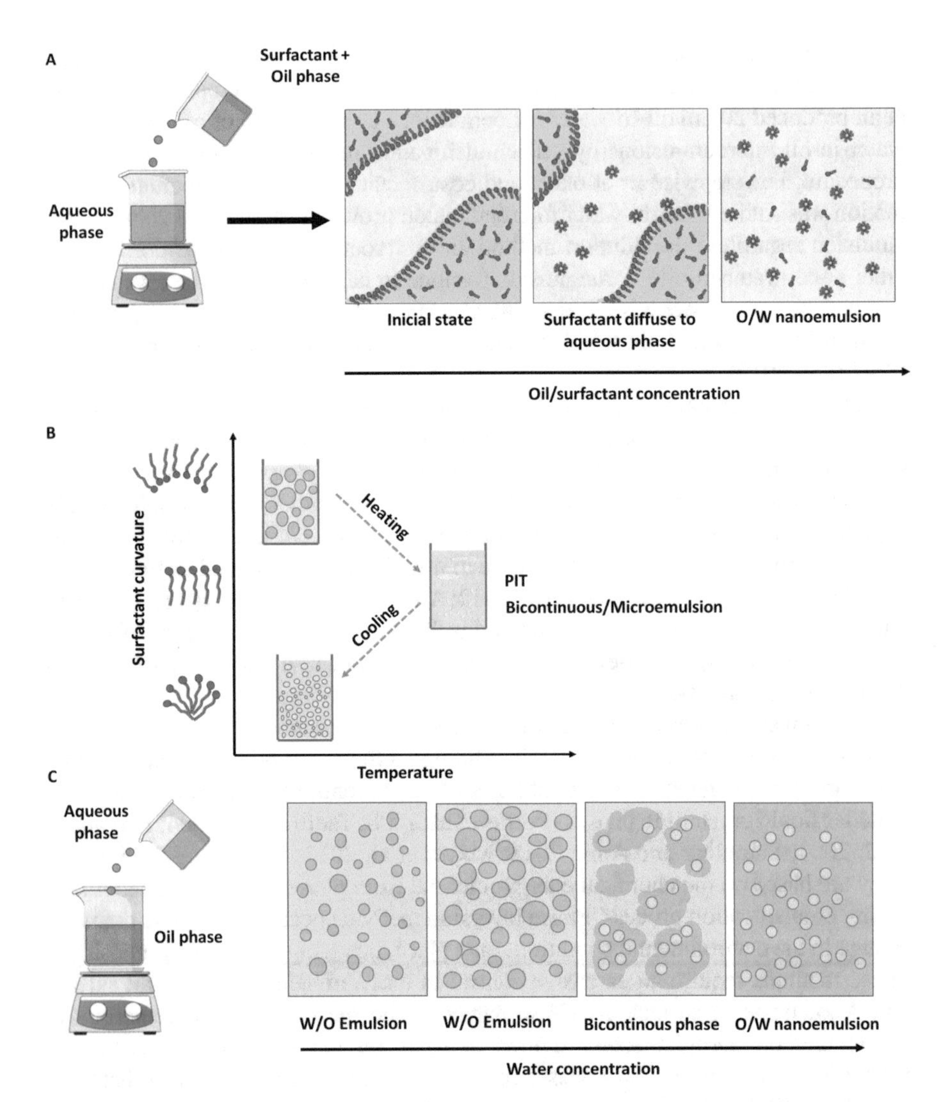

Fig. 4.6 Schematic diagram of low energy methods: The spontaneous emulsification (**a**), is based on the diffusion of a water-miscible component from the organic phase into aqueous phase when both phases are put into contact, and one of the phases contains a component miscible in both phases. In phase inversion temperature (**b**), there is an increase of the volumetric fraction of the dispersed phase that is added to a microemulsion and then, the curvature of the surfactant is altered. Emulsion inversion phase (**c**) is known to cause a catastrophic phase inversion, which occurs through the inversion in oil-to-water ratio. (Modified after Jin et al. 2016; Anton and Vandamme 2011; Rao and McClements 2011)

composition method using ionic surfactant, potassium oleate, for example. At low pH values, the carboxyl groups from fatty acids are non-ionized and, thus lipid soluble. Under this condition, these molecules can stabilize water-in-oil emulsion. However, this situation is inverted if pH is raised. At higher pH values, carboxyl groups become ionized which increases water-solubility and provides the stabilization of oil-in-water emulsion (Solè et al. 2006).

Phase inversion composition method is considered an interesting approach for large-scale production of nanoemulsions, since it relies only on the addition of one component to a mixture of components without requiring high temperatures of high-energy input, being also beneficial when temperature-sensitive components are used. Moreover, phase inversion composition method is not limited a specific type of surfactant (Solans and Solé 2012).

For temperature-sensitive surfactants, a phase inversion can be achieved by changing the temperature of the system, forcing the transition of an oil-in-water emulsion, prepared at low temperatures, for a water-in-oil emulsion, formed at higher temperatures, due to changes in the physicochemical properties of the surfactant. This is a typical phase inversion composition example of a transient phase inversion, known as phase inversion temperature, in which nanoemulsions are formed under a fixed composition by changing temperature, by a drastic change in surfactants water solubility, through an intermediate liquid crystalline or bicontinuous microemulsion phase (Fernandez et al. 2004; McClements and Rao 2011).

This process is triggered by the changes in physicochemical properties of surfactant as temperature changes. Polyethoxylated nonionic surfactants are highly temperature-dependent. They tend to become lipophilic with increasing temperature due to dehydration of the chain of the polar part of the surfactant, i.e., ethylene oxide groups. During heating, the ethylene oxide groups responsible for the hydrophilic characteristic of the surfactant are "hidden" and, consequently, there is modification of the affinity for the phases. As the system undergoes cooling, the surfactant passes through a point of zero curvature and with minimum surface tension, which provides conditions to the formation of nanoemulsions (Fernandez et al. 2004; Tadros et al. 2004).

The inversion point take place at a specific temperature, i.e., phase inversion temperature, when the solubility of the surfactant in water and in oil reach equivalent values. As temperature is continuously raised, the surfactant becomes more soluble in the oil phase than in the aqueous phase (McClements 2011). Nanoemulsions are obtained by rapidly breaking up microemulsions formed at phase inversion point by sudden cooling below phase inversion temperature (McClements 2011). Through the phase inversion temperature method very small droplet sizes and stable nanoemulsions are obtained. On the other hand, the systems are highly prone to coalescence, being unstable, which makes cooling step a critical aspect to obtain stable nanoemulsions (Tadros et al. 2004; Solans and Solé 2012).

In emulsion inversion point method (Fig. 4.6c), one solvent, i.e. water or oil, is continuously added under stirring to the dispersed phase until this solvent concentration becomes predominant, at constant temperature. Once a critical amount of solvent is achieved, droplets are so highly packed together that phase inversion occurs, and the emulsion reaches a phase inversion point from water-in-oil to oil-in-water or vice-versa. This causes the spontaneous surfactant curvature to change, i.e., micelles are assembled in reverse curvature, which disrupts them into smaller structures thereby obtaining emulsions with nanometric droplets (Anton et al. 2008).

Emulsion inversion point is considered to cause a catastrophic phase inversion, which occurs through the inversion in oil-to-water ratio. The surfactants used to produce the catastrophic phase inversion comprise small molecule synthetic emulsifiers able to stabilize both oil-in-water and water-in-oil emulsions. In this method, droplet size depends on the stirring speed and the solvent addition speed (McClements and Rao 2011).

Although the phase inversion temperature method facilitates the obtention of nanoemulsions with nanometric droplet sizes by the fact that it is possible to achieve very low values of interfacial tension, the dynamics of droplets coalescence can be extremely fast. In order to obtain stable nanoemulsions, the water-oil dispersion formed from phase inversion temperature method is readily cooled at a temperature just below it. Thus, the lamellar or the bicontinuous system collapse forming small droplets kinetically stable, with very small droplet size and narrow size droplet size. If the cooling process is not fast enough, the coalescence predominates and a polydisperse mixture is obtained (Fernandez et al. 2004).

According to the literature, not only the minimum interfacial tension produced during inversion of the curvature of the surfactant, but also the formation of the bicontinuous phase or liquid-crystalline phase prior to the inversion locus is closely related to the formation of nanoemulsions, in both emulsion inversion point and phase inversion temperature methods (Fernandez et al. 2004; Tadros et al. 2004). Moreover, complete solubilization of the oil in the bicontinuous phase or in the crystalline phase is an extremely important factor for the formation of fine nanoemulsions.

4.4.3 Comparison Between Obtention Methods

High energy emulsification processes are different regarding energy input and emulsification time: the high-pressure homogenizer, for example, is a high energy and low process time method; ultrasonication is a high energy and long process method; and rotor stator is a mixer that uses less energy, requiring longer process times. On the other hand, low-energy processes can provide smaller droplet sizes compared to high-energy methods. However, there are limited options for surfactants to be used in low-energy approaches, which exclude several natural surfactants as proteins or polysaccharides (Rao and McClements 2012). Besides, low-energy methods, as they cannot count on high shear energy devices, require large amounts of surfactants, which can limit the pharmaceutical application of the obtained nanoemulsions.

Jafari, He, and Bhandari compared different methods of high energy emulsification, it was observed that the average mechanical mixer produced emulsions with large droplet size, in a range above 10,000 nm, because, in this case, the main deformations forces are shear stresses in laminar flow that are not able to efficiently break the droplets. The stator rotor mixer proved to be more efficient than the average mechanical mixer in the formation of smaller droplets, of about 1000 nm, and with narrow size distribution, due to the incorporation of more forces on the shear that can form smaller droplet sizes, for example, inertial forces (Jafari et al. 2007).

Considering the other two emulsification systems, the high-pressure homogenization, and microfluidization, provided droplet sizes and size distribution smaller than the ultrasonication due to the greater efficiency in the rupture of the drops, or with cavitation along with shear and inertial forces, in addition to the higher energy input. From these results it can be shown that the difference in droplets size obtained is directly related to the energy input (Jafari et al. 2007).

Comparing both high and low energy emulsification methods, it was verified the variation of the droplet size as a function of time for nanoemulsions prepared using the phase inversion temperature method and high-pressure homogenization. In both cases, an increase in droplet hydrodynamic size after a certain time the preparation of nanoemulsions. However, when high pressure homogenization is used, the droplet size can be maintained at lower values for longer periods of time. On the other hand, phase inversion temperature method provides a faster destabilization of the system (Tadros et al. 2004).

Phase inversion composition method and high-pressure homogenization were compared in the production of oil-in-water efavirenz-loaded nanoemulsions. Phase inversion composition method was performed by dilution in water. Nanoemulsions were evaluated regarding the physical stability when submitted to heating-cooling cycles, centrifugation and freeze-thaw cycles, water-dispersibility, droplet size, rheology and dilution impact on droplet size. Both nanoemulsions showed good physical stability and water dispersibility. Regarding viscosity, there was no significant difference between nanoemulsions prepared by each method. Nanoemulsions prepared through phase inversion composition method showed to be more transparent than those prepared by high-pressure homogenization, this is due to the concentration of surfactant and oil/surfactant ratio used to prepare nanoemulsions by each method. Droplet size showed no significant difference between both methods. However, polydispersity Index was found to be significantly different between the methods. Phase inversion temperature method provided polydispersity index from 0.150 to 0.404, while high-pressure homogenization varied from 0.637 to 0.812. Both nanoemulsions proved to be stable under infinite dilution. Droplet size and polydispersity index remained unaltered after sample dilution. Drug release also showed differences between nanoemulsions made by high-energy and low-energy emulsification methods. Drug release from nanoemulsion prepared by high-pressure homogenization was smaller than that prepared by phase inversion composition method, which can be a consequence of polydispersity index (Kotta et al. 2015).

Yang et al. compared microfluidization with spontaneous emulsification in the production of food-grade nanoemulsions with 20% of oil phase and same type and concentration of surfactants. The influence of the surfactant used end surfactant-oil ratio, stability and droplet size were also evaluated. Microfluidization produced nanosized droplets with less than 100 nm, using lower surfactant-to-oil ratio than spontaneous emulsification. Besides, microfluidization showed a linear decrease in droplet radius as surfactant concentrations increased until it reached a value where no difference in droplet size is observed, suggesting that minimum droplet size was achieved. Spontaneous emulsification also produces droplets smaller 100 nm, even though to do so, it required considerably higher, i.e., ten-fold higher of surfactant concentration. Regarding the kinetic stability, microfluidization provided stable droplets against aggregation and gravitational separation phenomena for at least 30 days. Spontaneous emulsification, on the other hand, provided droplets sizes that increased after one-month storage. Moreover, nanoemulsions showed creaming instability signs (Yang et al. 2012).

4.5 Stability of Nanoemulsions

As previously mentioned, nanoemulsions are characterized for being kinetically stable. Nanoemulsions shelf life can be further improved with some stabilization strategies that aim at maintaining bulk and droplet properties stable for longer periods of time. In order to achieve stable nanoemulsions several stabilization methods can be used, such as electrostatic, steric and mechanical stabilization (Cardoso-Ugarte et al. 2016).

Repulsive electrostatic forces between droplets can ensure proper droplet separation and prevent coalescence and/or flocculation. This is related to surfactant superficial charge and is responsible for the high stability of ionically-charged nanoemulsions. Electrostatic stabilization is more important with smaller droplets, since they have an increased superficial area (Helgeson 2016).

Superficial charge is often provided by the surfactants, especially natural surfactants, as biopolymers with surface activity like proteins, polysaccharides, which can stablish electrostatical intermolecular interactions depending on concentration, pH and isoelectric point, and ionic strength of the solution (Cardoso-Ugarte et al. 2016). When pH of the surrounding solution is far from protein isoelectric point, proteins residues are charged and there is an electrostatic repulsion that hinders droplets to aggregate (Dickinson et al. 1991).

Non-ionic surfactants also can provide stabilization, even though they are not charged. Steric stabilization is preponderant in the case of non-ionic surfactants, particularly with amphiphilic non-ionic polymers, which can form voluminous interfacial films around droplets able to prevent coalescence. Amphiphilic block polymers are advantageous since less amounts of surfactants are required to stabilize droplets. On the other hand, in some cases, macromolecular surfactants can have difficult diffusion towards droplets interface due to their size and molecular mobility (Bouyer et al. 2012; T. Tadros 2009; Qian and McClements 2011).

Steric stabilization comprises three main mechanisms: (i) non-adsorbent macromolecules which provide an elastic film around droplets that avoid droplet collision and deformation, (ii) branched macromolecules that form a voluminous surface that prevent droplets from approaching, (iii) stabilization due to hydrophobic interactions between adsorbed macromolecular surfactants (Cardoso-Ugarte et al. 2016).

The addition of viscosity agents and gelling polymers in the outer phase of nanoemulsions is often called as mechanical stabilization, since they can reduce droplets mobility by providing a mechanical network that serves as a barrier to aggregation. Viscosity agents are often called as stabilizers and produce semi-solid or gel-like systems. However, stabilizers should be used with caution, since important properties as optical appearance, droplet size and encapsulation efficiency can be modified (Behrend et al. 2000; Dickinson 2009).

4.5.1 Instability Phenomena

As nanoemulsions are formed from non-spontaneous process they tend to undergo instability phenomena such as flocculation, coalescence and Ostwald ripening and gravitational phase separation as conventional emulsions. However, the small size of the droplets in a nanoemulsion confers enhanced kinetic stability, compared to ordinary emulsions (McClements 2011).

Thus, nanoemulsions can be destabilized by several different phenomena: (i) irreversible phenomena, related to the permanent modification of the droplet size and may lead to complete phase separation; (ii) reversible flocculation of droplets, which may be followed by creaming or sedimentation, according to the respective densities of the dispersed and continuous phases (Tadros et al. 2004; Sing et al. 1999; Abismail et al. 1999). In general, nanoemulsions tend to be more stable to gravitational separation, flocculation and coalescence and more susceptible to Ostwald ripening (McClements 2011).

Reversible phenomena involve aggregation and migration of droplets, as flocculation and creaming. Irreversible phenomena are related to the modification of droplet size, for examples, coalescence. Droplets can co-exist in nanoemulsions as individually separated entities or as flocs, i.e., droplets aggregates formed as consequence of attractive interactions among them, characterizing a flocculated system (McClements 2015; McClements and Jafari 2018).

Flocculation is a reversible phenomenon in which the droplets dispersed in an emulsion aggregate and migrate, aiming to reach the thermodynamic equilibrium by decreasing the chemical potential differences that exist throughout the system. During this process, the droplets collide randomly and can remain in contact after these shocks, producing aggregates or flocs (Katsumoto et al. 2001; Starov and Zhdanov 2003). In flocculation, droplets aggregate without the rupture of the interfacial surfactant film, being a reversible phenomenon. The flocculation rate depends on attractive forces between droplets, the frequency of collisions between droplets and how long they remain in contact (Wang et al. 2009). In a dilute system, flocculation may accelerate gravitational phase separation as a consequence of the increase in droplet size. On the contrary, in concentrated systems, flocculation can prevent gravitational phase separation, since the aggregates may produce a tree-dimensional network which may be a barrier to phase segregation (McClements and Jafari 2018).

According to the difference of density between both inner and outer phases, the aggregates formed in the flocculation phenomenon may show gravitational separation such as sedimentation or creaming. When less dense aggregates are formed, they can rise to the surface, characterizing the phenomenon of creaming. Contrarywise, when inner phase shows higher density, denser aggregates are formed and deposited at the bottom of the system, constituting the sedimentation process (Fig. 4.7). Both sedimentation and creaming phenomena are related to gravitational forces that can influence whether the aggregates will move upwards or downwards, depending on the difference of density between inner or outer phases. Hence, both phenomena are most prone to occur with droplets with increased droplet size, as larger objects are more susceptible to gravitational forces. Thus, gravitational sepa-

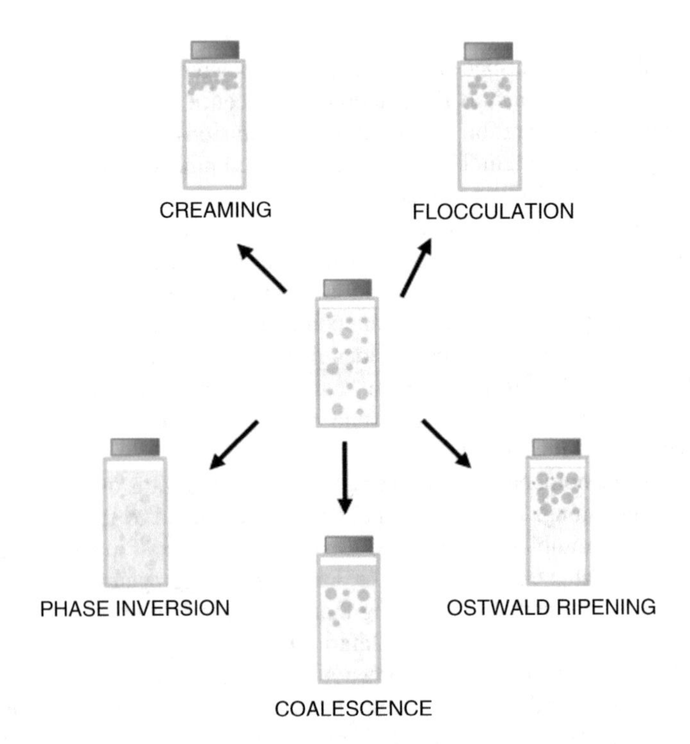

Fig. 4.7 Instability phenomena in emulsions and nanoemulsions**.** Reversible phenomena comprise aggregation and density-driven migration of droplets, as flocculation and creaming. Irreversible phenomena are related to a definitive increase of droplet size, for example: coalescence, Ostwald ripening and, the most drastic, phase inversion. (Modified after McClements and Jafari 2018; Taylor 1998)

ration can be reduced with smaller droplet sizes, increasing the viscosity of continuous phase or by reducing the density difference between dispersed and continuous phases (McClements 2011; McClements and Jafari 2018).

Some surfactants as proteins and polysaccharides can change droplet density by forming a shell layer on droplet surface, which can prevent gravitational separation phenomena such creaming, since it reduces density differences between inner and outer phases. Moreover, it can be possible to produce tunable-density droplets by controlling droplet size and thickness of surfactant layer (McClements 2011).

Irreversible phenomena include the Ostwald ripening, phase separation and coalescence, which may lead to the eventual complete phase separation (Sing et al. 1999). Coalescence can occur when droplets collide and merge, producing one larger droplet. Coalescence can lead to complete phase separation but can be avoided when repulsive interactions are provided by either electrostatic or steric effects, which is achievable by the right choice of surfactant (McClements and Jafari 2018).

Ostwald ripening is often considered as the main cause of instability of a nanoemulsion (Liu et al. 2006). This process is dependent on the polydispersity of the system, the solubility of dispersed phase in the continuous phase and the difference in solubility between droplets of different sizes. Ostwald ripening is the process by which the larger particles grow from the smaller droplets due to the greater solubility of the smaller droplets and by the molecular diffusion of the continuous phase.

In polydisperse nanoemulsions, there is a difference between the capillary pressure from different-sized droplets which makes the small droplets of the dispersed phase to diffuse into the large droplets (Chebil et al. 2013; Taylor 1998). As a consequence, the larger droplets grow from the smaller ones that have greater chemical potential. Thus, the droplets content diffuses through the dispersing phase due to the greater solubility of the smaller droplets, which does not require contact between the droplets. This phenomenon aims to decrease the total energy of the system by reducing the total interfacial area, which can be avoided with the use of insoluble oils and with the choice of suitable polymeric surfactants (T. Tadros et al. 2004; C. Solans et al. 2005; Constantinides et al. 2008).

The rate of droplet growth depends on the product of the solubility of the dispersed oil in the aqueous continuous phase and oil diffusion coefficient and is explained by the Lifshitz-Slezov-Wagner theory (Taylor 1998). It predicts that the Ostwald ripening rate presents a linear relationship between the dispersed phase droplet radius and time. Lifshitz-Slezov-Wagner theory assumes that the droplets of the dispersed phase are spherical and furthermore the distance between them is greater than the diameters of these droplets and the kinetics is controlled by molecular diffusion of the dispersed phase in the continuous phase. Also according to this theory, the Ostwald ripening rate in oil-in-water nanoemulsions, although is predicted to be lower than in conventional emulsions, is directly proportional to the solubility of the oil in the aqueous phase (Helgeson 2016).

However, the decrease in droplet size causes the increase of oil solubility in water, which is the driving force of Ostwald ripening. As the main factor for Ostwald ripening is the solubility of oil phase in water, it does not configure a real issue when it comes to poorly water-soluble oils. On the other hand, for nanoemulsions formulated with oils with some water-solubility, Ostwald ripening might represent the main instability event (McClements 2011).

Tadros et al. presented two methods that tended to delay the increase in droplet size. In the first, a further dispersed phase that is insoluble in the continuous phase is added to the nanoemulsion (Tadros et al. 2004). The fact that it is insoluble causes the droplets to remain almost unchanged in size, generating a balance in the chemical potential due to the partitioning of the dispersed phases. This method has limited application because of the difficulty of finding a phase that fully meets this process. From these results, reduction of the Ostwald ripening process can be observed by the addition to the system of a small amount of a second oil with low solubility in the aqueous phase.

The second method consists in the modification of the interfacial film between the oil phase and water, i.e. outer phase, by adding another surfactant, preferably a block copolymer A-B-A type. The effect would be the adsorption of this copolymer at the oil-in-water interface reducing interfacial tension and balancing curvature effects, which would reduce the diffusion rate of the dispersed phase, minimizing droplet ripening (T. Tadros et al. 2004). In addition, for a system containing a nonionic surfactant based on poly (ethylene oxide), the Ostwald ripening rate can also be decreased by adding a second surfactant having the same hydrophobic group and with a higher hydrophilic content, i.e., ethylene oxide chains, than the primary surfactant (C. Solans et al. 2005).

Several studies were performed to describe and to modulate the Ostwald ripening mechanism. They suggested that the Ostwald ripening can also be delayed or disrupted by size, interfacial viscosity or elasticity of the droplets. This occurs when the interfacial tension between the dispersed phase and the continuous phase equals zero. For a number of drops of the emulsion it was also shown that the Ostwald ripening could be disrupted by interfacial elasticity, even at finite interfacial tensions (Meinders and van Vliet 2004; Liu et al. 2006).

Repeated collisions might trigger coalescence, which is the mechanical fusion of two droplets (Helgeson 2016). Coalescence occurs after prolonged contact between the particles, when the adhesion energy between two droplets is greater than the turbulent energy that causes the dispersion. The mechanism is based on the rupture of the thin film between adjacent droplets, which leads to the joining of two droplets, forming one of larger size. The origin of the rupture of this film can occur due to a mechanical instability in the emulsion. When a large number of particles coalesce, the result is complete separation of the phases. The greater the extent of this phenomenon, the greater the tendency to complete phase separation (Sing et al. 1999).

Irreversible changes caused by Ostwald ripening lead to the formation of larger droplets, i.e., the formation of less stable emulsions, which may be responsible for phase separation. Phase inversion may occur due to temperature variation and/or change in composition of the emulsion.

The composition of nanoemulsions can be tuned towards to produce kinetically stable systems. The adequate selection of oil and aqueous phase and, most importantly, surfactants are essential to an ideal nanoemulsion design.

4.6 Characterization of Nanoemulsions

A wide range of analytical methodologies are required in order to fully characterize nanoemulsions droplet size, aggregation, charge and physical state and bulk properties as rheology and stability. The most used are described in more details here.

4.6.1 Dynamic Light Scattering

Droplet size distribution are some of the most important physical characteristics of nanoemulsions and can influence a great number of characteristics, since droplet size can influence bulk properties, such as optical properties, rheology and stability (McClements 2011) and also other functional properties as solubility, release profile and bioavailability (Bourbon et al. 2018). In general, when nanoemulsions are produced, a range of different droplet sizes are distributed all over the system. Hence, droplet size is often reported as mean droplet size and polydispersity index (PdI). PdI values below 0.3 is indicative of a narrow size distribution whereas values close to 1.0 are indicative of very high size heterogeneity (Klang and Valenta 2011).

Dynamic light scattering (DLS) ($\lambda = 500$ nm) is based on measurements of photon correlation spectroscopy caused by the Brownian motion of droplets. The diffusion of small isometric particles is rapid, causing faster fluctuations in the intensity of scattering light compared to large particles that diffuse more slowly. The particle/droplet size analyzer is based on the typical principle of DLS. This principle comprises four main components. The first of all is the laser beam, which is used to provide the light source to illuminate the particles inside the vial. Most of the laser beam passes rectilinearly through the sample, but another part of it is scattered by particles or droplets dispersed in the medium (Dorfmüller 1987).

A detector is used to measure the intensity of scattered light. As one single particle scatters light in all directions, it is possible to place a detector at any position and still detect the scattering. In DLS, the position of 90°, from the detector to the incident light beam, is a classical detection arrangement. However, it provides a narrower detectable size range and thus, can only be used for low sample concentrations (Dorfmüller 1987).

Another common device used for particle size analysis feature the position of the detector in 173° from the transmitted light beam. The scattered light intensity must be within a specific range for the detector to measure it successfully. When the amount of light detectable is out of this range attenuators should be applied to reduce the light intensity of the laser and thus reduce the scattered intensity (Dorfmüller 1987).

The intensity of the scattering signal is transmitted to the detector by a digital signal processor called correlator. The correlator compares the scattering intensity at successive time intervals to derive the rate at which the intensity varies. The detection optics of this equipment measure the information of the scattering near

180°, being known as backscatter detection, patented technology known as non-invasive backscatter (Porras et al. 2004).

Analysis of apparent hydrodynamic radius (r_h) are often performed under the dilute regimen, which makes it possible to obtain the diffusion coefficient D as the Eq. 4.3, where η is the viscosity of the medium (Wang et al. 2008).

$$r_h = KT / 6\pi\eta D \tag{4.3}$$

With very concentrated samples the apparent droplet size can be overestimated because of multiple scattering due to long-length colloidal interactions among close droplets. In this context DLS measurements can be often limited by the necessity of diluting the samples.

The mean droplet size obtained from DLS can be determined by different manner, for instance, number, intensity (Z-average) and volume values. Hence, it is fundamental to be clear about the type of determination applied to droplet size characterization, since these values can be very discrepant, depending on sample concentration, polydispersity and refraction index.

In order to characterize droplet size, techniques such as electron microscopy, X-Ray and Neutron Scattering of liquid emulsions, spectro-turbidity and dynamic light scattering can be used, the latter being the most used in the characterization of nanoemulsions (McClements 2015).

Droplet size can also be characterized by transmission electron microscopy, which can visually provide droplet size and distribution in nanoemulsions of nanometric droplet sizes. From this technique it is also possible to observe morphological spherical shape of droplets and evaluate the destabilization of the nanoemulsions, in which droplet coalescence can observed. However, transmission electron microscopy has also limitations, which are often related to sample preparation that can induce to coalescence upon solvent evaporation during drying step. Besides, contrast with heavy metals as drying can also alter droplets environment and induce to some instability phenomena. High-energy electron beams can ruin nanostructure of droplets, which can be overcome by Cryo- transmission electron microscopy techniques.

4.6.2 Zeta Potential

Nanoemulsions formulated with ionic surfactants feature superficial charge. The type and magnitude of the superficial charge of droplets are responsible for many characteristics of nanoemulsions, such as stability against aggregation phenomena and functional properties like mucoadhesivity and absorption (McClements 2015).

The electrical characteristics of nanoemulsion can range from strongly positive, to neutral and strongly negative depending on the surfactant charge and the surrounding conditions. Droplet superficial charge can be characterized by zeta potential. Zeta potential describes the electro-kinetic potential in a system. Charged nanodroplets dispersed in a liquid system attract opposite-charged ions close to surface, providing a double layer with a certain thickness that can vary depending on the type and concentration of counter-ions (Jin et al. 2016). With the distribution of counter-ions in the surrounding droplet layer, any movement between rigid and mobile phases provides an electrokinetic potential (Bourbon et al. 2018). The stern layer plays an important role in the stabilization of nanoemulsions, since stern layer is responsible for the repulsive forces that prevent droplets from aggregate.

Zeta potential measurements results are presented in millivolts (mV) and is carried in an electrophoretic cell, where two electrodes create an electrical field and migration of colloidal particles or droplets is measured. Zeta potential can be strongly affected by pH and ionic strength of the system (Bhattacharjee 2016). Other factors can greatly influence zeta potential as the state oh hydration and droplet morphology. According to literature, zeta potential values above 60 mV suggest excellent electrostatic stability, while results from 60 to 30 mV indicate good stability and from 5 to 15 mV suggest a region of limited flocculation and from 3 to 5 mV correspond to maximum flocculation (Bourbon et al. 2018). When zeta potential values are high in module, droplets would repulse each other and the system is less prone to coalescence (de Oliveira et al. 2014; Junyaprasert et al. 2009; Svetlichny et al. 2017).

Nanoemulsion stabilized by anionic surfactants show negative superficial charge and those stabilized by cationic surfactants present positive superficial charge. By controlling the amount and the type of surfactant used when formulating nanoemulsions, it is possible to tune the superficial charge according to the characteristics needed.

4.6.3 Rheological Behavior

Bulk physicochemical properties, as rheological behavior can deeply impact nanoemulsions applicability, particularly for some applications routes as topical or dermal application (Barradas et al. 2018). Rheology is an important tool to characterize nanoemulsion stability. In general, flocculated systems are more viscous than a non-flocculated nanoemulsion with the same droplet size and concentration (McClements and Rao 2011). Moreover, rheological viscoelastic properties reflect flow behavior and the deformation during nanoemulsions industrial production in processing steps as flow through pipes, mixing and packaging (Chung and McClements 2018).

Flow behavior of nanoemulsions are measured apparent viscosity values as a function of shear rate or shear stress. Viscoelastic properties are determined under dynamic oscillation modes and are presented as viscous (G') and elastic (G'') modulus as a function of deformation of frequency. Typically, dilute small-sized nanoemulsions can show low viscosity and a newtonian behavior as previously published by Barradas et al. (2017). The elastic modulus provides information about the elastic properties and the stored energy, being considered a solid-like property. On the other hand, the viscous modulus indicates the viscous properties and the energy dissipated as heat being a liquid-like property.

Nanoemulsions viscosity increases as droplet increases and when dispersed phase is more concentrated. As droplet concentration increases, there is a tendency to aggregation among them, which can increase both viscosity and viscoelastic behavior (Tadros 2004). Droplet aggregation can change flow behavior from newtonian to shear thinning and increase elastic modulus of concentrated nanoemulsions. Shift on rheological behavior occur once critical droplet concentration is achieved. Besides, rheological behavior of nanoemulsions can be altered by strong electrostatic attractive or repulsive interactions among the droplets, which can provide a three-dimensional network and increase viscoelastic behavior.

The viscosity of oil phase can also greatly influence rheology and viscosity of nanoemulsions. Longer fatty acids chains lengths are able to increase viscosity of vegetable oils. As oil degradation occur, the size of fatty acids chains can be reduced by hydrolysis, which often causes the reduction in nanoemulsions viscosity as reported by Barradas et al. 2017. Thus, rheology can be an important tool to access nanoemulsion stability.

Rheology can also be a useful approach to characterize nanoemulsions microstructure and classify them as viscous, viscoelastic or semi-solid materials. Transient non-destructive tests are often applied to study structural recovery time of nanoemulsions after a constant stress is applied. This measurement can provide important information regarding the strength of intermolecular bonds and microstructure of the sample. For example, for injectable nanoemulsions it is fundamental to predict formulation viscosity, rheological behavior and structural integrity after being submitted to high deformation upon passing through a syringe.

Dynamic oscillatory tests can determine viscoelastic properties of the samples, allowing to classify them into fluid-like, solid-like, gel-like and semi-solid materials (Helgeson 2016; Chung and McClements 2018). Both frequency and strain stress can be tuned to obtain information about the macrostructure of materials. Non-destructive experiments are often performed under small strains and low frequencies in order to maintain internal structure unaltered, within linear viscoelastic region. Largely destructive measurements are conducted to reproduce stresses to which nanoemulsions can be submitted during production or practical application. High stresses are applied to cause flow or large deformations. In these cases, for liquid-like fluid nanoemulsions, viscosity is reported as apparent viscosity at a particular shear stress (Chung and McClements 2018).

4.6.4 Conductivity

Calderò et al. produced nanoemulsions oil-in-water nanoemulsions from water-in-oil emulsions by phase inversion composition method as a template for polymeric nanoparticles by solvent evaporation. Conductivity determinations were used to confirm phase inversion from water-in-oil to oil-in-water (Calderó et al. 2011). As water concentration was increased, conductivity was significantly raised until a maximum, when it dropped. It was justified by the fact that at lower water concentrations, conductivity increases as the electrolytes in the system become gain mobility. However, with higher water concentrations, a dilution effect in the charged components takes place, which causes the conductivity (Calderó et al. 2011).

4.6.5 Phase Behavior

Small-angle neutron scattering (SANS) has been quite explored for structural characterization for phase behavior and surfactant curvature during low-energy emulsification methods (Solans and Solé 2012). Moreover, it is considered a valuable tool for characterizing bulk properties of nanoemulsions, since neutron wavelengths can probe nanosized materials (Graves et al. 2005). SANS provide the identification of ordered regions within the sample, which can be observed as spots in the scattering pattern. In contrast, amorphous and disordered regions show a diffuse pattern distributed all over the scattering profile. Moreover, a color-detector provide different intensity levels, which can emphasize scattering from ordered liquid crystalline regions in the presence of disordered phases (Sonneville-Aubrun et al. 2009).

Compared with DLS, SANS provides information of droplet size in a wide range of sizes and regarding the internal structure of nanosized droplets, by performing selective deuteration of the components. Besides droplet size information, SANS provide information regarding droplet structure, such as core-shell structure, globular drops or bicontinuous microemulsion (Wang et al. 2008).

Sonneville-Aubrun et al. produced 100 nm oil-in-water nanoemulsions by performing the addition water to a water-in-oil reverse microemulsions by phase transition methods and used SANS as a tool to study phase inversion stages. Measurements were performed immediately after the addition of water to the microemulsion. The SANS profiles showed that in the early stages of water addition, the initial formation of mesophase occurred, suggesting that a liquid crystalline region was formed as a transition phase. Moreover, SANS showed a strong change in surfactant curvature progressively or abruptly as the hydration of the surfactant polar group occurs, depending on the speed of water addition (Sonneville-Aubrun et al. 2009).

SANS has also been used to study silicone anionic nanoemulsions produced with different oil volume fractions, from 0.008 to 0.6. At dilute oil volume fractions, droplets are spherical. As silicon volume fractions increased above a critical jamming point, the primary peak increased, suggesting more frequent interactions and deformations among droplets due to electrostatic repulsion of neighboring droplets (Graves et al. 2005).

4.7 Application of Nanoemulsions

Nanoemulsions feature the most diverse applications in the cosmetics and pharmaceutical industry (Singh et al. 2017; Sutradhar and Amin 2013; Jaiswal et al. 2015). The small droplet size, increased surface area, high kinetic stability and optical transparency of nanoemulsions, compared to conventional emulsions, offer them advantages in many technological applications, some of them are explored below. The ability to encapsulate lipophilic bioactive molecules and improve solubility and stability of lipophilic components, makes oil-in-water nanoemulsions a useful tool to enhance the delivery of natural oils. Qian et al. produced β-lactoglobulin-stabilized nanoemulsion for encapsulation of carotenoids, natural antioxidant easily chemically degradable. The influence of ionic strength, temperature and pH on chemical stability of β-carotene and physical stability of nanoemulsions was studied. Color fading due to chemical degradation of β-carotene was significantly smaller to β-lactoglobulin-stabilized nanoemulsions, which showed to be an effective approach to increase chemical stability of bioactive molecules, such as β-carotene (Qian et al. 2012).

The encapsulation of lipophilic components, as vegetable oils provides easier handling or administration and incorporation into a pharmaceutical secondary formulation, for example, gels, lotions or creams. Besides, nanoencapsulation of vegetable oils can also increase bioavailability by increasing water solubility and due to small droplet size, promote rate and site-controlled delivery and protect from environmental degradation and prevent early evaporation (Dias et al. 2014; Sutradhar and Amin 2013).

The encapsulation of bioactive molecules and vegetable oils can be useful in developing novel pharmaceutical formulations for masking unpleasant taste or smell of some drugs, which is especially useful for pediatrics formulations (Amin et al. 2018). Moreover, nanoemulsions are useful tools to improve bioavailability of both synthetic drugs and biologically active lipids or probiotics, as polyphenols and oil-soluble vitamins, for example, which can improve the pharmacological effect (Chen et al. 2011; Salem and Ezzat 2018). Yen et al. developed nanoemulsions for improving bioavailability of andrographolide, a poorly water-soluble anti-inflammatory. The results indicated a significantly enhanced bioavailability of six-fold from nanoemulsions in comparison with conventional drug suspension. The ability of preventing gastrointestinal inflammatory disorders was much higher for nanoemulsions then for drug suspension (Yen et al. 2018). Besides, there is a general understanding that both solubility and bioavailability of poorly soluble drugs increase with reduced droplet size. The most plausible explanation is the enhancement of surface contact area, which can increase contact with solvents or cells.

Lane et al. studied the bioavailability of omega-3-rich algal oil encapsulated by nanoemulsions as a prophylactic strategy to prevent coronary heart disease and cerebrovascular disease in 11 subjects, which were fed with the formulations. This study investigated whether the ingestion of omega-3-loaded nanoemulsions increased bioavailability in comparison with free oil. The results showed that bioavailability of omega-3 and polyunsaturated oils was drastically enhanced in patients fed with nanoemulsions (Lane et al. 2014).

The solubility of oil encapsulated in nanoemulsions increases with the reduction of droplet size. According to McClements studies (McClements 2011), the solubility of a typical oil was increased by 2.24, 1.08, 1.01 and 1.00 when droplet sizes of 10, 100, 1000 and 10,000 nm, respectively, were obtained. It is due to modification in oil-water partition coefficient of the encapsulated substance (McClements 2011).

A great number of studies have reported that oil-in-water nanoemulsions can increase antimicrobial activity of essential oils against several microorganisms as bacteria and fungus (Salvia-Trujillo et al. 2015; Sonu et al. 2018; Chuesiang et al. 2019). The encapsulation of essential oils into nanosized droplets can lead to a great disruptive activity of essential oils on cell membranes of microorganisms (McClements and Jafari 2018). Chuesiang et al. produced cinnamon oil-loaded oil-in-water nanoemulsions by phase inversion temperature method and investigated water-miscibility and antimicrobial activity against *Escherichia coli, Salmonella enterica serova Triphimurim, Staphylococcus aureus* and *Vibrio parahaemolyticus*. Cinnamon oil features biological activities, which is related to cinnamaldehyde, cinnamon oil main constituent, that is able to interact with bacterial cell membrane (Chuesiang et al. 2019). Nanoemulsions increased antimicrobial activity probably due to the encapsulation of cinnamaldehyde, which is prone to chemical degradation in its free form. Besides, water dispersibility of cinnamon oil was enhanced by nanoencapsulation, which can allow the use as a natural preservative to be incorporated in food or beverages. Moreover, smaller droplets seemed to be more efficiently transported through bacterial membranes, and thus provided high antimicrobial activity in comparison with larger droplets, even though the latter contained higher amounts of cinnamon oil (Chuesiang et al. 2019).

A wide range of bioactive natural oils feature important antioxidant properties, however, as they are highly lipophilic molecules, the incorporation of these components into many aqueous pharmaceutical formulations can be limited. In this context nanoemulsions rise as an encapsulation approach to provide protection to the droplet content, while preserving bioactive molecules functional properties. Rinaldi et al. produced neem oil-loaded nanoemulsions by ultrasound sonication Tween 20® as surfactant. The antioxidant activity of neem oil alone and encapsulated into nanoemulsions were quite similar, suggesting that nanoemulsions are efficient in encapsulating bioactive molecules, while maintaining functional activity of neem oil. Moreover, cytotoxicity was significantly reduced when Neem oil was incorporated in droplets in comparison with free Neem oil (Rinaldi et al. 2017).

Nanoemulsions can be applied in the controlled release of bioactive molecules in the pharmaceutical and cosmetic area. This is due to the large surface area and low interfacial tension of droplets, which allows the effective penetration of active pharmaceutical ingredient. Because of the small size of nanosized oil-loaded droplets, they can penetrate the *stratum corneum* and can they also be applied in alcohol-free perfume formulations (Rai et al. 2018). The encapsulation of vegetable oils can be particularly beneficial for volatile components, such as essential oils and aromas. The encapsulation approach can control the release and evaporation rate of bioactive molecules and volatile components, which can bring important for aroma perception and duration (McClements 2015). Time-controlled oil release can be tuned by modulating the lipophilicity of the inner phase. High lipophilicity can lead to a more sustained oil release. Droplet size in the case of nanoemulsions is not a limiting parameter for release profile (McClements 2011).

Because the nanoemulsions are often transparent, they are related to freshness, purity and simplicity even when carrying great amounts of oil. This characteristic has been very much valued in both pharmaceutical and cosmetic industries. It is only achieved when droplet size is too small (<70 nm) to avoid strong light scattering and ensure optical transparency, which is also dependent on polydispersity. In that context, optically transparent products can be produced when small droplet size and narrow size distribution are obtained and maintained for a considerable period of time (Wooster et al. 2008). Transparent nanoemulsions can be prepared by both high-energy and low-energy methods by adjusting and optimizing oil and aqueous phase composition, surfactants and processing parameters to achieve small droplet sizes and prevent Ostwald ripening and aggregation phenomena (McClements 2011).

Nanoemulsions with small droplet size can be sterilized through filtration and lead to a wide variety of water-based pharmaceutical products. A wide variety of products are obtained with the use of nanoemulsions, for example: lotions, moisturizers and transparent gels, with different rheological behavior (Helgeson 2016). Parenteral, or injectable, administration of nanoemulsions is employed for a variety of purposes, i.e., nutrition, for example in the administration of vegetable oils, vitamins, among others, and topical or systemic drug release. Nanoemulsions are advantageous for intravenous administration because of the rigid requirements of this route of administration, particularly the need for a droplet size in the formulation below 1 μm (Hörmann and Zimmer 2016). The benefit of nanoemulsions in oral drug administration has also been reported in the absorption of the emulsion in the gastrointestinal tract which has been correlated with droplet size (Bali et al. 2011).

4.8 Conclusion

Nanoemulsions are unique nanocarriers for the delivery lipophilic components as they provide a more stable, bioavailable, readily manufacturable, and acceptable formulation. They also impart good protection to the entrapped bioactive molecules against the effects of external conditions, as they encapsulate bioactive molecules in the core of nanosized micelles. In addition, nanoemulsions exhibit high surface area, stability and tunable rheology, which can improve drug bioavailability, making many treatments less toxic and invasive. Recently, a growing interest in the use of natural oils has been taking place, since they are proving to feature antimicrobial, antioxidants and anti-inflammatory properties, among others. In this chapter the main aspects on nanoemulsion formulation, obtention methods, characterization techniques and applications were presented.

Both high-energy and low-energy emulsification methods provide nanoemulsions with small droplet size and stability. However, much research is still needed to the achieving scaling up of these processes and to understand the impact on size and stability aspects for nanoemulsions. The difficult to scale up nanoemulsions production is responsible for so few nanoemulsion-based products are commercially available in contrast with so much research being published in this field.

Before nanoemulsions become widespread in pharmaceutical field, other challenges must be overcome. First, pharmaceutical-grade excipients should be ideally chosen, such as synthetic polymers and surfactants. Next, there are some safety concerns involving nanotechnological products as nanoemulsion and nano-based products toxicological profile is different from conventional emulsion. In this context further research is needed to promote wide nanoemulsions production and utilization.

Acknowledgements This material is based upon the work supported by the Coordination for the Improvement of Higher Education Personnel (CAPES) and National Council for Scientific and Technological Development (CNPq).

References

Abismail B, Canselier JP, Wilhelm AM, Delmas H, Gourdon C (1999) Emulsification by ultrasound- drop size distribution and stability.Pdf. Ultrason Sonochem 6:75–83

Adjonu R, Doran G, Torley P, Agboola S (2014) Formation of whey protein isolate hydrolysate stabilised nanoemulsion. Food Hydrocoll 41:169–177. https://doi.org/10.1016/j.foodhyd.2014.04.007

Alam P, Shakeel F, Anwer MK, Foudah AI, Alqarni MH (2018) Wound healing study of eucalyptus essential oil containing nanoemulsion in rat model. J Oleo Sci 67(8):957–968. https://doi.org/10.5650/jos.ess18005

Alexandridis P, Holzwarth JF, Hatton TA (1994) Micellization of poly(ethylene oxide)-poly(propylene oxide)-poly(ethylene oxide) triblock copolymers in aqueous solutions: thermodynamics of copolymer association. Macromolecules 27(9):2414–2425. https://doi.org/10.1021/ma00087a009

Al-Subaie MM, Hosny KM, El-Say KM, Ahmed TA, Aljaeid BM (2015) Utilization of nanotechnology to enhance percutaneous absorption of acyclovir in the treatment of herpes simplex viral infections. Int J Nanomedicine. https://doi.org/10.2147/IJN.S83962

Amin F, Khan S, Shah SMH, Rahim H, Hussain Z, Sohail M, Ullah R, Alsaid MS, Shahat AA (2018) A new strategy for taste masking of azithromycin antibiotic: development, characterization, and evaluation of azithromycin titanium nanohybrid for masking of bitter taste using physisorption and panel testing studies. Drug Des Dev Ther 12:3855–3866. https://doi.org/10.2147/DDDT.S183534

Andreu V, Mendoza G, Arruebo M, Irusta S (2015) Smart dressings based on nanostructured fibers containing natural origin antimicrobial, anti-inflammatory, and regenerative compounds. Materials 8(8):5154–5193. https://doi.org/10.3390/ma8085154

Angioni A, Barra A, Cereti E, Barile D, Coïsson JD, Arlorio M, Dessi S, Coroneo V, Cabras P (2004) Chemical composition, plant genetic differences, antimicrobial and antifungal activity investigation of the essential oil of Rosmarinus Officinalis L. J Agric Food Chem 52(11):3530–3535. https://doi.org/10.1021/jf049913t

Anton N, Vandamme TF (2011) Nano-emulsions and micro-emulsions: clarifications of the critical differences. Pharm Res 28(5):978–985. https://doi.org/10.1007/s11095-010-0309-1

Anton N, Benoit J-P, Saulnier P (2008) Design and production of nanoparticles formulated from nano-emulsion templates – a review. J Control Release 128(3):185–199. https://doi.org/10.1016/j.jconrel.2008.02.007

Anwer MK, Jamil S, Ibnouf EO, Shakeel F (2014) Enhanced antibacterial effects of clove essential oil by nanoemulsion. J Oleo Sci. https://doi.org/10.5650/jos.ess13213

Asbahani A, El K, Miladi W, Badri M, Sala EH, Aït Addi H, Casabianca AEM et al (2015) Essential oils: from extraction to encapsulation. Int J Pharm 483(1–2):220–243. https://doi.org/10.1016/j.ijpharm.2014.12.069

Azeem A, Rizwan M, Ahmad FJ, Iqbal Z, Khar RK, Aqil M, Talegaonkar S (2009) Nanoemulsion components screening and selection: a technical note. AAPS PharmSciTech 10. https://doi.org/10.1208/s12249-008-9178-x

Babchin AJ, Schramm LL (2012) Osmotic repulsion force due to adsorbed surfactants. Colloids Surf B: Biointerfaces. https://doi.org/10.1016/j.colsurfb.2011.10.050

Badgujar SB, Patel VV, Bandivdekar AH (2014) Foeniculum Vulgare mill: a review of its botany, phytochemistry, pharmacology, contemporary application, and toxicology. Biomed Res Int. https://doi.org/10.1155/2014/842674

Bai L, Huan S, Jiyou G, McClements DJ (2016) Fabrication of oil-in-water nanoemulsions by dual-channel microfluidization using natural emulsifiers: saponins, phospholipids, proteins, and polysaccharides. Food Hydrocoll 61:703–711. https://doi.org/10.1016/j.foodhyd.2016.06.035

Bali V, Ali M, Ali J (2011) Nanocarrier for the enhanced bioavailability of a cardiovascular agent: in vitro, pharmacodynamic, pharmacokinetic and stability assessment. Int J Pharm. https://doi.org/10.1016/j.ijpharm.2010.10.018

Barradas TN, de Campos VEB, Senna JP, Coutinho CSC, Tebaldi BS, Silva KGH, Mansur CRE (2015) Development and characterization of promising o/w nanoemulsions containing sweet fennel essential oil and non-ionic sufactants. Colloids Surf A Physicochem Eng Asp 480. https://doi.org/10.1016/j.colsurfa.2014.12.001

Barradas TNTN, Juliana Perdiz Senna JP, Stephani Araujo Cardoso SA, Nicoli S, Padula C, Santi P, Rossi F, Gyselle de Holanda e Silva KG, Mansur CRE (2017) Hydrogel-thickened nanoemulsions based on essential oils for topical delivery of psoralen: permeation and stability studies. Eur J Pharm Biopharm 116:38–50. https://doi.org/10.1016/j.ejpb.2016.11.018

Barradas TN, Senna JP, Cardoso SA, de Holanda e Silva KG, Elias Mansur CR (2018) Formulation characterization and in vitro drug release of hydrogel-thickened nanoemulsions for topical delivery of 8-methoxypsoralen. Mater Sci Eng C 92. https://doi.org/10.1016/j.msec.2018.06.049

Behrend O, Ax K, Schubert H (2000) Influence of continuous phase viscosity on emulsification by ultrasound. Ultrason Sonochem. https://doi.org/10.1016/S1350-4177(99)00029-2

Bhattacharjee S (2016) DLS and zeta potential – what they are and what they are not? J Control Release. https://doi.org/10.1016/j.jconrel.2016.06.017

Bondi CA, Marks JL, Wroblewski LB, Raatikainen HS, Lenox SR, Gebhardt KE (2015) Human and environmental toxicity of sodium lauryl sulfate (SLS): evidence for safe use in household cleaning products. Environ Health Insights 9:27–32. https://doi.org/10.4137/EHI.S31765

Bonferoni MC, Rossi S, Cornaglia AI, Mannucci B, Grisoli P, Vigani B, Saporito F et al (2017) Essential oil-loaded lipid nanoparticles for wound healing. Int J Nanomedicine 13:175–186. https://doi.org/10.2147/IJN.S152529

Borges RS, Lima ES, Keita H, Ferreira IM, Fernandes CP, Cruz RAS, Duarte JL et al (2018) Anti-inflammatory and antialgic actions of a nanoemulsion of *Rosmarinus officinalis* L. essential oil and a molecular docking study of its major chemical constituents. Inflammopharmacology 26(1):183–195. https://doi.org/10.1007/s10787-017-0374-8

Botas G, Cruz R, de Almeida F, Duarte J, Araújo R, Souto R, Ferreira R et al (2017) Baccharis reticularia DC. and limonene nanoemulsions: promising larvicidal agents for *Aedes aegypti* (Diptera: Culicidae) control. Molecules 22(11):1990. https://doi.org/10.3390/molecules22111990

Bouchemal K, Briançon S, Perrier E, Fessi H (2004) Nano-emulsion formulation using spontaneous emulsification: solvent, oil and surfactant optimisation. Int J Pharm 280(1–2):241–251. https://doi.org/10.1016/j.ijpharm.2004.05.016

Bourbon AI, Gonçalves RFS, Vicente AA, Pinheiro AC (2018) Characterization of particle properties in nanoemulsions. Nanoemulsions (Elsevier):519–546. https://doi.org/10.1016/B978-0-12-811838-2.00016-3

Bouyer E, Mekhloufi G, Rosilio V, Grossiord JL, Agnely F (2012) Proteins, polysaccharides, and their complexes used as stabilizers for emulsions: alternatives to synthetic surfactants in the pharmaceutical field? Int J Pharm. https://doi.org/10.1016/j.ijpharm.2012.06.052

Calderó G, García-Celma MJ, Solans C (2011) Formation of polymeric nano-emulsions by a low-energy method and their use for nanoparticle preparation. J Colloid Interface Sci 353(2):406–411. https://doi.org/10.1016/j.jcis.2010.09.073

Cardoso-Ugarte GA, López-Malo A, Jiménez-Munguía MT (2016) Application of nanoemulsion technology for encapsulation and release of lipophilic bioactive compounds in food. Emulsions (Elsevier):227–255. https://doi.org/10.1016/B978-0-12-804306-6.00007-6

Carteau D, Bassani D, Pianet I (2008) The 'Ouzo Effect': following the spontaneous emulsification of trans-anethole in water by NMR. C R Chim. https://doi.org/10.1016/j.crci.2007.11.003

Cerqueira-Coutinho C, Santos-Oliveira R, dos Santos E, Mansur CR (2015) Development of a photoprotective and antioxidant nanoemulsion containing chitosan as an agent for improving skin retention. Eng Life Sci. https://doi.org/10.1002/elsc.201400154

Chaieb K, Hajlaoui H, Zmantar T, Kahla-Nakbi AB, Rouabhia M, Mahdouani K, Bakhrouf A (2007) The chemical composition and biological activity of clove essential oil, *Eugenia caryophyllata* (Syzigium Aromaticum L. Myrtaceae): a short review. Phytother Res. https://doi.org/10.1002/ptr.2124

Chebil A, Desbrières J, Nouvel C, Six J-L, Durand A (2013) Ostwald ripening of nanoemulsions stopped by combined interfacial adsorptions of molecular and macromolecular nonionic stabilizers. Colloids Surf A Physicochem Eng Asp 425:24–30. https://doi.org/10.1016/j.colsurfa.2013.02.028

Chen H, Khemtong C, Yang X, Chang X, Gao J (2011) Nanonization strategies for poorly water-soluble drugs. Drug Discov Today. https://doi.org/10.1016/j.drudis.2010.02.009

Chuesiang P, Siripatrawan U, Sanguandeekul R, Yang JS, McClements DJ, McLandsborough L (2019) Antimicrobial activity and chemical stability of cinnamon oil in oil-in-water nanoemulsions fabricated using the phase inversion temperature method. LWT 110:190–196. https://doi.org/10.1016/j.lwt.2019.03.012

Chung C, McClements DJ (2018) Characterization of physicochemical properties of nanoemulsions: appearance, stability, and rheology. Nanoemulsions (Elsevier):547–576. https://doi.org/10.1016/B978-0-12-811838-2.00017-5

Constantinides PP, Chaubal MV, Shorr R (2008) Advances in lipid nanodispersions for parenteral drug delivery and targeting. Adv Drug Deliv Rev 60(6):757–767. https://doi.org/10.1016/j.addr.2007.10.013

de Oliveira EF, Paula HCB, de Paula RCM (2014) Alginate/cashew gum nanoparticles for essential oil encapsulation. Colloids Surf B: Biointerfaces 113:146–151. https://doi.org/10.1016/j.colsurfb.2013.08.038

Dias D d O, Colombo M, Kelmann RG, Kaiser S, Lucca LG, Teixeira HF, Limberger RP, Veiga VF, Koester LS (2014) Optimization of copaiba oil-based nanoemulsions obtained by different preparation methods. Ind Crop Prod 59:154–162. https://doi.org/10.1016/j.indcrop.2014.05.007

Dickinson E (2003) Hydrocolloids at interfaces and the influence on the properties of dispersed systems. Food Hydrocoll. https://doi.org/10.1016/S0268-005X(01)00120-5

Dickinson E (2009) Hydrocolloids as emulsifiers and emulsion stabilizers. Food Hydrocoll 23(6):1473–1482. https://doi.org/10.1016/j.foodhyd.2008.08.005

Dickinson E, Galazka VB, Anderson DMW (1991) Emulsifying behaviour of gum arabic. Part 1: effect of the nature of the oil phase on the emulsion droplet-size distribution. Carbohydr Polym 14(4):373–383. https://doi.org/10.1016/0144-8617(91)90003-U

Dickinson E, Semenova MG, Antipova AS (1998) Salt stability of casein emulsions. Food Hydrocoll 12(2):227–235. https://doi.org/10.1016/S0268-005X(98)00035-6
Donsì F, Ferrari G (2016) Essential oil nanoemulsions as antimicrobial agents in food. J Biotechnol Elsevier BV. https://doi.org/10.1016/j.jbiotec.2016.07.005
Dorfmüller T (1987) R. Pecora (Ed.): dynamic light scattering – applications of photon correlation spectroscopy, Plenum Press, New York and London 1985. 420 Seiten, Preis: $ 59.90. Ber Bunsenges Phys Chem 91(4):498–499. https://doi.org/10.1002/bbpc.19870910455
Duarte JL, Amado JRR, Oliveira AEMFM, Cruz RAS, Ferreira AM, Souto RNP, Falcão DQ, Carvalho JCT, Fernandes CP (2015) Evaluation of larvicidal activity of a nanoemulsion of *Rosmarinus officinalis* essential oil. Rev Bras 25(2):189–192. https://doi.org/10.1016/j.bjp.2015.02.010
Elmahjoubi E, Frum Y, Eccleston GM, Wilkinson SC, Meidan VM (2009) Transepidermal water loss for probing full-thickness skin barrier function: correlation with tritiated water flux, sensitivity to punctures and diverse surfactant exposures. Toxicol in Vitro. https://doi.org/10.1016/j.tiv.2009.06.030
Farshi P, Tabibiazar M, Ghorbani M, Mohammadifar M, Amirkhiz MB, Hamishehkar H (2019) Whey protein isolate-guar gum stabilized cumin seed oil nanoemulsion. Food Biosci 28:49–56. https://doi.org/10.1016/j.fbio.2019.01.011
Fernandez P, André V, Rieger J, Kühnle A (2004) Nano-emulsion formation by emulsion phase inversion. Colloids Surf A Physicochem Eng Asp. https://doi.org/10.1016/j.colsurfa.2004.09.029
Goñi MG, Roura SI, Ponce AG, Moreira MR (2016) Clove (*Syzygium aromaticum*) oils. In: Preedy VR (ed) Essential oils in food preservation, flavor and safety. Academic, Amsterdam
Graves S, Meleson K, Wilking J, Lin MY, Mason TG (2005) Structure of concentrated nanoemulsions. J Chem Phys 122(13):134703. https://doi.org/10.1063/1.1874952
Gupta A, Eral HB, Hatton TA, Doyle PS (2016a) Controlling and predicting droplet size of nanoemulsions: scaling relations with experimental validation. Soft Matter. https://doi.org/10.1039/C5SM02051D
Gupta A, Eral HB, Hatton TA, Doyle PS (2016b) Nanoemulsions: formation, properties and applications. Soft Matter 12(11). https://doi.org/10.1039/C5SM02958A
Gutiérrez JM, González C, Maestro A, Solè I, Pey CM, Nolla J (2008) Nano-emulsions: new applications and optimization of their preparation. Curr Opin Colloid Interface Sci. https://doi.org/10.1016/j.cocis.2008.01.005
Hashem AS, Awadalla SS, Zayed GM, Maggi F, Benelli G (2018) Pimpinella Anisum essential oil nanoemulsions against Tribolium Castaneum—insecticidal activity and mode of action. Environ Sci Pollut Res 25(19):18802–18812. https://doi.org/10.1007/s11356-018-2068-1
Hashemi Gahruie H, Ziaee E, Eskandari MH, Hosseini SMH (2017) Characterization of basil seed gum-based edible films incorporated with Zataria Multiflora essential oil nanoemulsion. Carbohydr Polym 166:93–103. https://doi.org/10.1016/j.carbpol.2017.02.103
Hasssanzadeh H, Alizadeh M, Rezazad Bari M (2018) Formulation of garlic oil-in-water nanoemulsion: antimicrobial and physicochemical aspects. IET Nanobiotechnol 12(5):647–652. https://doi.org/10.1049/iet-nbt.2017.0104
Helgeson ME (2016) Colloidal behavior of nanoemulsions: interactions, structure, and rheology. Curr Opin Colloid Interface Sci. https://doi.org/10.1016/j.cocis.2016.06.006
Herman AAP, Herman AAP (2015) Essential oils and their constituents as skin penetration enhancer for transdermal drug delivery: a review. J Pharm Pharmacol. https://doi.org/10.1111/jphp.12334
Hernández MD, Sotomayor JA, Hernández Á, Jordán MJ (2016) Rosemary (*Rosmarinus officinalis* L.) oils. In: Preedy VR (ed) Essential oils in food preservation, flavor and safety. Elsevier, San Diego, pp 677–688
Hoeller S, Sperger A, Valenta C (2009) Lecithin based nanoemulsions: a comparative study of the influence of non-ionic surfactants and the cationic phytosphingosine on physicochemical behaviour and skin permeation. Int J Pharm 370(1–2):181–186. https://doi.org/10.1016/j.ijpharm.2008.11.014

Hörmann K, Zimmer A (2016) Drug delivery and drug targeting with parenteral lipid nanoemulsions – a review. J Control Release 223:85–98. https://doi.org/10.1016/j.jconrel.2015.12.016
Hudaib M, Speroni E, Di Pietra AM, Cavrini V (2002) GC/MS evaluation of thyme (*Thymus vulgaris* L.) oil composition and variations during the vegetative cycle. J Pharm Biomed Anal 29(4):691–700. http://www.ncbi.nlm.nih.gov/pubmed/12093498
Hunter RJ (2001) Foundations of colloid science, 2nd edn. Oxford University Press, New York
Hussain AI, Anwar F, Hussain Sherazi ST, Przybylski R (2008) Chemical composition, antioxidant and antimicrobial activities of basil (*Ocimum basilicum*) essential oils depends on seasonal variations. Food Chem 108(3):986–995. https://doi.org/10.1016/j.foodchem.2007.12.010
Jafari SM (2017) Nanoencapsulation technologies for the food and nutraceutical industries. Academic Print, London. https://doi.org/10.1016/B978-0-12-809436-5/00014-8
Jafari SM, He Y, Bhandari B (2007) Production of sub-micron emulsions by ultrasound and microfluidization techniques. J Food Eng 82(4):478–488. https://doi.org/10.1016/j.jfoodeng.2007.03.007
Jafari SM, Assadpoor E, He Y, Bhandari B (2008) Re-coalescence of emulsion droplets during high-energy emulsification. Food Hydrocoll. https://doi.org/10.1016/j.foodhyd.2007.09.006
Jafari SM, Paximada P, Mandala I, Assadpour E, Mehrnia MA (2017) Encapsulation by nanoemulsions. Nanoencapsulation Technol Food Nutraceut Ind 2017:36–73
Jaiswal M, Dudhe R, Sharma PK (2015) Nanoemulsion: an advanced mode of drug delivery system. 3 Biotech Springer Verlag. https://doi.org/10.1007/s13205-014-0214-0
Jin W, Xu W, Lian H, Li Y, Liu S, Li B (2016) Nanoemulsions for food: properties, production, characterization and applications. In: Grumezescu A (ed) Emulsions. Elsevier Inc, London, pp 1–29. https://doi.org/10.1016/B978-0-12-804306-6.00001-5
Junyaprasert VB, Teeranachaideekul V, Souto EB, Boonme P, Müller RH (2009) Q10-loaded NLC versus nanoemulsions: stability, rheology and in vitro skin permeation. Int J Pharm. https://doi.org/10.1016/j.ijpharm.2009.05.020
Jirovetz L, Buchbauer G, Stoilova I, Stoyanova A, Krastanov A, Schmidt E (2006) Chemical composition and antioxidant properties of clove leaf essential oil. J Agric Food Chem 54:6303–6307
Kataoka K, Harada A, Nagasaki Y (2001) Block copolymer micelles for drug delivery: design, characterization and biological significance. Adv Drug Deliv Rev 47(1):113–131
Katsumoto Y, Ushiki H, Lachaise J, Graciaa A (2001) Time evolution of the size distribution of nano-sphere droplets in the hexadecane-in-water miniemulsion stabilized by nonionic surfactants. Colloid Polym Sci 279(2):122–130. https://doi.org/10.1007/s003960000395
Khan MS, Krishnaraj K (2014) Phospholipids: a novel adjuvant in herbal drug delivery systems. Crit Rev Ther Drug Carrier Syst 31(5):407–428. https://doi.org/10.1615/CritRevTherDrugCarrierSyst.2014010634
Kiaei N, Hajimohammadi R, Hosseini M (2018) Investigation of the anti-inflammatory properties of calendula nanoemulsion on skin cells. Bioinspired, Biomimetic Nanobiomater 7(4):228–237. https://doi.org/10.1680/jbibn.17.00033
Klang V, Valenta C (2011) Lecithin-based nanoemulsions. J Drug Delivery Sci Technol 21(1):55–76. https://doi.org/10.1016/S1773-2247(11)50006-1
Kotta S, Khan AW, Ansari SH, Sharma RK, Ali J (2015) Formulation of nanoemulsion: a comparison between phase inversion composition method and high-pressure homogenization method. Drug Deliv 22(4):455–466. https://doi.org/10.3109/10717544.2013.866992
Kuhn KR, Cunha RL (2012) Flaxseed oil – whey protein isolate emulsions: effect of high pressure homogenization. J Food Eng 111(2):449–457. https://doi.org/10.1016/j.jfoodeng.2012.01.016
Lane KE, Li W, Smith C, Derbyshire E (2014) The bioavailability of an omega-3-rich algal oil is improved by nanoemulsion technology using yogurt as a food vehicle. Int J Food Sci Technol 49(5):1264–1271. https://doi.org/10.1111/ijfs.12455
Lawrence MJ, Rees GD (2000) Microemulsion-based media as novel drug delivery systems. Adv Drug Deliv Rev 45(1):89–121. https://doi.org/10.1016/S0169-409X(00)00103-4

Lee L, Norton IT (2013) Comparing droplet breakup for a high-pressure valve homogeniser and a microfluidizer for the potential production of food-grade nanoemulsions. J Food Eng 114(2):158–163. https://doi.org/10.1016/j.jfoodeng.2012.08.009

Lefebvre G, Riou J, Bastiat G, Roger E, Frombach K, Gimel J-C, Saulnier P, Calvignac B (2017) Spontaneous nano-emulsification: process optimization and modeling for the prediction of the nanoemulsion's size and polydispersity. Int J Pharm 534(1–2):220–228. https://doi.org/10.1016/j.ijpharm.2017.10.017

Lémery E, Briançon S, Chevalier Y, Bordes C, Oddos T, Gohier A, Bolzinger MA (2015) Skin toxicity of surfactants: structure/toxicity relationships. Colloids Surf A Physicochem Eng Asp. https://doi.org/10.1016/j.colsurfa.2015.01.019

Leong TSH, Wooster TJ, Kentish SE, Ashokkumar M (2009) Minimising oil droplet size using ultrasonic emulsification. Ultrason Sonochem. https://doi.org/10.1016/j.ultsonch.2009.02.008

Li QX, Chang CL (2016) Basil (*Ocimum basilicum* L.) oils. In: Preedy VR (ed) Essential oils in food preservation, flavor and safety. Academic, San Diego, pp 231–238

Li Z-h, Cai M, Liu Y-s, Sun P-l (2018) Development of finger citron (*Citrus medica* L. Var. Sarcodactylis) essential oil loaded nanoemulsion and its antimicrobial activity. Food Control 94:317–323. https://doi.org/10.1016/j.foodcont.2018.07.009

Liu W, Sun D, Li C, Liu Q, Xu J (2006) Formation and stability of paraffin oil-in-water nano-emulsions prepared by the emulsion inversion point method. J Colloid Interface Sci 303(2):557–563. https://doi.org/10.1016/j.jcis.2006.07.055

Lou Z, Chen J, Yu F, Wang H, Kou X, Ma C, Zhu S (2017) The antioxidant, antibacterial, antibiofilm activity of essential oil from *Citrus medica* L. Var. Sarcodactylis and its nanoemulsion. LWT 80:371–377. https://doi.org/10.1016/j.lwt.2017.02.037

Maestro A, Solè I, González C, Solans C, Gutiérrez JM (2008) Influence of the phase behavior on the properties of ionic nanoemulsions prepared by the phase inversion composition method. J Colloid Interface Sci 327(2):433–439. https://doi.org/10.1016/j.jcis.2008.07.059

Mandal S, DebManda M (2016) Thyme (*Thymus vulgaris* L.) oils. In: Preedy VR (ed) Essential oils in food preservation, flavor and safety. Academic, San Diego, pp 25–834

Mao L, Xu D, Yang J, Yuan F, Gao Y, Zhao J (2009) Effects of small and large molecule emulsifiers on the characteristics of B-carotene nanoemulsions prepared by high pressure homogenization. Food Technol Biotechnol 47(3):336–342

Marotti M, Piccaglia R, Giovanelli E (1996) Differences in essential oil composition of basil (*Ocimum basilicum* L.) Italian cultivars related to morphological characteristics. J Agric Food Chem 44(12):3926–3929. https://doi.org/10.1021/jf9601067

Mason TG, Wilking JN, Meleson K, Chang CB, Graves SM (2006) Nanoemulsions: formation, structure, and physical properties. J Phys Condens Matter 18(18):635–666. https://doi.org/10.1088/0953-8984/18/41/R01

Mayer S, Weiss J, McClements DJ (2013) Vitamin E-enriched nanoemulsions formed by emulsion phase inversion: factors influencing droplet size and stability. J Colloid Interface Sci. https://doi.org/10.1016/j.jcis.2013.04.016

Mazarei Z, Rafati H (2019) Nanoemulsification of *Satureja khuzestanica* essential oil and pure carvacrol; comparison of physicochemical properties and antimicrobial activity against food pathogens. LWT 100:328–334. https://doi.org/10.1016/j.lwt.2018.10.094

McClements DJ (2011) Edible nanoemulsions: fabrication, properties, and functional performance. Soft Matter 7(6):2297–2316. https://doi.org/10.1039/C0SM00549E

McClements DJ (2012) Nanoemulsions versus microemulsions: terminology, differences, and similarities. Soft Matter 8(6):1719–1729. https://doi.org/10.1039/C2SM06903B

McClements DJ (2015) Food emulsions: principles, practices, and techniques, 3rd edn. CRC Press, Boca Raton/London/New York

McClements DJ, Jafari SM (2018) General aspects of nanoemulsions and their formulation. Nanoemulsions (Elsevier):3–20. https://doi.org/10.1016/B978-0-12-811838-2.00001-1

McClements DJ, Rao J (2011) Food-grade nanoemulsions: formulation, fabrication, properties, performance, biological fate, and potential toxicity. Crit Rev Food Sci Nutr 51(4):285–330. https://doi.org/10.1080/10408398.2011.559558

McNamee BF, O'Riorda ED, O'Sullivan M (1998) Emulsification and microencapsulation properties of gum arabic. J Agric Food Chem. https://doi.org/10.1021/jf9803740

Meinders MBJ, van Vliet T (2004) The role of interfacial rheological properties on ostwald ripening in emulsions. Adv Colloid Interf Sci 108–109:119–126. https://doi.org/10.1016/j.cis.2003.10.005

Moghimi R, Ghaderi L, Rafati H, Aliahmadi A, McClements DJ (2016) Superior antibacterial activity of nanoemulsion of thymus daenensis essential oil against *E. Coli*. Food Chem 194:410–415. https://doi.org/10.1016/j.foodchem.2015.07.139

Myers D (2005a) Surfactants in solution. In: Surfactant science and technology, 3rd edn. Wiley, Hoboken, pp 107–142

Myers D (2005b) The organic chemistry of surfactants. In: Surfactant science and technology. Wiley, Hoboken, pp 1–448

Nirmal NP, Mereddy R, Li L, Sultanbawa Y (2018) Formulation, characterisation and antibacterial activity of lemon myrtle and anise myrtle essential oil in water nanoemulsion. Food Chem 254:1–7. https://doi.org/10.1016/j.foodchem.2018.01.173

Oliveira AEMFM, Bezerra DC, Duarte JL, Cruz RAS, Souto RNP, Ferreira RMA, Nogueira J et al (2017) Essential oil from *Pterodon emarginatus* as a promising natural raw material for larvicidal nanoemulsions against a tropical disease vector. Sustain Chem Pharm 6:1–9. https://doi.org/10.1016/j.scp.2017.06.001

Osanloo M, Sedaghat MM, Esmaeili F, Amani A (2018a) Larvicidal activity of essential oil of *Syzygium aromaticum* (clove) in comparison with its major constituent, eugenol, against *Anopheles stephensi*. J Arthropod Borne Dis 12(4):361–369. http://www.ncbi.nlm.nih.gov/pubmed/30918905

Osanloo M, Sereshti H, Sedaghat MM, Amani A (2018b) Nanoemulsion of dill essential oil as a green and potent larvicide against *Anopheles stephensi*. Environ Sci Pollut Res 25(7):6466–6473. https://doi.org/10.1007/s11356-017-0822-4

Ouzineb K, Lord C, Lesauze N, Graillat C, Tanguy PA, McKenna T (2006) Homogenisation devices for the production of miniemulsions. Chem Eng Sci 61(9):2994–3000. https://doi.org/10.1016/j.ces.2005.10.065

Pathania R, Khan H, Kaushik R, Khan MA (2018) Essential oil nanoemulsions and their antimicrobial and food applications. Curr Res Nutr Food Sci J. https://doi.org/10.12944/crnfsj.6.3.05

Pengon S, Chinatangkul N, Limmatvapirat C, Limmatvapirat S (2018) The effect of surfactant on the physical properties of coconut oil nanoemulsions. Asian J Pharm Sci 13(5):409–414. https://doi.org/10.1016/j.ajps.2018.02.005

Pérez-Recalde M, Ruiz Arias IE, Hermida ÉB (2018) Could essential oils enhance biopolymers performance for wound healing? a systematic review. Phytomedicine 38:57–65. https://doi.org/10.1016/j.phymed.2017.09.024

Politeo O, Jukic M, Milos M (2007) Chemical composition and antioxidant capacity of free volatile aglycones from basil (*Ocimum basilicum* L.) compared with its essential oil. Food Chem 101(1):379–385. https://doi.org/10.1016/j.foodchem.2006.01.045

Porras M, Solans C, González C, Martínez A, Guinart A, Gutiérrez JM (2004) Studies of formation of W/O nano-emulsions. Colloids Surf A Physicochem Eng Asp. https://doi.org/10.1016/j.colsurfa.2004.08.060

Porras M, Solans C, González C, Gutiérrez JM (2008) Properties of water-in-oil (W/O) nano-emulsions prepared by a low-energy emulsification method. Colloids Surf A Physicochem Eng Asp. https://doi.org/10.1016/j.colsurfa.2008.04.012

Prabaharan M (2011) Prospective of guar gum and its derivatives as controlled drug delivery systems. Int J Biol Macromol. https://doi.org/10.1016/j.ijbiomac.2011.04.022

Prajapati VD, Jani GK, Moradiya NG, Randeria NP (2013) Pharmaceutical applications of various natural gums, mucilages and their modified forms. Carbohydr Polym. https://doi.org/10.1016/j.carbpol.2012.11.021

Prud'homme RK, Guangwei W, Schneider DK (1996) Structure and rheology studies of poly(oxyethylene–oxypropylene–oxyethylene) aqueous solution. Langmuir 12(20):4651–4659. https://doi.org/10.1021/la951506b

Qian C, McClements DJ (2011) Formation of nanoemulsions stabilized by model food-grade emulsifiers using high-pressure homogenization: factors affecting particle size. Food Hydrocoll 25(5):1000–1008. https://www.sciencedirect.com/science/article/pii/S0268005X10002328

Qian C, Decker EA, Xiao H, McClements DJ (2012) Physical and chemical stability of β-carotene-enriched nanoemulsions: influence of PH, ionic strength, temperature, and emulsifier type. Food Chem 132(3):1221–1229. https://doi.org/10.1016/j.foodchem.2011.11.091

Rai VK, Mishra N, Yadav KS, Yadav NP (2018) Nanoemulsion as pharmaceutical carrier for dermal and transdermal drug delivery: formulation development, stability issues, basic considerations and applications. J Control Release. https://doi.org/10.1016/j.jconrel.2017.11.049

Rao J, McClements DJ (2011) Food-grade microemulsions, nanoemulsions and emulsions: fabrication from sucrose monopalmitate & lemon oil. Food Hydrocoll 25(6):1413–1423. https://doi.org/10.1016/j.foodhyd.2011.02.004

Rao J, McClements DJ (2012) Food-grade microemulsions and nanoemulsions: role of oil phase composition on formation and stability. Food Hydrocoll 29(2):326–334. https://doi.org/10.1016/j.foodhyd.2012.04.008

Rinaldi F, Hanieh PN, Longhi C, Carradori S, Secci D, Zengin G, Ammendolia MG et al (2017) Neem oil nanoemulsions: characterisation and antioxidant activity. J Enzyme Inhib Med Chem 32(1):1265–1273. https://doi.org/10.1080/14756366.2017.1378190

Rosety M, Ordóñez FJ, Rosety-Rodríguez M, Rosety JM, Rosety I, Carrasco C, Ribelles A (2001) Acute toxicity of anionic surfactants sodium dodecyl sulphate (SDS) and linear alkylbenzene sulphonate (LAS) on the fertilizing capability of gilthead (*Sparus aurata* L.) sperm. Histol Histopathol 16(3):839–843. https://doi.org/10.14670/HH-16.839

Saberi AH, Fang Y, McClements DJ (2014) Stabilization of vitamin E-enriched mini-emulsions: influence of organic and aqueous phase compositions. Colloids Surf A Physicochem Eng Asp. https://doi.org/10.1016/j.colsurfa.2014.02.042

Salem MA, Ezzat SM (2018) Nanoemulsions in food industry. In: Some new aspects of colloidal systems in foods. IntechOpen, London. https://doi.org/10.5772/intechopen.79447

Salvia-Trujillo L, Rojas-Graü A, Soliva-Fortuny R, Martín-Belloso O (2015) Physicochemical characterization and antimicrobial activity of food-grade emulsions and nanoemulsions incorporating essential oils. Food Hydrocoll. https://doi.org/10.1016/j.foodhyd.2014.07.012

Schwarz JC, Weixelbaum A, Pagitsch E, Löw M, Resch GP, Valenta C (2012) Nanocarriers for dermal drug delivery: influence of preparation method, carrier type and rheological properties. Int J Pharm. https://doi.org/10.1016/j.ijpharm.2012.08.003

Seibert JB, Rodrigues IV, Carneiro SP, Amparo TR, Lanza JS, Frézard FJG, de Souza GHB, dos Santos ODH (2019) Seasonality study of essential oil from leaves of *Cymbopogon densiflorus* and nanoemulsion development with antioxidant activity. Flavour Fragrance J 34(1):5–14. https://doi.org/10.1002/ffj.3472

Shahavi MH, Hosseini M, Jahanshahi M, Meyer RL, Darzi GN (2015) Evaluation of critical parameters for preparation of stable clove oil nanoemulsion. Arab J Chem. https://doi.org/10.1016/j.arabjc.2015.08.024

Shahavi MH, Hosseini M, Jahanshahi M, Meyer RL, Darzi GN (2016) Clove oil nanoemulsion as an effective antibacterial agent: Taguchi optimization method. Desalin Water Treat 57(39):18379–18390. https://doi.org/10.1080/19443994.2015.1092893

Shahba AA-W, Mohsin K, Alanazi FK (2012) Novel self-nanoemulsifying drug delivery systems (SNEDDS) for oral delivery of cinnarizine: design, optimization, and in-vitro assessment. AAPS PharmSciTech 13(3):967–977. https://doi.org/10.1208/s12249-012-9821-4

Sharma M, Mann B, Sharma R, Bajaj R, Athira S, Sarkar P, Pothuraju R (2017) Sodium caseinate stabilized clove oil nanoemulsion: physicochemical properties. J Food Eng 212:38–46. https://doi.org/10.1016/j.jfoodeng.2017.05.006

Sharma A, Sharma NK, Srivastava A, Kataria A, Dubey S, Sharma S, Kundu B (2018) Clove and lemongrass oil based non-ionic nanoemulsion for suppressing the growth of plant pathogenic Fusarium Oxysporum f.Sp. Lycopersici. Ind Crop Prod 123:353–362. https://doi.org/10.1016/j.indcrop.2018.06.077

Shaw DJ (1992) The colloidal state. In: Introduction to colloid and surface chemistry, 4th edn. Elsevier, Amsterdam, p 320. https://doi.org/10.1016/C2009-0-24070-0

Shchipunov YA (2015) Lecithin. In: Somasundaran P (ed) Encyclopedia of surface and colloid science. CRC Press, Boca Raton, pp 3674–3693. https://doi.org/10.1081/E-ESCS3

Sing AJF, Graciaa A, Lachaise J, Brochette P, Salager JL (1999) Interactions and coalescence of nanodroplets in translucent O/W emulsions. Colloids Surf A Physicochem Eng Asp 152(1–2):31–39. https://doi.org/10.1016/S0927-7757(98)00622-0

Singh Y, Meher JG, Raval K, Khan FA, Chaurasia M, Jain NK, Chourasia MK (2017) Nanoemulsion: concepts, development and applications in drug delivery. J Control Release. https://doi.org/10.1016/j.jconrel.2017.03.008

Sivakumar M, Tang SY, Tan KW (2014) Cavitation technology – a greener processing technique for the generation of pharmaceutical nanoemulsions. Ultrason Sonochem. https://doi.org/10.1016/j.ultsonch.2014.03.025

Solans C, Solé I (2012) Nano-emulsions: formation by low-energy methods. Curr Opin Colloid Interface Sci 17(5):246–254. https://doi.org/10.1016/j.cocis.2012.07.003

Solans C, Izquierdo P, Nolla J, Azemar N, Garcia-Celma MJ (2005) Nano-emulsions. Curr Opin Colloid Interface Sci. https://doi.org/10.1016/j.cocis.2005.06.004

Solè I, Maestro A, Pey CM, González C, Solans C, Gutiérrez JM (2006) Nano-emulsions preparation by low energy methods in an ionic surfactant system. Colloids Surf A Physicochem Eng Asp 288(1–3):138–143. https://doi.org/10.1016/j.colsurfa.2006.02.013

Solè I, Solans C, Maestro A, González C, Gutiérrez JM (2012) Study of nano-emulsion formation by dilution of microemulsions. J Colloid Interface Sci 376(1):133–139. https://doi.org/10.1016/j.jcis.2012.02.063

Sonneville-Aubrun O, Simonnet JT, L'Alloret F (2004) Nanoemulsions: a new vehicle for skincare products. Adv Colloid Interf Sci. https://doi.org/10.1016/j.cis.2003.10.026

Sonneville-Aubrun O, Babayan D, Bordeaux D, Lindner P, Rata G, Cabane B (2009) phase transition pathways for the production of 100 Nm oil-in-water emulsions. Phys Chem Chem Phys 11(1):101–110. https://doi.org/10.1039/B813502A

Sonu KS, Mann B, Sharma R, Kumar R, Singh R (2018) Physico-chemical and antimicrobial properties of d-limonene oil nanoemulsion stabilized by whey protein–maltodextrin conjugates. J Food Sci Technol 55(7):2749–2757. https://doi.org/10.1007/s13197-018-3198-7

Speranza A, Corradini MG, Hartman TG, Ribnicky D, Oren A, Rogers MA (2013) Influence of emulsifier structure on lipid bioaccessibility in oil–water nanoemulsions. J Agric Food Chem 61(26):6505–6515. https://doi.org/10.1021/jf401548r

Starov VM, Zhdanov VG (2003) Viscosity of emulsions: influence of flocculation. J Colloid Interface Sci 258(2):404–414. https://doi.org/10.1016/S0021-9797(02)00149-2

Strickley RG (2004) Solubilizing excipients in oral and injectable formulations. Pharm Res. https://doi.org/10.1023/B:PHAM.0000016235.32639.23

Sundararajan B, Moola AK, Vivek K, Kumari BDR (2018) Formulation of nanoemulsion from leaves essential oil of *Ocimum basilicum* L. and its antibacterial, antioxidant and larvicidal activities (*Culex quinquefasciatus*). Microb Pathog 125:475–485. https://doi.org/10.1016/j.micpath.2018.10.017

Sutradhar KB, Amin L (2013) Nanoemulsions: increasing possibilities in drug delivery. Eur J Nanomed. https://doi.org/10.1515/ejnm-2013-0001

Svetlichny G, Külkamp-Guerreiro IC, Dalla Lana DF, Bianchin MD, Pohlmann AR, Fuentefria AM, Guterres SS (2017) Assessing the performance of copaiba oil and allantoin nanoparticles on multidrug-resistant *Candida parapsilosis*. J Drug Delivery Sci Technol 40:59–65. https://doi.org/10.1016/j.jddst.2017.05.020

Sweedman MC, Tizzotti MJ, Schäfer C, Gilbert RG (2013) Structure and physicochemical properties of octenyl succinic anhydride modified starches: a review. Carbohydr Polym. https://doi.org/10.1016/j.carbpol.2012.09.040

Tadros TF (1994) Fundamental principles of emulsion applications. Colloids Surf A Physicochem Eng Asp 91:39–55

Tadros T (2004) Application of rheology for assessment and prediction of the long-term physical stability of emulsions. Adv Colloid Interf Sci. https://doi.org/10.1016/j.cis.2003.10.025

Tadros T (2009) Polymeric surfactants in disperse systems. Adv Colloid Interf Sci 147–148:281–299. https://doi.org/10.1016/j.cis.2008.10.005

Tadros T, Izquierdo P, Esquena J, Solans C (2004) Formation and stability of nano-emulsions. Adv Colloid Interf Sci. https://doi.org/10.1016/j.cis.2003.10.023

Taleb M, Abdeltawab N, Shamma R, Abdelgayed S, Mohamed S, Farag M, Ramadan M (2018) *Origanum vulgare* L. essential oil as a potential anti-acne topical nanoemulsion—in vitro and in vivo study. Molecules 23(9):2164. https://doi.org/10.3390/molecules23092164

Taylor P (1998) Ostwald ripening in emulsions. Adv Colloid Interf Sci 75(2):107–163. https://doi.org/10.1016/S0001-8686(98)00035-9

Tian Y, Chen L, Zhang W (2016) Influence of ionic surfactants on the properties of nanoemulsions emulsified by nonionic surfactants span 80/tween 80. J Dispers Sci Technol 37(10):1511–1517. https://doi.org/10.1080/01932691.2015.1048806

Tsujii K (1998) In: Tanaka T (ed) surface activity – principles, phenomena, and applications, 1st edn. Academic, San Diego

Wan J, Zhong S, Schwarz P, Chen B, Rao J (2019) Physical properties, antifungal and mycotoxin inhibitory activities of five essential oil nanoemulsions: impact of oil compositions and processing parameters. Food Chem 291:199–206. https://doi.org/10.1016/j.foodchem.2019.04.032

Wang L, Zhang Y (2017) Eugenol nanoemulsion stabilized with zein and sodium caseinate by self-assembly. J Agric Food Chem 65:14. https://doi.org/10.1021/acs.jafc.7b00194

Wang L, Mutch KJ, Eastoe J, Heenan RK, Dong J (2008) Nanoemulsions prepared by a two-step low-energy process. Langmuir 24(12):6092–6099. https://doi.org/10.1021/la800624z

Wang L, Tabor R, Eastoe J, Li X, Heenan RK, Dong J (2009) Formation and stability of nanoemulsions with mixed ionic–nonionic surfactants. Phys Chem Chem Phys 11(42):9772. https://doi.org/10.1039/b912460h

Wei Y, Ye X, Shang X, Peng X, Bao Q, Liu M, Guo M, Li F (2012) Enhanced oral bioavailability of silybin by a supersaturatable self-emulsifying drug delivery system (S-SEDDS). Colloids Surf A Physicochem Eng Asp. https://doi.org/10.1016/j.colsurfa.2011.12.025

Wiedmann T (2003) Surfactants and polymers in drug delivery. J Control Release 93:415. https://doi.org/10.1016/j.jconrel.2002.11.001

Wooster TJ, Golding M, Sanguansri P (2008) Impact of oil type on nanoemulsion formation and ostwald ripening stability. Langmuir 24(22):12758–12765. https://doi.org/10.1021/la801685v

Xing L, Mattice WL (1997) Strong solubilization of small molecules by triblock-copolymer micelles in selective solvents. Macromolecules 30(6):1711–1717. https://doi.org/10.1021/ma961175p

Yadav MP, Manuel Igartuburu J, Yan Y, Nothnagel EA (2007) Chemical investigation of the structural basis of the emulsifying activity of gum arabic. Food Hydrocoll. https://doi.org/10.1016/j.foodhyd.2006.05.001

Yang Y, Marshall-Breton C, Leser ME, Sher AA, McClements DJ (2012) Fabrication of ultrafine edible emulsions: comparison of high-energy and low-energy homogenization methods. Food Hydrocoll 29(2):398–406. https://doi.org/10.1016/j.foodhyd.2012.04.009

Yen C-C, Chen Y-C, Wu M-T, Wang C-C, Wu Y-T (2018) Nanoemulsion as a strategy for improving the oral bioavailability and anti-inflammatory activity of andrographolide. Int J Nanomedicine 13:669–680. https://doi.org/10.2147/IJN.S154824

Yerramilli M, Ghosh S (2017) Long-term stability of sodium caseinate-stabilized nanoemulsions. J Food Sci Technol 54(1):82–92. https://doi.org/10.1007/s13197-016-2438-y
Zhang J, Peppard TL, Reineccius GA (2015) Preparation and characterization of nanoemulsions stabilized by food biopolymers using microfluidization. Flavour Fragrance J 30(4):288–294. https://doi.org/10.1002/ffj.3244
Zhang S, Zhang M, Fang Z, Liu Y (2017) Preparation and characterization of blended cloves/cinnamon essential oil nanoemulsions. LWT 75:316–322. https://doi.org/10.1016/j.lwt.2016.08.046

Chapter 5
Lipid Nanoarchitectonics for Natural Products Delivery in Cancer Therapy

Vishal Sharad Chaudhari, Prakash Kishore Hazam, and Subham Banerjee

Abstract Therapeutic applications of natural products have been noteworthy. A large sum of the population thrives on these products for countering multiple diseases and ailments. However, poor pharmacokinetic and pharmacodynamic profile of these molecules halted their progression as potential drug candidates. Other than that, low yield values are some of the added limiting factors of natural products as well.

Therefore, various measures have been taken to maximize the potential of these molecules in terms of their drug delivery perspectives. After a plenty of efforts, researchers could direct themselves towards nanoformulations as it was showing promising outcomes in various reported instances. Hence, we attempted to review the lipid-based nanoformulations/nanoarchitectonics, as is was found to be promising in terms of their drug delivery approach. Lipids are natural molecules that are biocompatible, biodegradable and they are an amicable medium for natural products delivery systems. Lipid nanoarchitectonics includes both vesicular and particulate based systems like liposomes, lipid micelles, solid lipid nanoparticles; nanostructured lipid carriers, etc. Each of them has its own pros and cons, which makes them different from others. This chapter mainly covered the up-to-date information regarding various research work carried out on lipid nanoarchitectonics based natural product delivery for the treatment of cancer therapy.

Keywords Natural products · Lipid Nanoformulations · Cancer therapy · Improved delivery

---

V. S. Chaudhari · S. Banerjee (✉)
Department of Pharmaceutics, National Institute of Pharmaceutical Education and Research (NIPER), Guwahati, Assam, India
e-mail: subham.banerjee@niperguwahati.ac.in

P. K. Hazam
National Institute of Pharmaceutical Education and Research (NIPER), Guwahati, Assam, India

A. Saneja et al. (eds.), *Sustainable Agriculture Reviews 44*, Sustainable Agriculture Reviews 44, https://doi.org/10.1007/978-3-030-41842-7_5

Abbreviations

A2780	Human ovarian cancer cell line
A549 cells	Adenomic carcinomic human alveolar basal epithelial cells
B16F10	Mouse melanoma cells
BCRP	Breast cancer resistance protein
BCS	Biopharmaceutics Classification System
Caco-2	Human epithelial colorectal adenocarcinoma cell line
DSPE-PEG 2000	1,2-distearoyl-sn-glycero-3-phosphoethanolamine-N-amino(polyethylene glycol)-2000
NF	Nanoformulations
DMPC	1,2-dimyristoyl-sn-glycero-3-phosphocholine,
EE	Entrapment efficiency.
GMS	Glyceryl monostearate
HCT-116 cell line	Human colorectal carcinoma cell line
HSPC	Hydrogenated soybean phospholipid
HepG2 cells	Liver hepato-cellular carcinoma
HL-60	Human Leukemia cell line
HeLa and SiHa	Cervical cancer cell line
KHOS OS	Osteo sarcoma cancer cell line
LY1 cells	Lymphocyte cells
MCF-7	Michigan Cancer Foundation-7 breast cancer cell line
MDA-MB-231	M. D. Anderson Metastasis Breast cancer cell line
mV	Millivolt
μm	Micrometer
NCI-ADR-RES cells	National Cancer Institute Adriamycin resistant cells
nm	Nanometers
PS	Particle size
PDI	Polydispersity index
PC3 & LNCaP	Human prostate carcinomas
PSMA-positive PCa cells	Prostate Specific Membrane Antigen PCa cell lines
PEG	Polyethylene glycol
ROS	Reactive oxygen species
RAW 264.7 cell line	Mouse macrophage cell line
RGD	Arginine Glycine Aspartic acid
S no	Serial number
SKBR3 cell line	Human Breast Cancer cell line
SKOV-3	Ovarian cancer cell line
SW480	Colon cancer cell line
SLN	Solid Lipid Nanoparticles
SH-SY5Y	Neuroblastoma cells
TPGS	Tocopheryl Polyethylene Glycol Succinate
U87	Malignant glioma cells
ZP	Zeta potential

5.1 Introduction

Nature's baskets have been providing all sources of energy derivatives to all forms of life, since its inception (Da Poian et al. 2010). Nature acquires an arsenal of therapeutically significant molecules against multiple ailments (Newman and Cragg 2016). Age old traditional practices (Buenz et al. 2018) to the modern drug discovery arena are largely dependent or inspired by natural chemical moieties (Du 2018). Categorically, natural products are alkaloids, glycosides, saponins, terpenoids, phenolics, and these molecules carry out most of the therapeutic activities in our physiology (Springob and Kutchan 2009).

The global mortality load in the current scenario can be attributed to diseases like metabolic disorder (Saklayen 2018), microbial infections (Hazam et al. 2019), and genetic inadequacies (Wojcik et al. 2018), and few others. Among them, cancer is one of the predominant global killers that estimates up to 9.6 million death worldwide, with 18.1 million new cases (Bray et al. 2018). Broadly, anticancer therapeutics (Espinosa et al. 2003) can be classified as listed in Table 5.1. Many of these molecules are either synthetic, semi-synthetic or in terms of their origin.

In general, naturally occurring anticancer molecules are obtained by microbes, plants or marines sources (Demain and Vaishnav 2011)· (Tewari et al. 2019) (Table 5.2). The vast array of secondary metabolites through inherent combinatorial chemistry by nature itself has propelled the possibilities of therapeutic molecules with significant activity.

Natural molecules have shown promising anticancer properties in laboratory scale. Despite such considerable results, properties like poor solubility, narrow therapeutic window, inferior absorption rate, and lower bioavailability, as well as stability issues, have halted the use of bare molecules or common formulations. As per many studies, these formulations were also appeared to possess, inappropriate pharmacokinetic profile and reduced elimination half-life, resulting in compromised activity as its consequences (Xie et al. 2016). Therefore, nanoparticles were introduced to counter the shortcomings of already present preparations or formulations. Nano-formulations were found to possess amicable construct, surface property, lower side effects and improved pharmacokinetic property when compared with preceding formulations.

Nanoparticles are defined as particles ranging from 1 to 100 nm in its dimension (Watkins et al. 2015). Tailor-made approaches like nano-formulation constructs

Table 5.1 Broad classification of anticancer drugs

Serial number	Class	Subclass
1	Chemotherapy	Alkylators, antibiotics, antimetabolites, topoisomerase inhibitors, mitosis inhibitors
2	Hormonal therapy	Steroids, anti-estrogens, anti-androgens, anti-aromatase agents
3	Immunotherapy	Interferons, interleukins, vaccines

Table 5.2 Naturally occurring anticancer drugs

Microbial source	Plant source	Marine source
Daunorubicin, doxorubicin (adriamycin), epirubicin, pirirubicin, idarubicin, valrubicin, amrubicin Bleomycin, phleomycin Actinomycin D, actinomycin Mithramycin, streptozotecin, Pentostatin, Mitosanes mitomycin C, Enediynes calicheamycin, Glycosides rebeccamycin, Macrolide lactones epotihilones, Ixebepilone 2″-deoxycoformycin/ pentostatin Salinosporamide A Aclarubicin Epirubicin HCl	Vinblastine, Vincristine, Etoposide, Teniposide, Taxol, Navelbine, Taxotere, Camptothecin, Topotecan, Irinotecan, Arglabin, Homoharringtonine, Ingenolmebutate, Masoprocol/Nordihydroguaiaretic acid, Peplomycin, Solamargines, Alitretinoin, Elliptiniumacetate, Etoposide phosphate	Cytarabine, Pederin Theopederins, Annamides Trabectedin, Aplidine Ecteinascidin, Eribulin Plinabulin

Table 5.3 Broader classification nanoparticles

Organic Nanoparticles	Inorganic Nanoparticles
Lipid-based Nanoparticles	Carbon-based Nanoparticles
Polymeric Nanoparticles	Graphene
Hybrids (Lipid-polymer)	Gold based Nanoparticles
Dendrimeric Nanoparticles	Iron-based Nanoparticles
	Molybdenum disulfide

result in improvement of the amicable pharmacokinetic profile, reduced side effects and superior surface area to volume ratio as its consequences. Broadly, these formulations can be classified as organic and inorganic Nanoparticles (Enrico 2019) (Table 5.3).

5.2 Organic and Inorganic Nanoparticles

According to the nature of excipients or materialistic composition of nanoparticles used for formulation, nanoparticles are classified as organic and inorganic nanoparticles. Inorganic nanoparticulate systems utilize several metal ions, fluorescent dyes. Alteration in size, shape, structure of inorganic nanomaterial provides an amplified platform for generating biological interaction in our body (Kim et al. 2013). Nanoparticle bioengineering uses properties like super magnetism, quantum properties, optical properties and few others for the generation of better end product. In the basic construct, core part acts as a scaffold with dynamic structural and functional properties which may not be possible from a polymer or lipid-based system.

Inorganic nanoparticles can be handy as a theranostic or imaging purpose due to tunable opticophysical properties (Kim et al. 2013). Multifunctional inorganic nanoparticles can be used for non-invasive imaging purpose such as optical microscopy and Magnetic Resonance Imaging.

Basically, organic Nanoparticles comprises of organic excipients such as polymers, lipids, polysaccharides, proteins. It includes vesicular drug delivery systems like liposomes, lipid micelles, and particulate drug delivery systems like polymeric nanoparticles, solid lipid nanoparticles, lipid drug conjugates and few more (Heneweer et al. 2012). Organic nanoparticles having the potential of surface modification with various ligands can be attached on its surface. Functionalization is possible in terminal ends of dendrimers. Liposomes are bilayer systems having the ability to incorporate both hydrophilic and lipophilic drugs through it. Mostly these organic nanoparticles are having applications in drug delivery.

Inorganic nanoparticle comes with a few disadvantages which include the lesser drug entrapment as compare to organic nanoparticles. Therefore, inorganic nanoparticles are predominantly used for imaging purpose. Organic nanoparticles were developed to overcome the issues of chronic toxicity associated with inorganic nanoparticles. Use of biodegradable polymers and lipids make organic nanoparticles more advantageous than inorganic nanoparticles. Easy preparation methods of organic nanoparticles along with higher stability during storage and in biological fluids make them superior to inorganic nanoparticles (Jesus and Grazu 2012).

In the case of organic nanoparticles, polymers and lipids are used more frequently for formulation development. Lipids are generally obtained from the natural origin which makes them biodegradable and biocompatible. Hence, chances of toxicity induction from lipids vanish. This makes lipids more superior than polymers. More drug entrapment is observed in lipid-based nanoparticles as it provides a hydrophobic environment to drugs for solubilization. Dose dumping and unintentionally modified release pattern may observe with polymers make them secondary. Hence, the lipid-based organic nanoparticulate system provides a better platform for safe and effective drug delivery.

5.3 Lipid Nanoarchitectonics

The recent pharmaceutical scenario shows that drug development and drug delivery are very crucial aspects of the pharma industry. According to the Biopharmaceutics Classification System (BCS), most drugs fall into class II and III, where drugs are having either less solubility or less permeability. In the case of synthetic drugs, the modification in their physicochemical properties or formulation strategies can solve these issues easily. But in case of natural products, most of the drugs fall into BCS class IV and are hydrophobic in nature with very less solubility in water (Ghadi and Dand 2017). This decreases the oral bioavailability of natural products. Hence, nanotechnology comes into play to overcome these issues by formulating nanodimensional formulations such as polymeric nanoparticles, nanoemulsion and micelles.

Along with polymers, surfactants, co-surfactants, block copolymers are needed to be used to formulate nanoformulations. Use of synthetic excipients as a major component for formulation development may raise the concern of safety and toxicity. Therefore, the need of safe and nontoxic excipients occurs for nanoformulations development. Use of lipid excipients have solved the problems associated with other excipients which allows us to develop several advantageous drug delivery systems.

Lipids belong to a class of organic compounds with long chain fatty acids, which are lipophilic in nature, completely soluble in organic solvents (Porter et al. 2007). Lipids are highly water insoluble in nature. Some lipids are amphiphilic in nature, with some part of lipid is hydrophilic and remaining one is lipophilic in nature. With such type of lipids, both hydrophilic and lipophilic type of drugs can be incorporated. In lipid-based nanoformulations, the drug can be entrapped in lipid core or it can coat the drug without any leakage. Drug targeting can be specifically done with nanoparticles which can decrease the potential nanoparticle's toxicity. Hence lipid-based formulations are considered and developed as potential drug delivery systems (Shrestha et al. 2014).

Lipid nanoarchitectonics is a wide range term covering many types of lipid-based nanoparticulate as well as vesicular drug delivery systems like liposomes, lipid micelles, lipid emulsion, solid lipid nanoparticles, and nanostructured lipid carriers (Fig. 5.1) (Shrestha et al. 2014). These systems can easily dissolve lipophilic drugs in their lipid domain in dispersion medium which increases their drug loading capacity along with entrapment efficiency. The ultimate aim is to increase the bioavailability of drugs through lipid-based nanoarchitectonics (Fig. 5.2). These lipid nanoarchitectonics are explained well below with thorough literature review along with examples of different drugs delivered through them.

5.3.1 Liposomes

Liposome belongs to the vesicular type of novel drug delivery system, was being discovered in the 1960s by Bangham and his co-workers (Düzgüneş and Gregoriadis 2005). Liposomes consist of a bilayer structure of phospholipids which are concentric in nature that encloses hydrophilic region in the centre with lipophilic outer lipid covering. Both hydrophilic and lipophilic type of drugs can be incorporated into liposomes. Water-soluble drugs get solubilized in hydrophilic tail regions in the inner layer of micelles, while water-insoluble or lipophilic drugs are being solubilized in the outer layer of head regions of micelles. These are mostly best suited for the delivery of lipophilic drugs (Sharma and Sharma 1997). Being vesicular in nature, size of vesicles varies from 0.1 μm to 1 μm in diameter (Lian and Ho 2001; Sharma and Sharma 1997). These can be classified as small unilamellar vesicles (SUV), large unilamellar vesicles (LUV) and multilamellar vesicles (MLV), (Akbarzadeh et al. 2013) Liposomes can be modified according to their size, surface charges. Drug entrapment along with drug loading into liposomes can be specifically achieved with modification in liposome types.

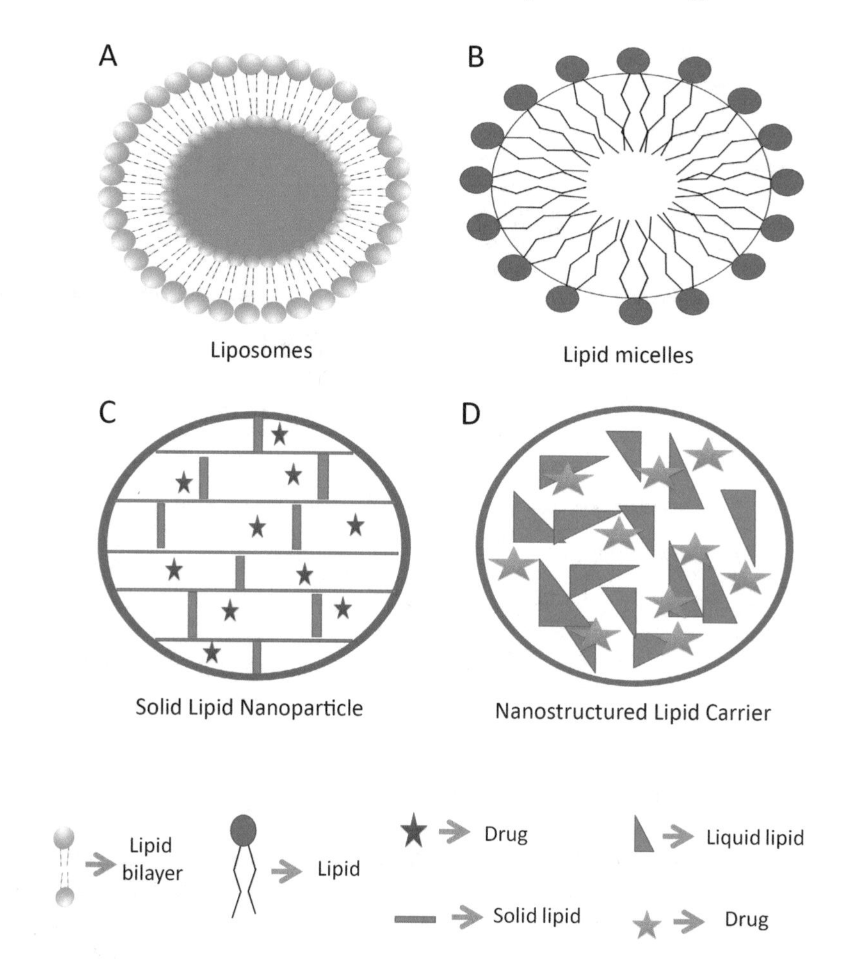

Fig. 5.1 Different varieties of lipid nanoarchitectonics

Liposomes itself having many advantages over other drug delivery systems like the increased capacity of drug loading, specific in tumour targeting, enhanced stability of drug by encapsulation, jeopardizes toxicity associated with drugs, improved pharmacokinetic properties and enhances skin penetration as well (Samad et al. 2007). Site-specific drug delivery can be easily achieved with liposomes especially in cancer. Several methods have been optimized for liposome preparation. These methods have been classified as mechanical dispersion methods, solvent removal methods and detergent removal methods (Akbarzadeh et al. 2013). Mechanical dispersion methods include thin film hydration, micro-emulsification, sonication, and freeze-thawing. Solvent removal methods include ether injection, ethanol injection, double emulsion, reverse phase evaporation method, and few others. And detergent removal method includes dialysis, dilution and reconstitution methods. Any of these methods can be used as per the requirement for formulation development and optimization. The final liposomal formulation has to be characterized for its properties

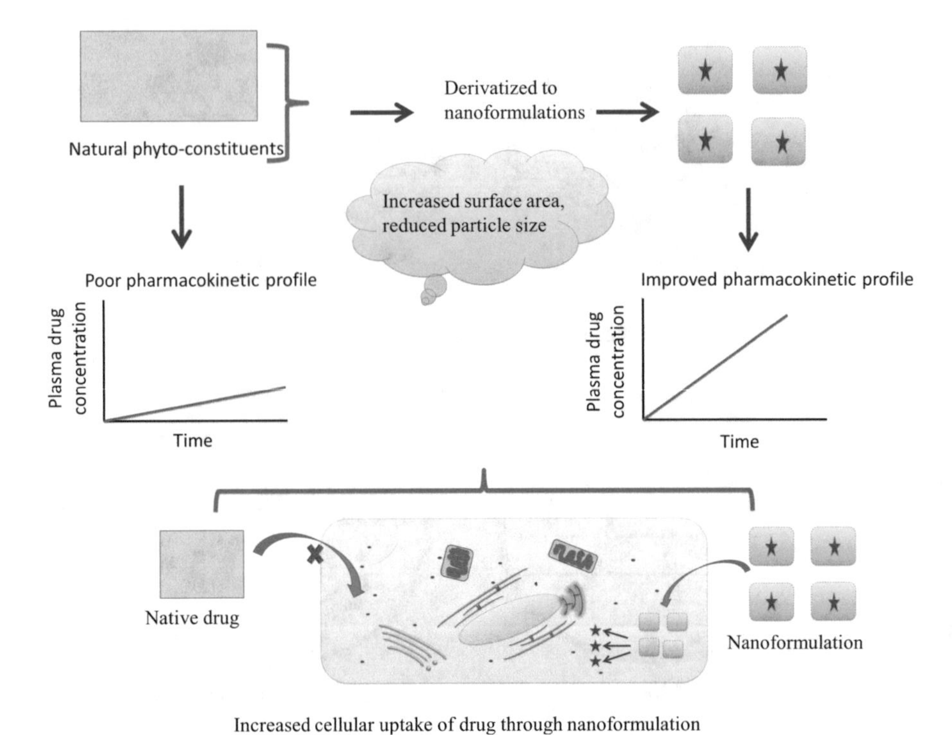

Fig. 5.2 Improved *in vitro/in vivo* performance exhibited by lipid nanoarchitectonics

with several instruments. It includes liposomal vesicle size analysis, polydispersity index (PDI) determination and zeta potential determination with particle size analyser or Zetasizer, surface morphology along with vesicle shape can be studied with Transmission electron microscopy (TEM), drug entrapment and release studies can be performed with dialysis (Samad et al. 2007).

Modifications can be done in liposomes according to the need or purpose of drug delivery. Conventional liposomes are able to deliver hydrophilic and hydrophobic drugs. Stealth liposomes were developed by polyethylene glycol (PEG) on the outer lipid bilayer membrane to avoid interaction with lipoproteins especially high-density lipoproteins (Immordino et al. 2006). It also increases negative charges on the surface along with stearic hindrance for serum proteins. Ultimately, hydrophobicity of liposome increases and can easily deliver lipophilic drugs with target specificity. Ligand-targeted liposomes can differ from conventional liposomes by the presence of a small molecule, peptide or polysaccharide on the surface of liposomes (Lamichhane et al. 2018; Olusanya et al. 2018). Theranostic liposomes are also called as multifunctional liposomes which are mostly used as an imaging agent or targeting antibodies (Lamichhane et al. 2018; Olusanya et al. 2018). Hence, the liposome is considered as versatile lipid-based nano-sized vesicular drug delivery system. Table 5.4 represents improved in vitro and in vivo profile of phytoconstituents after liposomal encapsulation.

Table 5.4 Anti-cancer phytoconstituents processed in the form of liposomes

Serial number	Natural product	Lipid components	Methodology	Particle properties			Targeting	Cancer cell lines	Findings	References
				Particle size (PS), Polydispersity index (PDI)	Zeta potential (ZP) mV	Entrapment efficiency (EE) %				
1.	Quercetin	Soy-phosphatidyl-choline, DSPE-PEG(2000) MAL-T7 conjugate, cholesterol	Film hydration	114.0 ± 1.80, (0.22)	–	95.19 ± 0.161	transferrin receptors targeting peptide surface-functionalized liposomes	A549 cells lung cancer	Higher fluorescence showed deeper penetration into cells, three fold increase in cell apoptosis	Riaz et al. (2019)
2.	Cisplatin and Curcumin	DSPE-PEG2000, Cholesterol, DMPC	Reverse microemulsion and film dispersion method	294.6 ± 14.8, (0.119)	10.8	99.5 Cisplatin and 93.4 Curcumin,	–	HepG2 cells Hepato-cellular carcinoma	Curcumin Synergises anti-cancer effect of cisplatin against HCC, Reduction in nephrotoxicity	Cheng et al. (2018)
3.	Quercetin & Adriamycin	Hydrogenated soybean phospholipid (HSPC):cholesterol: DSPE-PEG 2000	Thin-film evaporation and ultrasonic method	85	−14.9	97.49 Adriamycin and 95.50, Quercetin	–	Human Leukemia cell line (HL-60) and Human breast carcinoma cell strain Michigan Cancer Foundation-7 breast cancer cell line (MCF-7)	Reduction in IC50 value, synergistic effect of combination,	Riaz et al. (2019)

(continued)

Table 5.4 (continued)

Serial number	Natural product	Lipid components	Methodology	Particle properties			Targeting	Cancer cell lines	Findings	References
				Particle size (PS), Polydispersity index (PDI)	Zeta potential (ZP) mV	Entrapment efficiency (EE) %				
4.	Curcumin	Lecithin Octadecyl-amine,	Ethanol injection method	194 ± 0.25, (0.214)	31.9 ± 0.31	98.26 ± 1.33	Glycyrrhetinic acid modified Curcumin loaded cationic liposomes	HepG2 cells	Modification of curcumin with GA improved anti-cancer activity,	Chang et al. (2018)
5.	Quercetin	Soybean phosphatidylcholine, (DSPE)-PEG2000-RGD, cholesterol	Thin film hydration method	93.4 ± 7.2, (<0.14)	−20.4 ±0.6	89.2 ± 7.4	RGD-modified nanoliposomes	A549 Lung cancer cell lines	Targeting to Integrin αvβ3 with Arginine Glycine Aspartic acid (RGD) modified quercetin loaded nanoliposomes increased inhibition with tumor specificity	Zhou et al. (2018)
6.	Coix seed oil&β-carotene	Egg yolk phosphatidyl-choline, Cholesterol	The modified ethanol injection method	156.73 ± 5.93, (0.22)	−32.4 ±1.27	83.81 ±1.34	–	Caco-2 cell line	Reduced encapsulation drugs in combination than an individual one, but the enhanced anti-tumor activity of the combination	Bai et al. (2018)

7.	Curcumin	DOTAP, DOPE, C6 ceramide, and sodium cholate	Thin film hydration method followed by complexation	192.6 ± 9.0, (0.326)	56.4 ± 8.0	86.8 ± 6.0	–	Mouse melanoma cells (B16F10)	Faster growth inhibition, deeper penetration skin layers was observed for Curcumin liposomes complexed with siRNA	Jose et al. (2018)
8.	Brucea javanica oil	Soybean phosphatidyl-choline, cholesterol, (DSPE)-PEG2000	film-dispersion and biotin-streptavidin linkage methods	155.1 ±14.5, (0.227)	24.1 ± 0.54	92.2 ± 1.59	Luteinizing hormone releasing hormone receptor (LHRHR)	Human ovarian cancer A2780	Prolongation of circulation time, enhanced cellular uptake with cell specificity	Ye et al. (2016)
9.	Ginsenoside compound K (GCK)	TPGS, Egg yolk lecithin	Thin film hydration method	119.3 ± 1.4	1.9 ± 0.4	98.4 ± 2.3	TPGS modified liposomes	A549 Lung cancer cell lines	IC_{50} value reduction, Maximum fluorescence in 8 hr., effectively suppressed the tumor volume	Yang et al. (2016)
10.	Curcumin	Hydrogenated soya PC, cholesterol, DSPE-PEG2000	Extrusion technique	101 ± 1.2, (0.046)	−3.5 ± 0.2	96		pancreatic adenocarcinoma	Cell apoptosis with inhibition of ROS generation in time dependant manner,	Mahmud et al. (2016)

(continued)

Table 5.4 (continued)

Serial number	Natural product	Lipid components	Methodology	Particle properties			Targeting	Cancer cell lines	Findings	References
				Particle size (PS), Polydispersity index (PDI)	Zeta potential (ZP) mV	Entrapment efficiency (EE) %				
11.	Quercetin and Resveratrol	Lipoid S75, oleic acid	Sonication	79.0± 4.1, (0.12)	−40.0 ±6.7	71.2± 10.9(Q), & 72.1 ± 6.6 (R)	–	Human dermal Fibroblasts cultured in DMEM, Skin cancer	Improved cellular uptake of both drugs in combination along with wound healing with relevant ROS scavenging activity	Caddeo et al. (2016)
12.	Curcumin	Soya lecithin, cholesterol,	Thin film hydration method	219.5 ±9.3, (0.28)	27.7 ±0.9	68.9 ±0.6	–	HeLa and SiHa Cervical cancer	Cationic liposomes with enhanced cell apoptosis help in the improved anti-cancer potential of curcumin	Saengkrit et al. (2014)
13.	Epigallo-catechin and paclitaxel	L-α-Phosphatidylcholine and Cholesterol	Thin film hydration method	126.7 to 132.7, (0.4)	–	77.11 ± 2.30 PTX, 59.11 ± 3.51 EGCG	–	MDA-MB-231 Breast cancer	Synergistically induce tumour cell death	Ramadass et al. (2015)
14.	Silymarin	Hydrogenated soya PC, cholesterol, DSPE-PEG2000,	Thin film hydration method	145–168, (0.15–0.3)	–	70	β-sitosterolβ-D-glucoside (site-G) PEGylation	HepG2 cell	Low concentration of site-G showed more targeting along with PEGylated liposomes	Elmowafy et al. (2013)

15.	Wogonin	Soybean phospholipids (S100), cholesterol	Reverse evaporation method	90.5± 2.2	−16.8 ±4.5	97.39 ±5.0	Glycyrrhetinic acid-modified liposome	HepG2 cells	Highest uptake in liver tissues but gradually decreases in other tissues,	Tian et al. (2014)
16.	Curcumin	DMPG(1, 2-dimyristoylsn-glycero-3-(Phospho-rac-(1-glycerol)), DPPC (1, 2-dipalmitoylsn-glycero-3-Phosphocholine)	Thin-film evaporation	98.2	–	50	Curcumin- HPβ Cyclodextrin loaded into liposomes	KHOS OS cell line and MCF-7 breast cancer cell line	Reduction in IC50 of curcumin, Improved solubility followed by enhanced uptake into tumour cells	Dhule et al. (2012)
17.	Vincristine and quercetin	egg sphingomyelin cholesterol PEG2000 ceramide	film hydration method followed by extrusion	131.5 ± 13.8, (0.137)	–	93.6 ± 6.8	Estrogen receptor	MDA-MB-231 Breast cancer	Lower the dose of Vincristine when given in combination with Quercetin	Wong and Chiu (2010)

Abbreviations: *S. No* Serial Number, *PS* Particle size, *PDI* Polydispersity index, *ZP* Zeta potential, *mV* Millivolt, *EE* Entrapment efficiency, *DSPE-PEG* 2000–1,2-distearoyl-sn-glycero-3-phosphoethanolamine-N-amino (polyethylene glycol)-2000, *HSPC* Hydrogenated soybean phospholipid, *DMPC* 1,2-dimyristoyl-sn-glycero-3-phosphocholine

5.3.2 Lipid Micelles

Lipid-based drug delivery systems belong to the organic class of nanoparticle system. As we know lipids are amphiphilic molecules containing both hydrophilic head region (polar) and hydrophobic tail region (non-polar) (Torchilin 2007). A carboxylic group from fatty acid plays an important role in hydrophilic interactions while hydrocarbon chain in hydrophobic interactions. The tendency of surfactants followed by lipids in the aqueous environment exists as head region directing towards the water while hydrophobic tail region opposite to it facing towards the air. Above critical micellar concentration, micelles are formed inside the aqueous media. Lipid micelles are self-assembled spherical structures formed inside the solvent from lipids. It consists of a single layer of micelles enclosing hydrophobic tail region inside with hydrophilic heads forming a shell (Torchilin 2007). No aqueous environment is present in lipid micelles which allows only lipophilic drugs to get solubilize in it easily. Fatty acids have two long tails which consist of the hydrophobic part of micelles. Head region of micelles attaches to each other enclosing lipophilic tails inside. More the size of the hydrophilic head more is the steric hindrance. The size of micelles mostly depends upon the chain of alkyl moiety along with size of the hydrophilic head region. Mixed micelles are comprised of phospholipids along with polymeric micelles. It provides combined advantages of both lipid micelles and polymeric micelles (Almgren 2000; Zhang et al. 2008). Several advantages come with lipid micelles as drug delivery perspective. Micelles represent core-shell type structure which provides the lipophilic environment to water-insoluble drugs in aqueous solution, hence increases the solubility of such kind of drugs. Also, the toxicity from lipid micelles has been found to be very low due to the biocompatible and biodegradable nature of lipids. Improved penetration in tissues provides greater advantages according to drug delivery (Hanafy et al. 2018). On the other side, it shows the reduced drug loading capacities as the drug can easily ooze out from the micellar structure. Stability of maintaining the micellar structure in the body is very difficult as blood volume may dilute the critical micellar concentration (Hanafy et al. 2018). Table 5.5 represents the improved performance of phytoconstituents after of lipid micelles incorporation.

5.3.3 Solid Lipid Nanoparticles

The first form of lipid-based particulate system in nanoformulations comes to mind is solid lipid nanoparticles (SLNs). Solid lipid nanoparticles consist of solid lipids which remains solid at room temperature and at body temperature too (Das and Chaudhury 2011; Wissing et al. 2004). It is a very attractive approach for drug release in a controlled manner. Use of lipids in drug delivery minimizes various acute and chronic toxicities. It is also a type of nanoemulsion, in which the internal phase consists of solid particles. Various solid lipids have been used for the

Table 5.5 Reported anti-cancer phyto-molecules processed in the form of lipid micelles

S. no.	Natural product	Lipid components	Methodology	Particle properties			Targeting	Cancer cell lines	Findings	References
				PS, (PDI)	ZP mV	EE %				
1.	Narcilasine	lipid-modified cell-penetrating peptides, Cholesterol	probe sonication	96–138 (0.236)	96–138	90	–	HepG2 cells	The potent anti-tumor activity showed by narcilasine and siULK1 co-delivery through lipid micelles	Wang et al. (2018b)
2.	Paclitaxel	Cholesterol	film dispersion	167.90 ± 1.46, (0.196)	−21.91 ± 1.25	94.81 ± 1.37	BCRP protein targeting	MCF-7 cells	Reduction in adverse effects of paclitaxel with co-delivery of SiRNA	Zhu et al. (2017)
3.	Berberine	mPEG2000-DSPE, TPGS	solvent evaporation	4.1 ± 1.8 (0.320)	−23.5 ±0.9	95.7 ±3.6		PC3 &LNCaP human prostate carcinomas	TPGS enhances drug loading through hydrophobic interactions, Improved solubility of Berberine with increased cell apoptosis	Shen et al. (2016)

(continued)

Table 5.5 (continued)

S. no.	Natural product	Lipid components	Methodology	Particle properties			Targeting	Cancer cell lines	Findings	References
				PS, (PDI)	ZP mV	EE %				
4.	Wogonin	Cholesterol, PEG,2-(3-((S)-5-amino-1-carboxypentyl) ureido) pentanedioic acid (ACUPA)	self-assembly method	24 ± 5, (0.27 ± 0.03)	0.6 ± 1.4	95.8 ± 2.6	PSMA-positive PCa cells	LNCaP& PC3 Prostate cancer cells	PSMA-positive PCa cells showed Selectively improved cellular upatake of mixed micelles	Zhang et al. (2016)
5.	Baohuoside I	Lecithin and Solutol HS 15 (BLSM)	Thin film hydration method	103.47 ± 3.25		89.32	–	A549 small cell lung cancer	Solubility improved by 152.7 times through mixed micelles, Increased cytotoxicity	Yan et al. (2016)
6.	Curcumin	6sPCL-PMPC, PCL-PEG	solvent-evaporation method	211.3, (0.152)	−9.48	98.1 ± 3.1	–	HeLa cells	Phospholipid shielded micelles showed more cellular uptake, reduction in IC_{50} values of curcumin	Huang et al. (2014)

7.	Paclitaxel and curcumin	PEG_{2000}-PE	thin film hydration method	15 to 20	−18.9 ± 0.9	–	transferrin-receptor	NCI-ADR-RES cells Human MDR ovarian cancer	Targeted micelles found to be more effective than no-targeted micelles, Targeting doesn't increase cytotoxicity further.	Sarisozen et al. (2014)
8.	Curcumin	DSPE-PEG_{2000}	film rehydration	15.1 ± 2.7		90	vasoactive intestinal peptide (VIP) receptors	MCF-7 human breast cancer cell line	Cell apoptosis in a dose-dependent manner, Curcumin IC_{50} was almost reduced to half due to incorporating into micelles	Gülçür et al. (2013)

formulation of SLN are Glyceryl monostearate (GMS), glyceryl tristearate, Precirol, Compritol 888 ATO, etc. (Khosa et al. 2018) Lipid allows the drug to get encapsulated into its structure giving to solid lipid nanoparticles. There are various advantages come with this formulation system that helps in drug delivery. It can be used for controlled release of the drug along with site-specific targeting. Solid lipids which can be used in formulating SLNs are mostly biocompatible and biodegradable in nature (Table 5.6). Encapsulation of drugs which are photosensitive or moisture sensitive in nature gives protection against the degradation caused by light and moisture (Das and Chaudhury 2011; Sailaja et al. 2011; Shidhaye et al. 2008). But the system also has some limitations which include the presence of high water content leads to dilution of solid lipid nanoparticles. Lipids used for formulating SLNs need to be of high purity, which leads to reduced drug loading capacity as solid lipids having a tendency to crystallize in structure. As the crystallization of solid lipid may occur during storage, the expulsion of the encapsulated drug has been observed from SLNs (Das and Chaudhury 2011; Shidhaye et al. 2008). Aggregation of particles has been observed with several solid lipids which cause particle growth along with gelation due to formation linkages between lipids during storage. Hence, stability is the major concern in the case of solid lipid nanoparticles.

5.3.4 Nanostructured Lipid Carriers

The second generation of solid lipid nanoparticles is the nanostructured lipid carriers (NLCs) were introduced in order to overcome the side effects of SLNs as per the formulation aspects. The tendency of solid lipid to crystallize out in perfect lattice due to high purity gives less space to the drug for encapsulation. Hence, it decreases the loading capacity of SLNs also causes expulsion of the drug during storage (Das and Chaudhury 2011). Hence there was a need to tackle this issue of drug loading and expelling drug out. Incorporation liquid lipid in the system leads to improvement of drug loading capacity along with no drug expulsion from particles. This also avoids the crystallization of lipids making more space available for the drug to get encapsulate into it (Shidhaye et al. 2008). A mixture of solid lipid and liquid lipid provides a great medium to solubilize a large number of lipophilic drugs into it (Wissing et al. 2004). Various hydrophilic drugs can also be incorporated solubilizing into the aqueous phase of the primary emulsion during preparation method of NLCs. Drug release from NLCs can be controlled by varying the component percentages of solid lipid and liquid lipid. Core and shell-like structure have observed in case of NLCs which showed the immediate burst release from the outer shell part while slow release from the core of NLCs. Particles become solid after cooling which do not cause drug leakage during storage. Lipids used are biocompatible in nature which can easily cross the intestinal gut wall barrier with a reduced particle size up to 200 nm. NLCs have been explored in various diseases with different routes of administration (Ghasemiyeh and Mohammadi-Samani 2018). Natural product loaded nanostructured lipid carriers for the application in cancer is broadly discussed in Table 5.7.

Table 5.6 Reported anti-cancer phyto-molecules processed in the form of Solid lipid nanoparticles (SLNs)

S. no	Natural product	Lipid components	Surfactants	Methodology	Particle properties			Targeting	Cancer cell lines	Findings	References
					PS(nm), PDI	ZP (mV)	EE (%)				
1.	β-carotene	GMS, Gelucire50/13	Tween-80, Pluronic F68	Hot homogenization technique	203 ± 7.23, (0.185)	−7.21 ±0.82	68.3 ± 3.4	–	MCF-7 Breast cancer	Sustained release from lipid core by 1.92 times	Jain et al. (2019)
2.	Curcumin and Resveratrol	Gelucire50/13	Tween-80	Melt emulsification-probe sonication	127.8	−36.43 ± 7.7	–	–	HCT-116 cell line	HPβCD imparted Stability to SLN, potential anti-cancer activity in colorectal cancer	Gumireddy et al. (2019)
3.	Epigallocatechin gallate	GMS, stearic acid, soy lecithin	2% W/V Pluronic F68	Double emulsification evaporation	163 ± 3.2, (0.341)	−25.2 ±2.8	67.2 ± 3.5	Bioconjugation with Bombesin for gastrin-releasing peptide receptors(GRPR)	MDA-MB-231 Breast cancer	Improved anticancer activity with targeting, decrease in tumor volume, enhanced survivability	Radhakrishnan et al. (2019)
4.	Curcumin	Trilaurin	Pluronic F68	Cold dilution of microemulsion	206.2 ± 4.4 (0.220)	−26.8 ±5.8	90	–	CFPAC1,PANC-1 pancreatic adenocarcinoma	Feasible method of preparation, diverse bio distribution pattern	Chirio et al. (2019)
5.	Quercetin	Compritol	Tween 80	Microemulsification	85.5 (0.152)	−22.5	97.6	–	MCF-7 Breast cancer	Increased apoptotic and necrotic indexes, Unclear mechanism of induction of apoptosis	Niazvand et al. (2019)
6.	Annona muricata fruit extract	Stearic acid, soy lecithin	Poloxamer188	High-pressure homogenization followed by ultra-sonication	134.8	±30	83.26	–	MCF7 breast cancer	SLN showed more cytotoxicity than free extract	Sabapati et al. (2019)
7.	Curcumin	Stearic acid, lecithin, TPGS	Myrj59	Emulsion evaporation	125.2 (0.268)	−21.2 ±2.6	–	–	Hodgekin'slymo-homa	Increased efficacy with reduced toxicity and minimizing MDR	Guorgui et al. (2018)
8.	Citral	Imwittor® 900 k	Poloxamer 188	High pressure homogenization	97.7, (0.249)	−0.007	–	–	RAW 264.7 cell line	Higher NO inhibition by citral	Zielińska et al. (2018)

(continued)

Table 5.6 (continued)

S. no	Natural product	Lipid components	Surfactants	Methodology	Particle properties PS(nm), PDI	ZP (mV)	EE (%)	Targeting	Cancer cell lines	Findings	References
9.	Curcumin	NA	NA	NA	40	−25.3 ±1.3	72.47	–	SKBR3 breast cancer	Higher cellular uptake with cellular apoptosis by autophagy inhibition	Wang et al. (2018a)
10.	Curcumin	GMS	Poloxamer188	Emulsification	226.802 ± 3.92, (0.244)	–	67.88 ± 2.08	–	MDA-MB-231 Breast cancer,	Reduced cell viability with higher cellular apoptosis of Cur-SLN compared to native free curcumin	Bhatt et al. (2018)
11.	Pomegranate extract	Stearic acid, soy lecithin	Tween 80	Hot homogenization followed by the ultra-sonication technique	407.5 to 651.9, (407.5 to 651.9)	−29.8 to −45.5	56.02 to 65.23	–	MCF-7 Breast cancer	Studied the effect of several variables on the formulation, Reduction in IC50 values by 47 folds with enhanced bioefficacy	Badawi et al. (2018)
12	Wogonin	Stearic acid,	Poloxamer 188, Lecithin	Hot-melted evaporation technique	225.0 ± 8.3, (0.18)	−41.7 ± 3.2	93.2 ± 4.7	–	MCF-7 breast cancer	Increased bioavailability with enhanced cell apoptosis by cell cycle arrest	Baek et al. (2018)
13.	Capsaicin	Phosphatidylcholine, Cholesterol, Cholesteryl hemisuccinate	–	Thin film hydration	50–200	–	95.5 ± 2.4	Folic acid(FA) conjugation, PEG decoration	SKOv-3 ovarian cancer	FA increases cellular uptake, PEG prolongs the circulation of SLNs	Lv et al. (2017)
14.	Linalool	Myristyl myristate	Pluronic F68	Sonication	90–130	−4.0	80	–	HepG2, A549	Cellular uptake enhancement, dose dependant cell apoptosis	Rodenak-Kladniew et al. (2017)
15.	Curcumin	Cholesterol	Poloxamer 188	High shear homogenization	166.4 ± 3.5, (0.236)	−22.4 ±1.46	76.9 ± 1.9	–	MDA-MB-231 Breast cancer	Enhanced BA along with cytotoxicity	Rompicharla et al. (2017)

16.	Resveratrol	Stearic acid, GMS	Myrj56	Emulsification and low-temperature solidification method	168 ± 10.7(0.26)	−23.5 ± 1.6	25.2 ± 1.7	–	MDA-MB-231 Breast cancer,	Cell cycle arrest at G0/G1 phase, dose dependant cell apoptosis	Wang et al. (2017b)
17.	Rhein	Precirol ATO5, Lecinol	2.5% (w/v) of poloxamer188	Hot homogenization followed by the ultra-sonication technique	120.8 ± 7.9, (0.1)	−16.9 ± 2.3	90.2 ± 2.1	–	SW480 colon cancer	Improved poor solubility and tissue permeability, Increased cellular partitioning of R-SLNs with decreased cell viability	Feng et al. (2017)
18.	Lycopene	Compritol ATO 888 and gelucire	Pluronic F 68 (0.1%, w/w)	Homogenization-evaporation technique	185.1 ±9.3, (0.192)	−9.48 ±1.29	79.6 ±2.9	–	MCF-7 breast cancer cells	Retention of anti-cancer and antioxidant potential of LYC in SLN along with lowered IC_{50}	Wang et al. (2017a)
19.	Epigallocatechin gallate	GMS, Stearic acid, soy lecithin	2% W/V Pluronic F68	Emulsion solvent evaporation	< 200, (<0.2)	−30.3	67.2	–	MDA-MB-231 Breast cancer, DU-145 human prostate cancer cells	EGCG-SLNs showed 8.1 and 3.8 times increased cytotoxicity against MDA-MB-231 and DU-145 cells as compared to free EGCG	Radhakrishnan et al. (2016)
20.	Curcumin	Stearic acid, lecithin	Myrj52	Emulsification and low-temperature solidification technique	56.2	−26.2 ±1.3	37 ± 2.5	–	A549 cells non-small cell lung carcinoma	More cell apoptosis with Cur-SLNs than free drug, improved BA	Jiang et al. (2017)
21.	Curcumin and Piperine	GMS, Lecithin, Oleic acid	TPGS and Brij 78	Emulsification and low-temperature solidification technique	~130.8, (0.152)	~20	87.4 ± 0.6 (C), 14.7 ± 0.2 (P)	–	A2780/Taxol cells	Significant increase in cytotoxicity, Combination showed positive results against MDR Taxol cells	Tang et al. (2017)
22.	Vincristine and temozolomide	Stearic acid, soya lecithin	DDAB	solvent displacement technique	179.1 ±5.2, (0.19)	+35.7 ± 3.8	80.8 ± 3.2(T), 84.5 ± 4.6(V)	–	U87 malignant glioma cells	Higher cell growth inhibition	Wu et al. (2016)

(continued)

Table 5.6 (continued)

S. no	Natural product	Lipid components	Surfactants	Methodology	Particle properties PS(nm), PDI	ZP (mV)	EE (%)	Targeting	Cancer cell lines	Findings	References
23.	Sesamol	GMS or cetyl alcohol	Egg lecithin, Tween 80	Microemulsification	127.9 ± 1.4, (0.256)	–	88.21 ± 0.096	–	HL-60 cell lines	Controlled release system with high accumulation in skin, Higher cell apoptosis	Geetha et al. (2015)
24.	Aloe-emodin	Glyceryl monostearatev(GMS)	Poloxamer 188, Poloxamer 407. Tween 20	High pressure homogenization	88.9 ± 5.2, (0.298)	−42.8	97.71	–	MCF-7, HepG2	Sustained release over 48 h, Increased cellular uptake	Chen et al. (2015)
25.	Curcumin	Dynasan 114®, Sefsol-218®	Pluronic F-68	Hot high-pressure homogenization followed by high shear dispersion method	149.4 ±5.3, (0.187)	21.0 ±2.8	66.22 ±3.89	–	MCF-7 cells	Liquid lipid reduced the particle size significantly, helps in cell uptake and apoptosis	Sun et al. (2013)
26.	Camptothecin	Compritol 888 ATO		Supercritical fluid technology	–	–	–	–	MCF-7 cells	Uptake and retention of Cap-SLN in cancer cells with higher cell death	Acevedo-Morantes et al. (2013)
27.	Andrographolide	Cetyl alcohol	Tween 80	Solvent injection technique	154.1 ±10.7, (0.172)	−40.3 ±0.8	91.4 ±0.4	–	MCF-7 cells	Bioavailability was improved with higher cell apoptosis at lower concentrations	Parveen et al. (2014)
28.	Emodin	Glycerol monostearate (GMS), stearic acid	Poloxamer 188 and Tween 80	High-pressure homogenization	28.6 ±3.1	−31 ±1.1	60.37 ±2.2	–	MCF-7 and MDA-MB-231 cells	The biphasic drug release pattern of drug with time-dependent cell apoptosis	Wang et al. (2012)
29.	Curcumin, aspirin, sulforaphane	Stearic acid	Poloxamer 188	Hot melt emulsion technique	150 (A), 249 (C)	–	85 (A), 69 (C)	–	MIA PaCa-2, Panc-1 cells	The synergistic effect observed with the combination of SLNs with free Sulforaphane drug,	Sutaria et al. (2012)

30.	Curcumin	Hydrogenated soy phosphatidylcholine DSPE, Cholesterol, Triolein	Poloxamer 188	Hot homogenization method	194 ± 0.89	12.43 ± 0.26	84.99 ± 0.73	Transferrin receptor-mediated SLN	SH-SY5Y neuroblastoma cells	Tf-C-SLN decreased the dose, protects the drug from MPS, higher cellular apoptosis	Mulik et al. (2012)
31.	Curcumin	Compritol 888 ATO,	Polysorbate 80, soy lecithin	Microemulsification	134.6	–	92.33 ±1.63	–	PC3, HL-60, A549	Apoptotic potential of Cur-SLN at low doses, involved in many molecular pathways of cell growth	Vandita et al. (2012)
32.	Curcumin	Hydrogenated soy phosphatidylcholine DSPE, Cholesterol, Triolein	Poloxamer 188	Homogenization method	206 ± 3.2	12.43 ± 1.58	77.27 ± 2.34	Transferrin receptor-mediated SLN	MCF-7 breast cancer cells	Improved photostability of Cur along with apoptosis of tumor cells	Mulik et al. (2010)

Table 5.7 Reported anticancer phyto-molecules processed in the form of Nanostructured lipid carriers (NLCs)

S. no.	Natural product	Lipid components	Surfactants	Methodology	Particle properties			Targeting	Cancer cell lines	Findings	References
					PS(nm), PDI	ZP (mV)	EE (%)				
1.	β-carotene α-Tocopherol	Murumuru butter (A. murumuru),	Cremophor RH40, Span 80	Low emulsification technique	33.4 ± 0.81, (0.20)	–	73	–	HEPG cells	Fast permeation through cells was observed with murumuru butter	Gomes et al. (2019)
2.	Doxorubicin and β-element	Compritol® 888 ATO, Lecithin, Miglyol® 812	Tween® 80	Hot homogenization and ultra-sonication method	190, (<0.2)	−30.9 to −41.3	89.3 ± 3.9 (Dox)87.7 ± 4.2 (EL)	P[H] sensitive targeting	A549/ADR, MRC5 cells	Effective targeting and cellular uptake with EPR effect	Cao et al. (2019)
3.	Alpha Lipoic–Ellagic Acid, Fluvastatin (FLV)	Gelucire® 44/14, Compritol® 888 ATO, Almond oil	Soya lecithin	Hot emulsification–ultra-sonication method	85.2 ± 4.1	−25.1 ± 3.4	98.2 ± 1.1	–	PC3 Cell line prostate cancer	Synergistic effect of ALA-EA on FLV cytotoxicity through NLCs with enhanced cellular uptake	Fahmy (2018)
4.	Lapachone and doxorubicin	Compritol® 888 ATO, GMS, oleic acid	soybean phosphatidylcholine (S100)	Melted ultrasonic dispersion method	100.2 ± 6.8, (0.123)	−23.5 ± 3.6	83.8 (La), 92.3 (Dox)	–	MCF-7 ADR cell line	NLCs showed high cytotoxicity, reduction dose of Dox,	Li et al. (2018)
5.	Gambogic acid	Lecithin, Compritol 888 ATO,	Myrj52	Emulsification and solvent evaporation	20.96 ± 1.13 (0.23)	−5.86 ± 0.64	99.46 ± 0.4	cell penetrating peptides (cRGD and RGERPPR)	MDA-MB-231 Breast cancer	More cellular uptake of NLCs, RGD decorated GA-NLCs showed higher accumulation in tumor cells	Huang et al. (2018)
6.	Curcumin	Lecithin, Hydrogenated soyabean oil	lysophosphatidylcholine,	Melt emulsification	340.6 ± 33.64, (0.172)	–	–	Ginsenoside modified Cur-NLC	HCT116 and HT29 human colon cancer lines	Increased cellular uptake but no evidence of an increase in the anticancer activity of curcumin, Modification can help to target colon cancer	Vijayakumar et al. (2019)

7.	Epigallocatechin gallate	Precirol, Miglyol	Poloxamer 407	Hot homogenization	85	−21	83	Arginyl-glycyl-aspartic acid (RGD)for adhesion to αvβ3 integrin	MDA-MB-231 Breast cancer	Due to adhesion to αvβ3 integrin, effective cellular uptake, and accumulation in tumor cells	Hajipour et al. (2018)
8.	Orcinol glucoside	Tristearin, oleic acid	Tween 80	Hot homogenization ultra-sonication	160–230	−8 to −20	99.62 ± 1.5	polyethylene glycol-25-stearate (PEG-25-SA) and polyethylene glycol-55-stearate (PEG-55-SA)	HepG2 and Huh-7Hepatocell-ular Carcinoma HCT-116 and AGS cells of GIT	PEGylation avoids RES recognition enhances the intracellular concentration of OG-NLC	Nahak et al. (2018)
9.	Curcumin	Compritol, Captex, Migloyl, Soyalecithin	Sodium taurocholate	Precipitation technique	< 250 (<0.3)	−10 - -35	50–100	–	A2780S and A2780CP cells in ovarian cancer	Cytotoxicity observed with nuclear condensation	Bondì et al. (2017)
10.	Artesunate	Compritol, labrafil 1944, lecithin	Tween® 80	Hot homogenization and ultra-sonication method	117.5 ± 6.1	−19.4 ± 0.9	92.93 ± 1.47	–	MCF-7, MDA-MB-231 cells	Improved solubility, cellular uptake and effective anti-cancer activity	Tran et al. (2016)
11.	Ursolic acid	Tribehenin, trierucin, behenic acid, oleic acid	Tween 80	Hot homogenization sonication	208 ± 1 (0.34)	−17 ± 0.6	78.38 ± 1.9	–	human leukemic cell line K562 and melanoma cell line B16	UA-NLCs with higher penetration in the cell membrane, improved anticancer activity of Ursolic acid	Nahak et al. (2016)
12.	Etoposide and curcumin	GMS, oleic acid	Soyabeanphosphatidyl-choline	solvent injection technique	114.3 ±4.1, (0.26)	+33.7 ± 3.8	83.1 ±3.5 (ETP) 81.5 ±2.7 (Cur)	–	SGC7901 cell line Human gastric cancer	Highest cytotoxicity achieved with lower IC50 value, high distribution in tumor cells	Jiang et al. (2016)

(continued)

Table 5.7 (continued)

S. no.	Natural product	Lipid components	Surfactants	Methodology	Particle properties			Targeting	Cancer cell lines	Findings	References
					PS(nm), PDI	ZP (mV)	EE (%)				
13.	Baicalein and doxorubicin	Stearic acid, Precirol ATO 5, Cremophor ELP,	Soyabeanphosphatidyl-choline	Emulsion evaporation–solidification at the low temperature	103.5 ±2.2	+12.6 ± 1.2	90.8 (Bac) 91.5 (Dox)	Hyaluronic acid decorated NLC	MCF-7/ADR cells Breast cancer	Targeting along with co-delivery of herbal drug with Dox enhances cytotoxicity, Reduction in systemic toxicity	Liu et al. (2016)
14.	Curcumin	Tripalmitin, oleic acid,	polysorbate 80	Hot high-pressure homogenization (HPH) technique	214.0	–	88.6	–	A172 cells Brain cancer	Time-dependent cellular uptake of Cur-NLCS, Lowers the IC_{50} of Cur significantly, able to induce apoptosis in A172 brain cancer cells	Chen et al. (2016)
15.	Curcumin	Precirol ATO-5, oleic acid, soybean phosphatidyl-choline	Tween-80	Solvent diffusion method	126.8 ±3.4, (0.16)	+12.6 ± 1.8	82.7 ±2.9	Folate targeting	MCF-7 cells Breast cancer	Particle size reduction enhances cellular uptake and growth inhibition too	Lin et al. (2016)
16.	Doxorubicin and Vincristine	Compritol 888ATO, Stearic acid, Cremophor ELP	Didecyldimethyla-mmonium bromide (DDAB)	solvent injection technique	95.8 ± 2.1	+21.3 ± 2.8	87.9 ± 2.4 (Dox) 83.3 ± 2.1(VCR)	–	LY1 cells	Synergistic effect by co-delivery of herbal drug, NLCs reduced the IC_{50} values	Dong et al. (2016)
17.	Vincristine and temozolomide	Compritol 888 ATO, Cremophor ELP, Soyabeanphosphatidyl-choline	DDAB	solvent diffusion method	117.4 ±2.8, (0.009)	+29.8 ± 3.2	88.9 ± 3.6 (TMZ), 85.4 ± 2.8 (VCR)	--	U87 malignant glioma cells	The decrease in IC50 of drugs, NLCs showed higher inhibition rate than a free drug solution	Wu et al. (2016)

18.	Curcumin and Genistein	GMS, Lecithin, Oleic acid	Tween 80	High-speed homogenizer and ultrasonic probe	122 ± 6, (0.29)	−47 ± 2	93 ± 1 (Cur), 82 ± 1 (Gen)	–	PC3 cells prostate cancer	Improved bioavailability of both drugs with synergistic anti-prostate cancer potential	Aditya et al. (2013)
19.	Isoliquiriti-genin	GMS, Miglyol812	Poloxamer 188, Tween80	solvent injection technique	160.73 ± 6.08	–	96.74 ± 1.81	–	H22-bearing mice	Higher bio-distribution of drugs loaded NLCs in tumour bearing mice	Zhang et al. (2013)
20.	Quercetin	Glyceryl tridecanoate, glyceryl tripalmitate, α-tocopheryl acetate, Kolliphor HS15	1% of NaCl in deionized water	Phase inversion method	30, (0.059)	−15	95	–	MCF-7&MDA-MB-231 cells	Pronounced cell apoptosis in breast cancer cell line,	Sun et al. (2014)

5.4 Conclusion

Natural products comprise as a primary source of therapeutics to treat multiple disorders. Even today, a large sum of rural populations and ethnic tribes relies majorly on natural products for their therapeutic uses. These source are also vital for modern science in terms of drug discovery perspectives. However, factors like low solubility, bitter taste, and lower absorption rate limits their justifiable application. Therefore, formulation modification approaches have been applied to these molecules for improved therapeutic activity. In one such approach, nanotechnology has been reported to be helpful in countering unlikely characteristics of natural products. There have been reports of improved bioavailability, solubility, absorption, lower dosage regimen, taste masking, etc. due to the advent of nanotechnology. In particular, this is significantly related to retained in-vivo activities, unlike compromised outcomes in case of native molecules.

Among all of the nano-constructs, a specialized form of lipid-based nanoformulations is one such prominent example that has been fairly recent and advantageous. This class of formulations has made significant contributions to the delivery of therapeutically important natural products. The formulations like liposomes, lipid micelles, SLNs, and NLCs are reported to improve the pharmacokinetic profile of the derivatized natural anticancer products. Therefore, the use of improvised technology can always be handy to overcome noted shortcomings, in the pursuit of translational drug delivery approach.

Acknowledgments Authors acknowledge National Mission on Himalayan Studies (**GBPI/NMHS-2017-18/HSF-02**), Ministry of Environment Forest and Climate Change, Govt. of India for providing necessary funding support. Authors are also thankful to the National Institute of Pharmaceutical Education and Research (NIPER)-Guwahati for providing access to necessary literature resources and essential library facilities for writing this book chapter.

Disclosure The authors declare no competing financial interest.

References

Acevedo-Morantes CY, Acevedo-Morantes MT, Suleiman-Rosado D, Ramírez-Vick JE (2013) Evaluation of the cytotoxic effect of camptothecin solid lipid nanoparticles on MCF7 cells. Drug Deliv 20(8):338–348. https://doi.org/10.3109/10717544.2013.834412

Aditya NP, Shim M, Lee I, Lee Y, Im MH, Ko S (2013) Curcumin and genistein coloaded nanostructured lipid carriers: in vitro digestion and antiprostate cancer activity. J Agric Food Chem 61(8):1878–1883. https://doi.org/10.1021/jf305143k

Akbarzadeh A, Rezaei-Sadabady R, Davaran S, Joo SW, Zarghami N, Hanifehpour Y, Samiei M, Kouhi M, Nejati-Koshki K (2013) Liposome: classification, preparation, and applications. Nanoscale Res Lett 8(1):102. https://doi.org/10.1186/1556-276X-8-102

Almgren M (2000) Mixed micelles and other structures in the solubilization of bilayer lipid membranes by surfactants. Biochim Biophys Acta 1508(1–2):146–163. https://doi.org/10.1016/s0005-2736(00)00309-6

Badawi NM, Teaima MH, El-Say KM, Attia DA, El-Nabarawi MA, Elmazar MM (2018) Pomegranate extract-loaded solid lipid nanoparticles: design, optimization, and in vitro cytotoxicity study. Int J Nanomedicine 13:1313. https://doi.org/10.2147/IJN.S154033

Baek J-S, Na Y-G, Cho C-W (2018) Sustained cytotoxicity of wogonin on breast cancer cells by encapsulation in solid lipid nanoparticles. Nano 8(3):159. https://doi.org/10.3390/nano8030159

Bai C, Zheng J, Zhao L, Chen L, Xiong H, McClements DJ (2018) Development of Oral delivery systems with enhanced antioxidant and anticancer activity: coix seed oil and β-carotene coloaded liposomes. J Agric Food Chem 67(1):406–414. https://doi.org/10.1021/acs.jafc.8b04879

Bhatt H, Rompicharla SV, Komanduri N, Aashma S, Paradkar S, Ghosh B, Biswas S (2018) Development of curcumin-loaded solid lipid nanoparticles utilizing glyceryl Monostearate as single lipid using QbD approach: characterization and evaluation of anticancer activity against human breast cancer cell line. Curr Drug Deliv 15(9):1271–1283. https://doi.org/10.2174/1567201815666180503120113

Bondì ML, Emma MR, Botto C, Augello G, Azzolina A, Di Gaudio F, Craparo EF, Cavallaro G, Bachvarov D, Cervello M (2017) Biocompatible lipid nanoparticles as carriers to improve curcumin efficacy in ovarian cancer treatment. J Agric Food Chem 65(7):1342–1352. https://doi.org/10.1021/acs.jafc.6b04409

Bray F, Ferlay J, Soerjomataram I, Siegel RL, Torre LA, Jemal A (2018) Global cancer statistics 2018: GLOBOCAN estimates of incidence and mortality worldwide for 36 cancers in 185 countries. CA Cancer J Clin 68(6):394–424. https://doi.org/10.3322/caac.21492

Buenz EJ, Verpoorte R, Bauer BA (2018) The ethnopharmacologic contribution to bioprospecting natural products. Annu Rev Pharmacol Toxicol 58:509–530. https://doi.org/10.1146/annurev-pharmtox-010617-052703

Caddeo C, Nacher A, Vassallo A, Armentano MF, Pons R, Fernàndez-Busquets X, Carbone C, Valenti D, Fadda AM, Manconi M (2016) Effect of quercetin and resveratrol co-incorporated in liposomes against inflammatory/oxidative response associated with skin cancer. Int J Pharm 513(1–2):153–163. https://doi.org/10.1016/j.ijpharm.2016.09.014

Cao C, Wang Q, Liu Y (2019) Lung cancer combination therapy: doxorubicin and beta-elemene co-loaded, pH-sensitive nanostructured lipid carriers. Drug Des Devel Ther 13:1087–1098. https://doi.org/10.2147/DDDT.S198003. eCollection 2019

Chang M, Wu M, Li H (2018) Antitumor activities of novel glycyrrhetinic acid-modified curcumin-loaded cationic liposomes in vitro and in H22 tumor-bearing mice. Drug Deliv 25(1):1984–1995. https://doi.org/10.1080/10717544.2018.1526227

Chen R, Wang S, Zhang J, Chen M, Wang Y (2015) Aloe-emodin loaded solid lipid nanoparticles: formulation design and in vitro anti-cancer study. Drug Deliv 22(5):666–674. https://doi.org/10.3109/10717544.2014.882446

Chen Y, Pan L, Jiang M, Li D, Jin L (2016) Nanostructured lipid carriers enhance the bioavailability and brain cancer inhibitory efficacy of curcumin both in vitro and in vivo. Drug Deliv 23(4):1383–1392. https://doi.org/10.3109/10717544.2015.1049719

Cheng Y, Zhao P, Wu S, Yang T, Chen Y, Zhang X, He C, Zheng C, Li K, Ma X (2018) Cisplatin and curcumin co-loaded nano-liposomes for the treatment of hepatocellular carcinoma. Int J Pharm 545(1–2):261–273. https://doi.org/10.1016/j.ijpharm.2018.05.007

Chirio D, Peira E, Dianzani C, Muntoni E, Gigliotti LC, Ferrara B, Sapino S, Chindamo G, Gallarate M (2019) Development of solid lipid nanoparticles by cold dilution of microemulsions: Curcumin loading, preliminary in vitro studies, and biodistribution. Nano 9(2):230. https://doi.org/10.3390/nano9020230

Da Poian AT, El-Bacha T, MRMP L (2010) Nutrient utilization in humans: metabolism pathways. Nat Educ 3(9):1

Das S, Chaudhury A (2011) Recent advances in lipid nanoparticle formulations with solid matrix for oral drug delivery. AAPS Pharm Sci Tech 12(1):62–76. https://doi.org/10.1208/s12249-010-9563-0

Demain AL, Vaishnav P (2011) Natural products for cancer chemotherapy. Microb Biotechnol 4(6):687–699. https://doi.org/10.1111/j.1751-7915.2010.00221.x

Dhule SS, Penfornis P, Frazier T, Walker R, Feldman J, Tan G, He J, Alb A, John V, Pochampally R (2012) Curcumin-loaded γ-cyclodextrin liposomal nanoparticles as delivery vehicles for osteosarcoma. Nanomedicine 8(4):440–451. https://doi.org/10.1016/j.nano.2011.07.011

Dong X, Wang W, Qu H, Han D, Zheng J, Sun G (2016) Targeted delivery of doxorubicin and vincristine to lymph cancer: evaluation of novel nanostructured lipid carriers in vitro and in vivo. Drug Deliv 23(4):1374–1378. https://doi.org/10.3109/10717544.2015.1041580

Du G-H (2018) Natural small molecule drugs from plants. Springer, Singapore: https://doi.org/10.1007/978-981-10-8022-7

Düzgüneş N, Gregoriadis G (2005) Introduction: the origins of liposomes: Alec Bangham at Babraham methods in enzymology, vol 391. Elsevier, London, pp 1–3. https://doi.org/10.1016/S0076-6879(05)91029-X

Elmowafy M, Viitala T, Ibrahim HM, Abu-Elyazid SK, Samy A, Kassem A, Yliperttula M (2013) Silymarin loaded liposomes for hepatic targeting: in vitro evaluation and HepG2 drug uptake. Eur J Pharm Sci 50(2):161–171. https://doi.org/10.1016/j.ejps.2013.06.012

Enrico C (2019) Nanotechnology-based drug delivery of natural compounds and phytochemicals for the treatment of Cancer and other diseases studies in natural products chemistry, vol 62. Elsevier, Amsterdam, pp 91–123. https://doi.org/10.1016/B978-0-444-64185-4.00003-4

Espinosa E, Zamora P, Feliu J, Gonzalez Baron M (2003) Classification of anticancer drugs–a new system based on therapeutic targets. Cancer Treat Rev 29(6):515–523. https://doi.org/10.1016/S0305-7372(03)00116-6

Fahmy UA (2018) Augmentation of Fluvastatin cytotoxicity against prostate carcinoma PC3 cell line utilizing alpha Lipoic-Ellagic acid nanostructured lipid carrier formula. AAPS Pharm Sci Tech 19(8):3454–3461. https://doi.org/10.1208/s12249-018-1199-5

Feng H, Zhu Y, Fu Z, Li D (2017) Preparation, characterization, and in vivo study of rhein solid lipid nanoparticles for oral delivery. Chem Biol Drug Des 90(5):867–872. https://doi.org/10.1111/cbdd.13007

Geetha T, Kapila M, Prakash O, Deol PK, Kakkar V, Kaur IP (2015) Sesamol-loaded solid lipid nanoparticles for treatment of skin cancer. J Drug Target 23(2):159–169. https://doi.org/10.3109/1061186X.2014.965717

Ghadi R, Dand N (2017) BCS class IV drugs: highly notorious candidates for formulation development. J Control Release 248:71–95. https://doi.org/10.1016/j.jconrel.2017.01.014

Ghasemiyeh P, Mohammadi-Samani S (2018) Solid lipid nanoparticles and nanostructured lipid carriers as novel drug delivery systems: applications, advantages and disadvantages. Res Pharm Sci 13(4):288. https://doi.org/10.4103/1735-5362.235156

Gomes G, Sola M, Rochetti A, Fukumasu H, Vicente A, Pinho S (2019) β-carotene and α-tocopherol coencapsulated in nanostructured lipid carriers of murumuru (Astrocaryum murumuru) butter produced by phase inversion temperature method: characterization, dynamic in vitro digestion and cell viability study. J Microencapsul 36(1):43–52. https://doi.org/10.1080/02652048.2019

Gülçür E, Thaqi M, Khaja F, Kuzmis A, Önyüksel H (2013) Curcumin in VIP-targeted sterically stabilized phospholipid nanomicelles: a novel therapeutic approach for breast cancer and breast cancer stem cells. Drug Deliv Transl Res 3(6):562–574. https://doi.org/10.1007/s13346-013-0167-6

Gumireddy A, Christman R, Kumari D, Tiwari A, North EJ, Chauhan H (2019) Preparation, characterization, and in vitro evaluation of Curcumin-and resveratrol-loaded solid lipid nanoparticles. AAPS PharmSciTech 20(4):145. https://doi.org/10.1208/s12249-019-1349-4

Guorgui J, Wang R, Mattheolabakis G, Mackenzie GG (2018) Curcumin formulated in solid lipid nanoparticles has enhanced efficacy in Hodgkin's lymphoma in mice. Arch Biochem Biophys 648:12–19. https://doi.org/10.1016/j.abb.2018.04.012

Hajipour H, Hamishehkar H, Nazari Soltan Ahmad S, Barghi S, Maroufi NF, Taheri RA (2018) Improved anticancer effects of epigallocatechin gallate using RGD-containing nanostructured lipid carriers. Artif Cells Nanomed Biotechnol 46(sup1):283–292. https://doi.org/10.1080/21691401.2017.1423493

Hanafy NAN, El-Kemary M, Leporatti S (2018) Micelles structure development as a strategy to improve smart cancer therapy. Cancers 10(7). https://doi.org/10.3390/cancers10070238

Hazam PK, Goyal R, Ramakrishnan V (2019) Peptide based antimicrobials: design strategies and therapeutic potential. Prog Biophys Mol Biol 142:10–22. https://doi.org/10.1016/j.pbiomolbio.2018.08.006

Heneweer C, Gendy SE, Penate-Medina O (2012) Liposomes and inorganic nanoparticles for drug delivery and cancer imaging. Ther Deliv 3(5):645–656. https://doi.org/10.4155/tde.12.38

Huang L, Cai M, Xie X, Chen Y, Luo X (2014) Uptake enhancement of curcumin encapsulated into phosphatidylcholine-shielding micelles by cancer cells. J Biomater Sci Polym Ed 25(13):1407–1424. https://doi.org/10.1080/09205063.2014.941261

Huang R, Li J, Kebebe D, Wu Y, Zhang B, Liu Z (2018) Cell penetrating peptides functionalized gambogic acid-nanostructured lipid carrier for cancer treatment. Drug Deliv 25(1):757–765. https://doi.org/10.1080/10717544.2018.1446474

Immordino ML, Dosio F, Cattel L (2006) Stealth liposomes: review of the basic science, rationale, and clinical applications, existing and potential. Int J Nanomedicine 1(3):297

Jain A, Sharma G, Thakur K, Raza K, Shivhare U, Ghoshal G, Katare OP (2019) Beta-carotene-encapsulated solid lipid nanoparticles (BC-SLNs) as promising vehicle for Cancer: an investigative assessment. AAPS PharmSciTech 20(3):100. https://doi.org/10.1208/s12249-019-1301-7

Jesus M, Grazu V (2012) Nanobiotechnology: inorganic nanoparticles vs organic nanoparticles, vol 4. Elsevier, Amsterdam

Jiang H, Geng D, Liu H, Li Z, Cao J (2016) Co-delivery of etoposide and curcumin by lipid nanoparticulate drug delivery system for the treatment of gastric tumors. Drug Deliv 23(9):3665–3673. https://doi.org/10.1080/10717544.2016.1217954

Jiang S, Zhu R, He X, Wang J, Wang M, Qian Y, Wang S (2017) Enhanced photocytotoxicity of curcumin delivered by solid lipid nanoparticles. Int J Nanomedicine 12:167. https://doi.org/10.2147/IJN.S123107

Jose A, Labala S, Ninave KM, Gade SK, Venuganti VVK (2018) Effective skin cancer treatment by topical co-delivery of curcumin and STAT3 siRNA using cationic liposomes. AAPS PharmSciTech 19(1):166–175. https://doi.org/10.1208/s12249-017-0833-y

Khosa A, Reddi S, Saha RN (2018) Nanostructured lipid carriers for site-specific drug delivery. Biomed Pharmacother 103:598–613. https://doi.org/10.1016/j.biopha.2018.04.055

Kim CS, Tonga GY, Solfiell D, Rotello VM (2013) Inorganic nanosystems for therapeutic delivery: status and prospects. Adv Drug Deliv Rev 65(1):93–99. https://doi.org/10.1016/j.addr.2012.08.011

Lamichhane N, Udayakumar T, D'Souza W, Simone I, Raghavan S, Polf J, Mahmood J (2018) Liposomes: clinical applications and potential for image-guided drug delivery. Molecules 23(2):288. https://doi.org/10.3390/molecules23020288

Li X, Jia X, Niu H (2018) Nanostructured lipid carriers co-delivering lapachone and doxorubicin for overcoming multidrug resistance in breast cancer therapy. Int J Nanomedicine 13:4107–4119. https://doi.org/10.2147/ijn.s163929

Lian T, Ho RJ (2001) Trends and developments in liposome drug delivery systems. J Pharm Sci 90(6):667–680. https://doi.org/10.1002/jps.1023

Lin M, Teng L, Wang Y, Zhang J, Sun X (2016) Curcumin-guided nanotherapy: a lipid-based nanomedicine for targeted drug delivery in breast cancer therapy. Drug Deliv 23(4):1420–1425. https://doi.org/10.3109/10717544.2015.1066902

Liu Q, Li J, Pu G, Zhang F, Liu H, Zhang Y (2016) Co-delivery of baicalein and doxorubicin by hyaluronic acid decorated nanostructured lipid carriers for breast cancer therapy. Drug Deliv 23(4):1364–1368. https://doi.org/10.3109/10717544.2015.1031295

Lv L, Zhuang Y-x, Zhang H-w, Tian N-n, Dang W-z, Wu S-y (2017) Capsaicin-loaded folic acid-conjugated lipid nanoparticles for enhanced therapeutic efficacy in ovarian cancers. Biomed Pharmacother 91:999–1005. https://doi.org/10.1016/j.biopha.2017.04.097

Mahmud M, Piwoni A, Filiczak N, Janicka M, Gubernator J (2016) Long-circulating curcumin-loaded liposome formulations with high incorporation efficiency, stability and anticancer activity towards pancreatic adenocarcinoma cell lines in vitro. PLoS One 11(12):e0167787. https://doi.org/10.1371/journal.pone.0167787

Mulik RS, Mönkkönen J, Juvonen RO, Mahadik KR, Paradkar AR (2010) Transferrin mediated solid lipid nanoparticles containing curcumin: enhanced in vitro anticancer activity by induction of apoptosis. Int J Pharm 398(1–2):190–203. https://doi.org/10.1016/j.ijpharm.2010.07.021

Mulik RS, Mönkkönen J, Juvonen RO, Mahadik KR, Paradkar AR (2012) Apoptosis-induced anticancer effect of transferrin-conjugated solid lipid nanoparticles of curcumin. Cancer Nanotechnol 3(1):65. https://doi.org/10.1007/s12645-012-0031-2

Nahak P, Karmakar G, Chettri P, Roy B, Guha P, Besra SE, Soren A, Bykov AG, Akentiev AV, Noskov BA (2016) Influence of lipid core material on physicochemical characteristics of an ursolic acid-loaded nanostructured lipid carrier: an attempt to enhance anticancer activity. Langmuir 32(38):9816–9825. https://doi.org/10.1021/acs.langmuir.6b02402

Nahak P, Gajbhiye RL, Karmakar G, Guha P, Roy B, Besra SE, Bikov AG, Akentiev AV, Noskov BA, Nag K (2018) Orcinol Glucoside loaded polymer-lipid hybrid nanostructured lipid carriers: potential cytotoxic agents against gastric, Colon and Hepatoma carcinoma cell lines. Pharm Res 35(10):198. https://doi.org/10.1007/s11095-018-2469-3

Newman DJ, Cragg GM (2016) Natural products as sources of new drugs from 1981 to 2014. J Nat Prod 79(3):629–661. https://doi.org/10.1021/acs.jnatprod.5b01055

Niazvand F, Orazizadeh M, Khorsandi L, Abbaspour M, Mansouri E, Khodadadi A (2019) Effects of Quercetin-loaded nanoparticles on MCF-7 human breast Cancer cells. Medicina 55(4):114. https://doi.org/10.3390/medicina55040114

Olusanya T, Haj Ahmad R, Ibegbu D, Smith J, Elkordy A (2018) Liposomal drug delivery systems and anticancer drugs. Molecules 23(4):907. https://doi.org/10.3390/molecules23040907

Parveen R, Ahmad F, Iqbal Z, Samim M, Ahmad S (2014) Solid lipid nanoparticles of anticancer drug andrographolide: formulation, in vitro and in vivo studies. Drug Dev Ind Pharm 40(9):1206–1212. https://doi.org/10.3109/03639045.2013.810636

Porter CJH, Trevaskis NL, Charman WN (2007) Lipids and lipid-based formulations: optimizing the oral delivery of lipophilic drugs. Nat Rev Drug Discov 6:231. https://doi.org/10.1038/nrd2197

Radhakrishnan R, Kulhari H, Pooja D, Gudem S, Bhargava S, Shukla R, Sistla R (2016) Encapsulation of biophenolic phytochemical EGCG within lipid nanoparticles enhances its stability and cytotoxicity against cancer. Chem Phys Lipids 198:51–60. https://doi.org/10.1016/j.chemphyslip.2016.05.006

Radhakrishnan R, Pooja D, Kulhari H, Gudem S, Ravuri HG, Bhargava S, Ramakrishna S (2019) Bombesin conjugated solid lipid nanoparticles for improved delivery of epigallocatechin gallate for breast cancer treatment. Chem Phys Lipids 224. https://doi.org/10.1016/j.chemphyslip.2019.04.005

Ramadass SK, Anantharaman NV, Subramanian S, Sivasubramanian S, Madhan B (2015) Paclitaxel/Epigallocatechin gallate coloaded liposome: a synergistic delivery to control the invasiveness of MDA-MB-231 breast cancer cells. Colloids Surf B: Biointerfaces 125:65–72. https://doi.org/10.1016/j.colsurfb.2014.11.005

Riaz MK, Zhang X, Wong KH, Chen H, Liu Q, Chen X, Zhang G, Lu A, Yang Z (2019) Pulmonary delivery of transferrin receptors targeting peptide surface-functionalized liposomes augments the chemotherapeutic effect of quercetin in lung cancer therapy. Int J Nanomedicine 14:2879. https://doi.org/10.2147/IJN.S192219

Rodenak-Kladniew B, Islan GA, de Bravo MG, Durán N, Castro GR (2017) Design, characterization and in vitro evaluation of linalool-loaded solid lipid nanoparticles as potent tool in cancer therapy. Colloids Surf B: Biointerfaces 154:123–132. https://doi.org/10.1016/j.colsurfb.2017.03.021

Rompicharla SVK, Bhatt H, Shah A, Komanduri N, Vijayasarathy D, Ghosh B, Biswas S (2017) Formulation optimization, characterization, and evaluation of in vitro cytotoxic potential of curcumin loaded solid lipid nanoparticles for improved anticancer activity. Chem Phys Lipids 208:10–18. https://doi.org/10.1016/j.chemphyslip.2017.08.009

Sabapati M, Palei NN, Molakpogu RB (2019) Solid lipid nanoparticles of Annona muricata fruit extract: formulation, optimization and in vitro cytotoxicity studies. Drug Dev Ind Pharm 45(4):577–586. https://doi.org/10.1080/03639045.2019.1569027

Saengkrit N, Saesoo S, Srinuanchai W, Phunpee S, Ruktanonchai UR (2014) Influence of curcumin-loaded cationic liposome on anticancer activity for cervical cancer therapy. Colloids Surf B: Biointerfaces 114:349–356. https://doi.org/10.1016/j.colsurfb.2013.10.005

Sailaja AK, Amareshwar P, Chakravarty P (2011) Formulation of solid lipid nanoparticles and their applications. Curr Pharma Res 1(2):197

Saklayen MG (2018) The global epidemic of the metabolic syndrome. Curr Hypertens Rep 20(2):12–12. https://doi.org/10.1007/s11906-018-0812-z

Samad A, Sultana Y, Aqil M (2007) Liposomal drug delivery systems: an update review. Curr Drug Deliv 4(4):297–305. https://doi.org/10.2174/156720107782151269

Sarisozen C, Abouzeid AH, Torchilin VP (2014) The effect of co-delivery of paclitaxel and curcumin by transferrin-targeted PEG-PE-based mixed micelles on resistant ovarian cancer in 3-D spheroids and in vivo tumors. Eur J Pharm Biopharm 88(2):539–550. https://doi.org/10.1016/j.ejpb.2014.07.001

Sharma A, Sharma US (1997) Liposomes in drug delivery: progress and limitations. Int J Pharm 154(2):123–140. https://doi.org/10.1016/s0378-5173(97)00135-x

Shen R, Kim JJ, Yao M, Elbayoumi TA (2016) Development and evaluation of vitamin E d-α-tocopheryl polyethylene glycol 1000 succinate-mixed polymeric phospholipid micelles of berberine as an anticancer nanopharmaceutical. Int J Nanomedicine 11:1687. https://doi.org/10.2147/IJN.S103332

Shidhaye S, Vaidya R, Sutar S, Patwardhan A, Kadam V (2008) Solid lipid nanoparticles and nanostructured lipid carriers-innovative generations of solid lipid carriers. Curr Drug Deliv 5(4):324–331. https://doi.org/10.2174/156720108785915087

Shrestha H, Bala R, Arora S (2014) Lipid-based drug delivery systems. J Pharm 2014:10. https://doi.org/10.1155/2014/801820

Springob K, Kutchan TM (2009) Introduction to the different classes of natural products plant-derived natural products. Springer, pp 3–50. https://doi.org/10.1007/978-0-387-85498-4_1

Sun J, Bi C, Chan HM, Sun S, Zhang Q, Zheng Y (2013) Curcumin-loaded solid lipid nanoparticles have prolonged in vitro antitumour activity, cellular uptake and improved in vivo bioavailability. Colloids Surf B: Biointerfaces 111:367–375. https://doi.org/10.1016/j.colsurfb.2013.06.032

Sun M, Nie S, Pan X, Zhang R, Fan Z, Wang S (2014) Quercetin-nanostructured lipid carriers: characteristics and anti-breast cancer activities in vitro. Colloids Surf B Biointerfaces 113:15–24. https://doi.org/10.1016/j.colsurfb.2013.08.032

Sutaria D, Grandhi BK, Thakkar A, Wang J, Prabhu S (2012) Chemoprevention of pancreatic cancer using solid-lipid nanoparticulate delivery of a novel aspirin, curcumin and sulforaphane drug combination regimen. Int J Oncol 41(6):2260–2268. https://doi.org/10.3892/ijo.2012.1636

Tang J, Ji H, Ren J, Li M, Zheng N, Wu L (2017) Solid lipid nanoparticles with TPGS and Brij 78: a co-delivery vehicle of curcumin and piperine for reversing P-glycoprotein-mediated multi-drug resistance in vitro. Oncol Lett 13(1):389–395. https://doi.org/10.3892/ol.2016.5421

Tewari D, Rawat P, Singh PK (2019) Adverse drug reactions of anticancer drugs derived from natural sources. Food Chem Toxicol 123:522–535. https://doi.org/10.1016/j.fct.2018.11.041

Tian J, Wang L, Wang L, Ke X (2014) A wogonin-loaded glycyrrhetinic acid-modified liposome for hepatic targeting with anti-tumor effects. Drug Deliv 21(7):553–559. https://doi.org/10.3109/10717544.2013.853850

Torchilin VP (2007) Micellar nanocarriers: pharmaceutical perspectives. Pharm Res 24(1):1–16. https://doi.org/10.1007/s11095-006-9132-0

Tran TH, Nguyen AN, Kim JO, Yong CS, Nguyen CN (2016) Enhancing activity of artesunate against breast cancer cells via induced-apoptosis pathway by loading into lipid carriers. Artif Cells Nanomed Biotechnol 44(8):1979–1987. https://doi.org/10.3109/21691401.2015.1129616

Vandita K, Shashi B, Santosh KG, Pal KI (2012) Enhanced apoptotic effect of curcumin loaded solid lipid nanoparticles. Mol Pharm 9(12):3411–3421. https://doi.org/10.1021/mp300209k

Vijayakumar A, Baskaran R, Baek J-H, Sundaramoorthy P, Yoo BK (2019) In vitro cytotoxicity and bioavailability of Ginsenoside-modified nanostructured lipid carrier containing Curcumin. AAPS PharmSciTech 20(2):88. https://doi.org/10.1208/s12249-019-1295-1

Wang S, Chen T, Chen R, Hu Y, Chen M, Wang Y (2012) Emodin loaded solid lipid nanoparticles: preparation, characterization and antitumor activity studies. Int J Pharm 430(1–2):238–246. https://doi.org/10.1016/j.ijpharm.2012.03.027

Wang H, Sun G, Zhang Z, Ou Y (2017a) Transcription activator, hyaluronic acid and tocopheryl succinate multi-functionalized novel lipid carriers encapsulating etoposide for lymphoma therapy. Biomed Pharmacother 91:241–250. https://doi.org/10.1016/j.biopha.2017.04.104

Wang W, Zhang L, Chen T, Guo W, Bao X, Wang D, Ren B, Wang H, Li Y, Wang Y (2017b) Anticancer effects of resveratrol-loaded solid lipid nanoparticles on human breast cancer cells. Molecules 22(11):1814. https://doi.org/10.3390/molecules22111814

Wang W, Chen T, Xu H, Ren B, Cheng X, Qi R, Liu H, Wang Y, Yan L, Chen S (2018a) Curcumin-loaded solid lipid nanoparticles enhanced anticancer efficiency in breast Cancer. Molecules 23(7):1578. https://doi.org/10.3390/molecules23071578

Wang X, Wu F, Li G, Zhang N, Song X, Zheng Y, Gong C, Han B, He G (2018b) Lipid-modified cell-penetrating peptide-based self-assembly micelles for co-delivery of narciclasine and siULK1 in hepatocellular carcinoma therapy. Acta Biomater 74:414–429. https://doi.org/10.1016/j.actbio.2018.05.030

Watkins R, Wu L, Zhang C, Davis RM, Xu B (2015) Natural product-based nanomedicine: recent advances and issues. Int J Nanomedicine 10:6055. https://doi.org/10.2147/IJN.S92162

Wissing S, Kayser O, Müller R (2004) Solid lipid nanoparticles for parenteral drug delivery. Adv Drug Deliv Rev 56(9):1257–1272. https://doi.org/10.1016/j.addr.2003.12.002

Wojcik MH, Schwartz TS, Yamin I, Edward HL, Genetti CA, Towne MC, Agrawal PB (2018) Genetic disorders and mortality in infancy and early childhood: delayed diagnoses and missed opportunities. Genet Med 20(11):1396–1404. https://doi.org/10.1038/gim.2018.17

Wong M-Y, Chiu GN (2010) Simultaneous liposomal delivery of quercetin and vincristine for enhanced estrogen-receptor-negative breast cancer treatment. Anti-Cancer Drugs 21(4):401–410. https://doi.org/10.1097/CAD.0b013e328336e940

Wu M, Fan Y, Lv S, Xiao B, Ye M, Zhu X (2016) Vincristine and temozolomide combined chemotherapy for the treatment of glioma: a comparison of solid lipid nanoparticles and nanostructured lipid carriers for dual drugs delivery. Drug Deliv 23(8):2720–2725. https://doi.org/10.3109/10717544.2015.1058434

Xie J, Yang Z, Zhou C, Zhu J, Lee RJ, Teng L (2016) Nanotechnology for the delivery of phytochemicals in cancer therapy. Biotechnol Adv 34(4):343–353. https://doi.org/10.1016/j.biotechadv.2016.04.002

Yan H-m, Song J, Zhang Z-h, Jia X-b (2016) Optimization and anticancer activity in vitro and in vivo of baohuoside I incorporated into mixed micelles based on lecithin and Solutol HS 15. Drug Deliv 23(8):2911–2918. https://doi.org/10.3109/10717544.2015.1120365

Yang L, Xin J, Zhang Z, Yan H, Wang J, Sun E, Hou J, Jia X, Lv H (2016) TPGS-modified liposomes for the delivery of ginsenoside compound K against non-small cell lung cancer: formulation design and its evaluation in vitro and in vivo. J Pharm Pharmacol 68(9):1109–1118. https://doi.org/10.1111/jphp.12590

Ye H, Liu X, Sun J, Zhu S, Zhu Y, Chang S (2016) Enhanced therapeutic efficacy of LHRHa-targeted brucea javanica oil liposomes for ovarian cancer. BMC Cancer 16(1):831. https://doi.org/10.1186/s12885-016-2870-4

Zhang L, Chan JM, Gu FX, Rhee JW, Wang AZ, Radovic-Moreno AF, Alexis F, Langer R, Farokhzad OC (2008) Self-assembled lipid–polymer hybrid nanoparticles: a robust drug delivery platform. ACS Nano 2(8):1696–1702. https://doi.org/10.1021/nn800275r

Zhang XY, Qiao H, Ni JM, Shi YB, Qiang Y (2013) Preparation of isoliquiritigenin-loaded nanostructured lipid carrier and the in vivo evaluation in tumor-bearing mice. Eur J Pharm Sci: Off J Eur Fed Pharm Sci 49(3):411–422. https://doi.org/10.1016/j.ejps.2013.04.020

Zhang H, Liu X, Wu F, Qin F, Feng P, Xu T, Li X, Yang L (2016) A novel prostate-specific membrane-antigen (PSMA) targeted micelle-encapsulating wogonin inhibits prostate cancer cell proliferation via inducing intrinsic apoptotic pathway. Int J Mol Sci 17(5):676. https://doi.org/10.3390/ijms17050676

Zhou X, Liu H-Y, Zhao H, Wang T (2018) RGD-modified nanoliposomes containing quercetin for lung cancer targeted treatment. OncoTargets Therapy 11:5397. https://doi.org/10.2147/OTT.S169555

Zhu W-j, Shu-di Yang C-xQ, Zhu Q-l, Chen W-l, Li F, Yuan Z-q, Liu Y, You B-g, Zhang X-n (2017) Low-density lipoprotein-coupled micelles with reduction and pH dual sensitivity for intelligent co-delivery of paclitaxel and siRNA to breast tumor. Int J Nanomed 12:3375. https://doi.org/10.2147/IJN.S126310

Zielińska A, Martins-Gomes C, Ferreira NR, Silva AM, Nowak I, Souto EB (2018) Anti-inflammatory and anti-cancer activity of citral: optimization of citral-loaded solid lipid nanoparticles (SLN) using experimental factorial design and LUMiSizer®. Int J Pharm 553(1–2):428–440. https://doi.org/10.1016/j.ijpharm.2018.10.065

Chapter 6
Inorganic Particles for Delivering Natural Products

Jairam Meena, Anuradha Gupta, Rahul Ahuja, Amulya K. Panda, and Sangeeta Bhaskar

Abstract Natural products are complex molecules that have been widely used in traditional medicine for therapeutics and diagnostics applications. Despite their long history of use, some challenges are associated with many natural product derived pharmaceuticals, like inadequate stability, poor absorption, distribution, metabolism and excretion. Medicinal chemists have been successful in addressing many of these challenges through structural modifications of the parent compound, but even so, analysis suggests that up to 20% of natural product leads are taken through unchanged as the final drug product. Even modified compounds are a challenge to administer, requiring the use of novel formulations and delivery strategies to enable the launch of an effective natural product derived drug into the market. To outwit these concerns, formulation of these natural product derived bioactive compounds using nanotechnology has been used as a potential tool in diagnostic and therapeutic applications. Compounds of organic or inorganic origin that are prepared from different metals, metal oxides, chitosan, sodium alginate, poly lactic acid, poly lactic co-glycolic acid, synthetic as well as natural origin polymers are amongst commonly used materials for development of natural product nanoformulations.

This book chapter deals in detail with the properties, synthesis, advantages and toxicity of inorganic particles like those of silver, gold, iron oxide and silica with the aim to shed light on the delivery of natural products for therapeutic and diagnostic purposes. Adjustable size and shape, large surface area, ease of functionalization and additional bioactivities associated with inorganic nanoparticles are some of the properties that give them an edge over other delivery methods. Apart from enhancing the stability of molecule, high-density surface ligands attachment enables the targeted delivery with enhanced therapeutic efficacy. Among the inorganic nanoparticles, metallic nanoparticles made up of silver or gold are increasingly being used for biomedical purposes because of their biocompatibility, versatility, broad antimicrobial activity as well as visible light extinction property. Silver and gold possesses

J. Meena (✉) · A. Gupta · R. Ahuja · A. K. Panda · S. Bhaskar
National Institute of Immunology, New Delhi, India
e-mail: jairam.meena20@gmail.com

A. Saneja et al. (eds.), *Sustainable Agriculture Reviews 44*, Sustainable Agriculture Reviews 44, https://doi.org/10.1007/978-3-030-41842-7_6

peculiar properties such as Surface Plasmon Resonance associated which are not associated with other delivery vehicles like liposomes, dendrimers or micelles. Metal oxides such as Iron oxide (Fe_2O_3, Fe_3O_4) and silica (SiO_2) with various surface modifications and as hybrid are now the popular choices for delivering natural products for a variety of applications.

Keywords Natural product delivery · Gold nanoparticles · Silver nanoparticles · Iron oxide nanoparticles · Silica nanoparticles · Curcumin · Paclitaxel · Flavonoids · Quercetin

Abbreviations

FDA	Food and Drug Administration
MCM-41	Mobil composition of matter no 41
PEG	Polyethylene glycol
PEI	Polyethylenimine

6.1 Introduction

Natural products are amongst the widely used medicinal compounds for targeting and treating various bourgeoning and emerging diseases. Plants, animals and minerals are common natural source for the extraction these complex chemical molecules by various extraction processes, and/or by chemical synthesis and chemo-enzymatic synthesis. These compounds serve as the source of potential drug lead molecules. However, low bioavailability and stability limits the therapeutic application of natural products. The average calculated octanol-water partition coefficient (log P) of natural products is 2.9 while of drugs is 2.1 suggesting that natural products are more lipophilic than drugs. Natural products are more rigid than drugs because they contain larger fused ring system and on an average contain two rotatable bonds less than drugs (Wetzel et al. 2011). To circumvent these concerns, encapsulation of these bioactive compounds via nanotechnology (nanomaterials are substances having at least one dimension, 100 nm in size) has been materialized with potential applications in diagnostic and therapeutic developments. Nanomaterials can be of inorganic as well as organic origin and could be synthesized from different polymers, metals, metal oxides, chitosan, sodium alginate, poly lactic co-glycolic acid (PLGA), poly lactic acid (PLA), synthetic polymers etc. These nanomaterials protect the drug/bioactive molecule from the hazardous environment, pH and temperature variation and target the molecule to desired organ with an increase in therapeutic efficacy. Inorganic particles draw much attention due to adjustable size and shape, high surface area, ease of functionalization, high-density surface ligands attachment as well as additional bioactivities associated with them for desired applications.

Inorganic nanoparticles of non-metallic origin ($Al(OH)_3$, Fe_2O_3, ZnO, SiO_2, Fe_3O_4, CeO_2, ITO, CaO, ATO, ZrO_2) and metals and metal alloys (Pt, Cu, Pd, Fe, Ni, Co, Ag, Au, Al, Mn) have been attempted for various purposes. Metallic nanoparticles show strong plasma absorption and enhanced rayleigh as well as surface raman scattering for imaging applications. However, most of these are thermodynamically unstable and may contain some impurities such as oxides and nitrides. Most of these impurities come during synthesis and may cause irritation or toxicity or both. Among metallic nanoparticles silver or gold (Ag or Au) particles are most preferably and increasingly being used for different biomedical purposes either because of their visible light extinction behavior, versatility, antimicrobial property and good biocompatibility (Kasthuri et al. 2009a, b). Silver and gold possess specific properties such as surface plasmon resonance which dendrimers, liposomes and micelles don't. Metal oxides such as Iron oxide (Fe_2O_3, Fe_3O_4) and silica (SiO_2) with various surface modifications and as hybrid are popularly used to deliver natural products. Iron oxides exhibits super paramagnetic properties at the size range of 8–10 nm, which not only helps in targeting under the influence of magnetic field, but also governs easy synthesis, separation and hyperthermia derived cancer cell killing. On the other hand silica show high payload due to mesoporous character and easy surface functionalization. This book chapter deals in details with the properties, synthesis, advantages and toxicity of silver, gold, iron oxide and silica particles citing some examples to show delivery of natural products. Figure 6.1 shows the schematic representation of natural products delivery via inorganic nanoparticles.

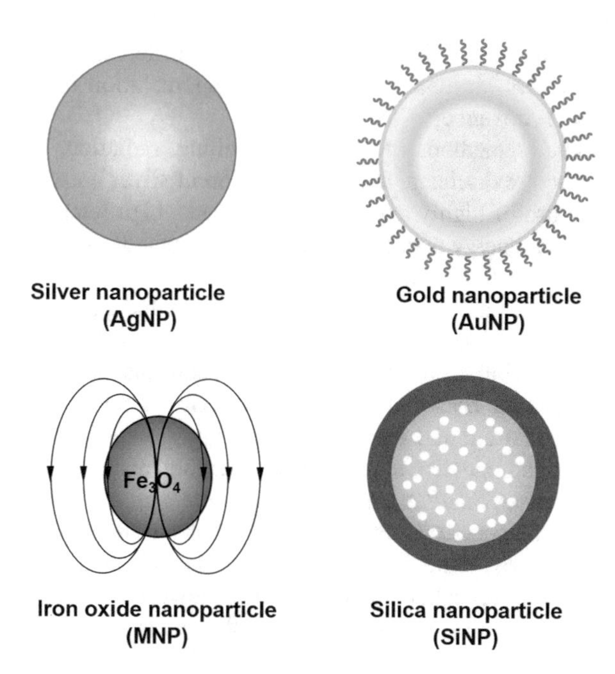

Fig. 6.1 Natural products delivery via silver, gold, iron oxide and silica nanoparticles

6.2 Inorganic Nanoparticles: Properties, Synthesis, Advantages and Toxicity

6.2.1 Metallic Nanoparticles

6.2.1.1 Silver Nanoparticles

Silver nanoparticles, due to their attractive physiochemical properties, easy to fabricate in desired size range with biological functionality, and non-toxic nature governs its extensive use in the biomedical applications. Silver based nanosystem and nanomaterials have shown antibacterial, antiviral, antifungal, antioxidant properties and used for drug/natural products delivery, tissue scaffold fabrication, wound dressing and protective coating applications. The antimicrobial properties exhibited by silver nanoparticles are related to their physiochemical properties such as shape, size, concentration, colloidal state and surface charge. Furthermore, silver nanoparticles allow the coordination of many ligands, and enabling easy surface functionalization (Tran and Le 2013; Le Ouay and Stellacci 2015; Phull et al. 2016).

Synthesis

Different methods such as chemical, physical and biological are used for the synthesis of silver nanoparticles. Using chemical method, Ag + ions are reduced to Ag (metallic silver) which agglomerates to form oligomeric clusters. Chemical synthesis is carried out using different inorganic and organic reducing agents such as sodium citrate, sodium borohydride, sodium ascorbate, elemental hydrogen, N, N-dimethylformamide and tollens method. However, different, coating agents, surfactants and polymers are used to inhibit the agglomeration of particles during chemical synthesis (Iravani et al. 2014).

Physical synthesis methods involves, ball milling, radiation, flame pyrolysis, sonication, electric arc discharge and laser ablation of silver bulk material in solution. Despite the use of highly expensive equipments, high temperature and pressure, as well as high energy consumption, physical methods are more eco-friendly as compared to chemical methods.

Biological synthesis are known as "green" methods and gained immense popularity due to devoid of any toxic chemical use. Using green synthesis; silver nanoparticles are prepared via bio-reduction of Ag + ions in aqueous medium using bacteria, fungi, algae, and plants. Bacterial synthesis comprises of selection and cultivation of suitable bacterial strain and maintenance of highly aseptic conditions. Culture supernatant of nonpathogenic and pathogenic microorganisms like *E. coli*, *B. indicus*, *A. flavus*, *B. cereus*, *Bacillus strain CS 11*, *P. proteolytica*, *P. meridian*, *S. aureus* etc. are used to synthesize silver nanoparticles, and can be achieved either intracellularly or extracellularly. Extracellular synthesis is easy, cheaper and required lesser time hence preferred over intracellular synthesis. However, the mechanism behind the nanoparticles formation using the microbial organisms is poorly understood which hinders the scaling laboratory technique for the industrial synthesis. (Nanda and Saravanan 2009; Shivaji et al. 2011; Das et al. 2014; Ahmed et al. 2016). Plant based green synthesis of silver nanoparticles is gaining immense popularity because of its easy accessibility, economic feasibility, environment friendly nature, formulation simplicity along with the possibility of large scale production. Different plant extracts such as *Azadirecta Indica, Madhuca longifolia, Crocus sativus L., Calliandra haematocephala*, Grape seed, Andean blackberry fruit, Granium leaf extract and Marigold flower etc. have been reported for the green synthesis of silver nanoparticles (Patil et al. 2018; Li et al. 2007; Ahmed et al. 2016; Bagherzade et al. 2017; Raja et al. 2017; Rivera-Rangel et al. 2018). The phytochemical extracts includes flavonoids, terpenoids, saponins, phenolics, tannins, catechins, enzymes, proteins, polysaccharides and rich complex compounds which acts as reducing, stabilizing and capping agents are extensively used during "green" method of silver nanoparticles synthesis. Due to such added advantages, silver nanoparticles have been extensively investigated for targeted drug delivery and diagnostic applications.

The phytochemical composition varies significantly from plant to plant and in different parts of plants and has admissible effect on the quality of nanoparticles. For example, spherical silver nanoparticles were prepared using *Syzygium cumini* fruit extract and it has been suggested that flavonoids present in the plant extract were responsible for the reduction of Ag + to Ag which is very crucial for particle synthesis (Mittal et al. 2014). In another study, spherical silver nanoparticles were synthesized in 3–20 nm size range with leaf broth of *Arbutus unedo* which contains both reducing and stabilizing agents which are responsible for particle synthesis (Kouvaris et al. 2012), similarly, *Erythrina indica* root extract was used for silver nanoparticles synthesis in size range of 20–118 nm (Sre et al. 2015). Shail et al. reported that extract of *Origanum vulgare L.* contains mixture of flavonoids, alkaloid and terpenoids which imparts reducing and capping properties during silver nanoparticles synthesis (Shaik et al. 2018). Nanoparticle synthesis parameters such as; concentration of silver ions, plant extract concentration and phytochemical composition, temperature, pH, microwave assistance, mechanical stirring speed and time had a significant effect, it was observed that synthesis of plant extract capped silver nanoparticles were higher at 90 °C as compared to room temperature with yielding of larger size particles at higher temperature. It was also reported that, antimicrobial properties of silver nanoparticles were linearly increases with use of higher concentrations of plant extracts during synthesis as the half maximal inhibitory concentration values were decreased from 4% to 21% with increase in solubility of nanoparticles (Khan et al. 2013, 2014). After nanoparticles synthesis, characterization can be done using physicochemical properties such as shape, size, surface area, purity and coating along with the associated electrochemical properties namely; charge, zeta potential, redox, conductivity and surface plasmon resonance for the synthesis variability assessment (Li et al. 2017). Silver nanoparticles have been widely and frequently used for tumor specific drug delivery (Kajani et al. 2016). It is reported that light scattering cross section of silver nanoparticles is ~10 times greater than similar size gold nanoparticles, and are strongest light scatterers amongst the nobel metal particles. Because of these characteristics, silver nanoparticles have also been employed as photoactivated drug delivery vector and biological sensors (Brown et al. 2013). Silver nanoparticle based delivery systems facilitate intracellular detection and controlled release of natural products through chemical or photothermal of photochemical triggers e.g. Silver nanoparticles in the size range of 60–80 nm were developed, which contains anti-sense oligonucleotides for ICAM-1 (intracellular adhesion molecule-1) silencing. Light induced release of anti-sense oligonucleotide demonstrated wound healing and therefore suggested their therapeutic potential for the treatment of Crohn's disease (Brown et al. 2013).

Toxicity Assessment of Silver Nanoparticles
Bioactive molecule incorporated surface functionalized silver nanoparticles have been exploited as promising oral formulation for improving solubility and bioavailability, reducing toxicity enhancing bioactive compound release. These formulations provide better therapeutic tool against cancer, wound healing and pathogenic microbial diseases etc. There is little information regarding exposure of silver nanoparticles to environment, animals and humans but their potential risks reporting contradictory results are available with short and long term exposure (Kittler et al. 2010; Bouwmeester et al. 2011; Loeschner et al. 2011).

Daily approximate silver intake by humans through food and/or water is 0.4–30 μg, which indicates that in the given range silver nanoparticles can be used without any sever short term toxicities for therapeutic applications (Hadrup and Lam 2014). Silver nanoparticles have demonstrated excellent antimicrobial properties and non-toxicity towards healthy mammalian cells (Stensberg et al. 2011). However, there are some reports where silver nanoparticle related toxic effects have been observed in rat hepatocytes, neuronal cells (El Mahdy et al. 2015), murine stem cells and human lung epithelial cells (Pinzaru et al. 2018). Wen H et al. has investigated the silver nanoparticles mediated acute toxicity and genotoxicity in Sprague-Dawley rats where 61.1% of nanoparticles were 27.3–106.2 nm sizes. Studies also reported that silver nanoparticles are mainly concentrated in lungs followed by spleen, liver, kidney, thymus and heart. These particles also enhance the blood urea, total bilirubin, alanine aminotransferase and creatinine which indicate abnormal liver functions. Further histopathological examination of liver, kidney, thymus and spleen after exposure with silver nanoparticles showed extensive organ damages which were also complementing the abnormal liver function tests. At the cellular level significant chromosomal breakage and polyploidy were observed which indicates silver nanoparticles derived genotoxicity (Wen et al. 2017). In an *in vivo* toxicity study using rat ear model, silver nanoparticles mediated permanent or temporary hearing loss and significant mitochondrial dysfunction was also observed. A Lower silver nanoparticles concentration has also been detected in retinal cells where these particles induced structural disruption and oxidative stress (Antony et al. 2015). Also, silver nanoparticles surface charge might modulate the particle uptake, its translocation to various tissues and organs and thereby cytotoxicity (Franci et al. 2015; Wu et al. 2017).

6.2.1.2 Gold Nanoparticles

Gold nanoparticles has garnered attention in recent biomedical field especially for ultrasensitive biomolecular detection, selective cancer cell killing by photothermal therapy, specific cells and protein labeling and cellular therapeutic delivery. Gold nanoparticles permits easy tailoring into different size, shape, and attachment of different functional groups, chemically biocompatible and have intrinsic tunable optical properties (Kumar et al. 2012). The gold nanoparticle size is related to surface plasmon resonance, by changing the thiol/gold ratio during synthesis, size of conjugated gold nanoparticles can be controlled and therefore surface plasmon resonance could be fine tune according to the application requirements. Such as particles with smaller size can be synthesized using higher amount of thiol (SH) and vice versa (Bhattacharya and Srivastava 2003). Characteristic UV absorbance at 520 nm can be observed using 10 nm gold nanoparticles, however, a blue or red shift can be observed with size variation. Figure 6.2 shows the schematic representation of different types of gold nanoparticles and effect of size on the optical properties. Gold nanorod shows the characteristic absorbance towards near infra-red range (690 nm–900 nm), these intrinsic optical properties of gold nanoparticles can be exploited for its use as composite theranostic agents in clinic (Tong et al. 2009). Gold nanoparticles of 1–100 nm, when dispersed in water are also designated as colloidal gold, these nanoparticles comprises; gold nanorods, gold nanoshells, gold nanocages, gold nanosphere and gold nanoparticles with stimuli-responsive surface enhanced raman scattering.

Synthesis

For the synthesis of gold nanoparticles, chemical method is most commonly and widely used which encompasses the reduction of gold salt (HAuCl4) with the help of reducing agents. In 1857, Michael Faraday, first reported gold nanosphere synthesis by two phase synthesis using tetraoctyl ammonium bromide having high air and thermal stability. Citrate reduction method has been used for gold nanoparticles

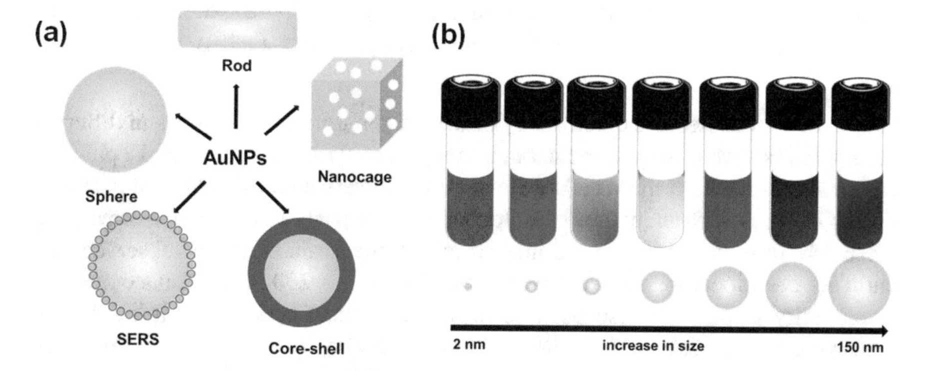

Fig. 6.2 (**a**) Types of gold nanoparticles (**b**) effect of size on the optical properties of gold nanospheres

synthesis, where the amount of citrate, a gold reducing agent, governed the yield of monodisperse gold nanospheres and it is reported that with lower amount of citrate higher amount of gold nanospheres can be obtained. In another method, in an aqueous medium, gold nanoparticles of about 20 nm were synthesized by Turkevich J et al. by single phase reduction of gold tetrachloroauric acid by sodium citrate (Turkevich et al. 1951). Bastus et al. have synthesized the gold nanoparticles of 200 nm size through kinetically controlled seed growth using same reducing agent i.e. sodium citrate (Bastus et al. 2011).

Green Synthesis Plant based materials such as; stem, root, seed, latex and leafs have been attempted for the synthesis of gold nanoparticles. Ankamwar B et al. used lemon grass extract and tamarind leaf extract for gold nanotriangle synthesis (Ankamwar et al. 2005). Spherical shaped gold nanoparticles have been synthesized using *Piper longum* extract (Yu et al. 2016). Similarly, aqueous extracts of *Citrus maxima* (Nakkala et al. 2016), aqueous extracts of neem (*Azadirachta indica*) (Anuradha et al. 2010), extract of *Allium cepa* (Parida et al. 2011) and dilute extract of *Phyllanthus amarus* (Kasthuri et al. 2009a, b), aqueous extract of *Terminalia chebula* (Edison and Sethuraman 2012), *Cassia fistula* extract (Daisy and Saipriya 2012), leaves and bark extracts of *Ficus caricaz* (Teimuri-Mofrad et al. 2017), *Plumeria Alba* (Frangipani flower) (Nagaraj et al. 2012), fruit peel extract of *Momordica charantia* (Pandey et al. 2012), extract of *Benincasa hispida* seed (Aromal and Philip 2012) and many more extracts have been used for gold nanoparticles synthesis.

Narayanan and Sakthive et al. reported gold nanoparticles of the size range 7–58 nm with different shapes such as triangular, decahaderal or spherical using *Coriandrum sativum* leaf extract (Narayanan and Sakthivel 2008). Similarly, Apiin, an ingredient present in banana leaf extract was also capable of reduction of ions to gold for nanoparticles synthesis (Kasthuri et al. 2009a, b). Nagajyothi et al. generated 8.02 nm gold nanoparticle using *Lonicera japonica* flower aqueous extract (Nagajyothi et al. 2012). Gold nanoparticles of 2.5–27.5 and 1.25–17.5 nm with an average size of 10 and 3 nm respectively were also produced using ethanolic extract of black tea and its free ethanol tannin extract, where these extracts worked as reducing and stabilizing agents during synthesis (Banoee et al. 2010).

Toxicity

Pan, Y. et al. assessed the cytotoxicity of 0.8–15 nm gold nanoparticles in different cells such as macrophages, melanoma, connective tissue fibroblast and epithelial cells and reported that these cells were sensitive to very small size of gold nanoparticles (1.4 nm) with induction of mitochondrial damage and oxidative stress, but when 15 nm particles were used instead of 1.4 nm particles, it was observed that particles are biocompatible and non-toxic (Pan et al. 2007). Gold nanoparticles cytotoxicity studies has also been performed in human cells where it has shown nontoxicity up to 250 mM, while ionic gold showed cytotoxicity at a relatively lower concentration of 25 mM (Connor et al. 2005). With surface functionalized particles it was observed that the chemical groups present at the surface of gold nanoparticles are responsible for its effect on cells. Takahashi et al. reported the reduced cytotoxicity of gold nanorods (6.5 nm × 11 nm) in HeLa cells when cetyl trimethyl ammonium bromide (CTAB) at surface was replaced with polyethylene glycol (Niidome et al. 2006). Similarly Shukla et al. reported that 3.7 nm PEGylated gold nanoparticles of spherical size were able to enter into HeLa cell's nucleus without any toxicity due to its neutral surface (Shukla et al. 2005). Hetero biofunctionalization of gold nanoparticles with thiol polyethylene glycol acid (HS-PEG-COOH) or capping with bovine serum albumin did not produce any mortality or behavioral change in mice at the studied doses suggesting their biocompatibility (Nghiem et al. 2012). Gold nanoparticles have demonstrated size-dependent organ distributions where smaller particles of 5–15 nm showed wider organ distribution compared to large size particles of 50–100 nm. These gold nanoparticles were mainly found in liver and spleen (De Jong et al. 2008; Semmler-Behnke et al. 2008; Sonavane et al. 2008; Chen et al. 2009; Cho et al. 2009a, b; Kim et al. 2009). In another study, it was observed that coating of 20 nm gold nanoparticles with TA-terminated PEG5000 stabilized the particles and reduces the toxicity (Zhang et al. 2009; Lipka et al. 2010). Chen, Y. S. et al. conducted in vivo studies in mice by injecting different sized gold nanoparticles of 5 and 50 nm and observed lethal effects (Chen et al. 2009). A study by Cho, W.S. et al. has also reported the similar results where 13 nm PEGylated gold nanoparticles when injected intravenously into mice, were predominantly concentrated in the liver, kupffer cells, spleen macrophages and induces an acute inflammation in liver as observed in biopsy study (Cho et al. 2009a, b). Effect of route of administration on the toxicity of gold nanoparticles has also been studied and it was observed that oral administration of gold nanoparticles significantly reduces the body weight, spleen index, and red blood cells. Also, when oral, intra-peritoneal and intravenous routes were compared, it was observed that oral and intra-peritoneal dosing is more toxic than intravenous injections. Hence targeted delivery of nanoparticles using tail vain injections will be ideal for radiotherapy, photothermal therapy, and natural product or drug delivery (Shukla et al. 2005).

6.2.2 *Non-Metallic Nanoparticles*

6.2.2.1 Iron Oxide Nanoparticles

Recently, magnetic nanoparticles of iron oxide, also called magnetite or Fe_3O_4, have approved by U.S. Food and Drug Administration (FDA) for clinical imaging and drug delivery applications. These particles are extensively used for imaging purposes as preclinical and clinical studies proven its safety and biocompatibility. Easy synthesis, multifunctional surface modification, magnetic bio-separation along with magnetic fluid hyperthermia revolutionize the use of Fe_3O_4 magnetic nanoparticles for loading a variety of agents which are highly useful in current therapeutic and diagnostic applications. Magnetic nanoparticles allow easy tailoring of different physical and chemical parameters such as chemical group modification, magnetism, shape, size, crystallinity and surface charge for targeting purposes. As next generation vehicles for targeted delivery of natural products and drugs, magnetic nanoparticles are amongst the preferred one due to devoid of undesired toxicities. These particles can be selectively targeted and concentrated to an organ by application of an external magnetic field (Verma et al. 2013; Ali et al. 2016). Oxidation state of iron governs its magnetic properties and hence has profound impact on magnetic properties of iron oxide nanoparticles. Use of magnetite (Fe_3O_4) is preferred over hematite, as magnetite contains both Fe^{3+} and Fe^{2+} ions and also known as ferrous-ferric oxide and has a brownish-black color with a metallic luster is known for strong magnetic properties compared to hematite (Fe_2O_3) which appears as reddish brown solid and the oxidation state of iron is (+III), for drug delivery and targeted therapeutic applications. Figure 6.3 shows the effect of oxidation state on the magnetic properties of Iron oxide nanoparticles.

Synthesis

Synthesis of iron nanoparticles can be done by thermal decomposition of Fe(N-nitrosophenyl hydroxylamine)$_3$ / Fe(acetylacetone)$_3$/ $Fe(CO)_5$ followed by oxidation. Facile decomposition of iron pentacarbonyl [$Fe(CO)_5$], which is a metastable organometallic compound, also denoted as iron carbonyl, at 140–160 °C under an inert atmospheric conditions using decalin as solvent and one of the three polymer as a surfactant and catalyst; polybutadiene, poly(styrene co-butadiene) and poly(styrene-co-4 vinylpyradine) yields magnetic iron nanoparticles (Lide 2004). Synthesis of monodispersed magnetite nanoparticles at high temperature (265 °C) using Fe(acetylacetonate)$_3$ in phenyl ether in the presence of alcohol, oleic acid, and oleylamine have been reported. In an another study, using microemulsion method, iron nanoparticles with an average size of ∽3 nm has been synthesized with the help

Fig. 6.3 Magnetite and hematite iron oxide nanoparticles

of trioctyl phosphine oxide as a stabilizing agent (Guo et al. 2001). Magnetic nanoparticles of monodisperse maghemite in a size range of 3.5 ± 0.6 nm with high magnetization saturation values along with monolayer oleic acid capping have been reported by Vidal et al. using one-pot microemulsion method (Vidal-Vidal et al. 2006), although this method is rarely used as a matters of convenience. Ionic surfactant such as quaternary ammonium compound (Martino et al. 1997; Seip and O'Connor 1999; Li et al. 2003) and nonionic e.g., polyether-based (Martino et al. 1997) long-chain surfactants are commonly used in micellar synthesis. Reduction of iron salts and its oxides using widely used reducing agents such as hydrazide (Seip and O'Connor 1999), sodium borohydride (Luborsky and Paine 1960; Glavee et al. 1995; Li et al. 2003), and lithium borohydride for the iron nanoparticles precipitation during synthesis is also reported. Use of surfactant during nanoparticles precipitation helps to prevent particle agglomeration. Likewise, co-precipitation is the other conventional method used for Fe_3O_4 or γ-Fe_2O_3 nanoparticles fabrication; which involves 1:2 molar ratios of ferrous and ferric ion salts. The above mentioned molar ratios are first mixed and then precipitated at room temperature or higher temperature using the highly basic solutions such as sodium hydroxide. Different synthesis conditions; temperature, stirring rate, precipitating agents, type of salts used, the ratio of ferrous and ferric ions, pH value affects the size and shape of magnetic nanoparticles (Wu et al. 2008).

Besides the above mention methods, *hydrothermal synthesis* is used for the synthesis of iron oxide nanoparticles with controlled shape and size. The method use high vapor pressure (0.3–4 MPa) and temperature (130–250 °C) for nanoparticles crystallization. Wang et al. obtained 40 nm particles with saturation magnetization (85.8 emu·g^{-1}) and high crystallinity using Fe_3O_4 at high temperature for 6 h (Wang et al. 2003). Similarly, with help of sodium bis(2-ethylhexyl) sulfosuccinate surfactant, Zheng et al. synthesized 27 nm magnetite nanoparticles using the hydrothermal synthesis (Zheng et al. 2006) which exhibited a superparamagnetic behavior at room temperature. Magnetic nanoparticles can also be synthesized using mechanical methods such as high-energy milling (Kerekes et al. 2002). But this method is not generally employed due to production of high poly dispersity and irregular shape and size.

Green biosynthesis is generally preferred over other method because the magnetite nanoparticles generated using this method possesses higher biocompatibility and biodegradability, due to the special non-toxic surface coating of green materials. Phumying et al. prepared magnetite nanoparticles using *Aloe vera* extract at different conditions (Phumying et al. 2013). In another study brown seaweed (*Sargassum muticum*) were used to synthesize magnetite nanoparticles (El-Kassas et al. 2016), similarly Rajendran et al. described the synthesis of magnetite nanoparticles using *Sesbania grandiflora* leaf extract as a photocatalyst (Rajendran and Sengodan 2017). Similar attempts were made by different groups using green biosynthesis such as, magnetite nanoparticles of 40–70 nm were synthesized using *Rubus glacus Benth* or *Andean blackberry* leaves (Kumar et al. 2016), nanoparticles has also been synthesized using *Syzygium cumini* seed extract (Venkateswarlu et al. 2014, 2019), *Punica Granatum* rind extract (Venkateswarlu et al. 2014, 2019), plantain peel extract (Venkateswarlu et al. 2013) and lemon juice (Bahadur et al. 2017). In many green synthesis approaches Lemon juice was used where it serves as the source of citric acid for surface capping and particle size control of magnetic nanoparticles. Other natural products such as potato (Buazar et al. 2016), *Mimosa pudica* root (Niraimathee et al. 2016), arabic gum (Horst et al. 2017) has also been attempted for iron nanoparticles synthesis.

Magnetic nanoparticles has also been prepared using waste materials such as tea residue, coffee waste hydrochar and aqueous corn leaf extract and used for arsenic removal (Lunge et al. 2014), acid red 17 (azo dye) removal (Khataee et al. 2017) and in drug delivery due to antioxidant and antibacterial activities (Khataee et al. 2017). Iron nanoparticles has not only been synthesized using plants but also other green materials such as natural polymer (Sodium alginate (Gao et al. 2008), chitosan (Shrifian-Esfahni et al. 2015), polysaccharides (pectin) (Namanga et al. 2013), glucose (D-maltose) (Demir et al. 2013), amino acid (arginine) (Wang et al. 2009), vitamin (Nicotinic acid) (Attallah et al. 2016), enzyme (Urease) (Shi et al. 2014) and fungi (yeast) (Zhou et al. 2009) were also used.

Toxicity

Use of iron oxide nanoparticles were increasing day by day for various biomedical applications. U.S. Food and Drug Administration (FDA) has approved Feridex® (ferumoxides), a dextran-coated magnetic nanoparticles as a imaging contrast agent for the detection of liver lesions (Bulte 2009) and Feraheme® (ferumoxytol) for the treatment of anemia i.e. iron deficiency in adult patients with chronic kidney disease (Wang 2015). Therefore, concerns have been raised regarding its safety, distribution and clearance. Feng Q et al. has analyzed the *in vitro* cellular uptake, bio-distribution, clearance and toxicity of iron oxide nanoparticles which are well characterized and commercially available. The average size, hydrodynamic size, method of preparation, type of coating, surface functionalization affects the efficacy and toxicity of iron oxide nanoparticles (Nel et al. 2009). Similarly effect of particle size and different surface coating on blood half-life and magnetic behavior of iron oxide nanoparticles have been reported by Roohi F et al. They have found the inverse relationship of pharmacokinetic properties with particle size, as the hydrodynamic radius of nanoparticles gets decreased, a significant increase in blood half-life time and biodistribution of iron oxide nanoparticles were observed, thus affecting uptake of these particles by different organs and ultimately toxicity. In an another study, effect of different coating materials, polyacrylic acid, carboxy dextran, polyethylene glycol (PEG) and starch on 50 nm magnetic nanoparticles were observed and it was reported that polyacrylic acid-iron oxide nanoparticles displayed the shortest blood half-life, followed by carboxy dextran-iron oxide nanoparticles, starch-iron oxide nanoparticles, and PEG-iron oxide nanoparticles. These differences were ascribed by different ionic characteristics of the coating agents (Roohi et al. 2012). Feng Q et al. has compared the toxicity and uptake of Polyethylenimine-coated iron oxide nanoparticles (PEI-iron oxide nanoparticles) with the PEGylated one (PEG-iron oxide nanoparticles). PEI-iron oxide nanoparticles displayed significantly higher uptake by macrophages and cancer cells than PEG- iron oxide nanoparticles.

PEI-iron oxide nanoparticles displayed severe cytotoxicity than PEG-iron oxide nanoparticles also. This may be due to reactive oxygen species production and apoptosis. Whereas PEG- iron oxide nanoparticles did not exhibited toxicity but it induces autophagy, which imparts biocompatibility and reduces the cytotoxicity. The bio-distribution studies signifies that, iron oxide nanoparticles larger than 100 nm in size are rapidly concentrated in liver and spleen via macrophage phagocytosis whereas particles smaller than 10 nm are likely to be eliminated through renal clearance. (Kumar et al. 2011). *In vivo* studies in BALB/c mice displayed no obvious toxicity with PEG-iron oxide nanoparticles, whereas dose dependent toxicity was observed in PEI-iron oxide nanoparticles which signify the importance of coating materials (Feng et al. 2018). Gokduman K et al. has performed the dose (0–400 μg/ml) treatment (single dosing vs. cumulative dosing) and time-dependent toxicity of super-paramagnetic iron oxide nanoparticles of 10 nm size on primary rat hepatocytes, which state that toxicity was increases significantly with increasing concentration and treatment duration (Gokduman et al. 2018). Apart from biomedical applications, iron oxide nanoparticles can be used for removal of metal contamination from aqueous solutions (Ge et al. 2012). Therefore, toxicity assessment studies were also performed in Zebrafish to study the effect of these particles on developmental toxicity in fish in aquatic environments where developmental toxicity were observed as presence of mortality, hatching delay, and malformation at higher concentrations (≥10 mg/L) of iron oxide nanoparticles (Zhu et al. 2012).

6.2.2.2 Silica Nanoparticles

Mesoporous silica nanoparticles, because of high surface area, well ordered mesopores with larger pore volume have been used as multifunctional drug delivery system. The particle size, shape and surface can be easily modified depending upon the desired purpose (Vallet-Regi et al. 2007; Fernandez-Fernandez et al. 2011; Rosenholm et al. 2011; Baeza et al. 2015). Functionalization is generally performed over silanol-containing surface for controlled drug release and drug loading. Since the first use of silica based material, MCM-41 for drug delivery by Vallet-Regí et al. in 2001 (Vallet-Regi et al. 2001), other silica materials, such as SBA-15 or MCM-48, and some metal-organic hybrid and their nanoparticles with modifications have been developed for loading of different biologically active compounds. Silica oxide is widely used in dentist industry for many purposes such as tooth and bone implants, ceramic plates (Lührs and Geurtsen 2009), orthopedics (scaffolds) (Zhu et al. 2014; Zhou et al. 2017) and for specialized medical device manufacturing (Brunner et al. 2009; Cao and Zhu 2014). Additional advantages associated with the use of silica are: high thermal stability, chemical inactivity, resistance to microbial attack, high hydrophilicity and biocompatibility, high loading capacity making it an attractive material for biomedical purposes (Gonçalves 2018). Furthermore, U.S. Food and Drug Administration (FDA) have recommended amorphous silica and silicates as safe at the oral intake up to 1500 mg per day. Amorphous silica is widely used in cosmetic product manufacturing, for foods and oral delivery of drugs (Kasaai 2015; Go et al. 2017; Gonçalves 2018).

Synthesis

Silica nanoparticles are generally synthesized by sol-gel, flame decomposition and reverse micro-emulsion methods. Reverse micelles are formed when organic solvent containing surfactant is added to aqueous phase, polar head group orient themselves to form microcavities containing water using reverse micro-emulsion method (Tan et al. 2011). Silica nanoparticles are also formed in the same manner when silicon alkoxides and catalyst addition results in particle synthesis inside the microcavities. Like the other methods which have several limitation, major challenge associated with this method is high cost and difficulties in surfactant removal (Bagwe et al. 2006; Liu and Han 2010; Wang et al. 2014). Naka Y et al. has reported one-pot water-oil-microemulsion technique for manufacturing of organo-unctionalized silica nanoparticles. Use of mixture of organosilane (3-aminopropyltriethoxysilane), phenyltrimethoxysilane, 3-mercaptopropyltrimethoxysilane, vinyltriethoxysilane, 3-cyanoethyltriethoxysilane and polyoxyethylene nonylphenol ether in the solution of tetraethyl orthosilicate is employed to synthesize monodisperse silica nanoparticles in the size range 25–200 nm (Naka et al. 2010). Kozlecki et al. has reported silica nanoparticle synthesis using modified oil-in-water microemulsion. Silica nanoparticles in the size range of 170–422 nm with high surface area (>300 m^2/g) were reported using heptanes, 2-ethylhexanole, Tween®85 non-ionic surfactant, tetraethyl orthosilicate and ammonium hydroxide. Similarly, precipitation of β-cyclodextrin and hydroxyethylcellulose have been used to obtain particles with similar characteristics (Koźlecki et al. 2016).

Silica nanoparticles can also be synthesized from metal organic precursors through high temperature flame decomposition technique, which is also known as chemical vapor decomposition (Silva 2004). Chemical reaction of silicon tetrachloride with hydrogen and oxygen is also used for silica nanoparticles synthesis, but desired particles size, shape and morphology are major limitations which are difficult to control, also this method is mainly used to produce particles in powder form. Yan F et al. have also prepared silica nanoparticles in the same manner where water vapors at ~150 °C were used to hydrolyzed silicon tetrachloride vapors. The obtained porous amorphous silica were having a large specific surface area (342.44 m^2/g), with broad size distribution of 162.8 ± 41.0 nm (Yan et al. 2014).

Sol-gel process which involves hydrolysis and condensation of metal alkoxide such as tetraethyl orthosilicate or inorganic salts e.g. sodium silicate in the presence of mineral acid e.g. hydrochloric acid or base e.g. ammonia as catalyst under mild conditions, is/are commonly used for the synthesis of silica and silica nanoparticles for different therapeutic and diagnostic applications (Stöber et al. 1968; Hench and West 1990; Klabunde et al. 1996). In the course of sol-gel based synthesis, in the first step, silanol groups were generated using tetraethyl orthosilicate hydrolysis, in the second step, condensation/polymerization is carried out between silanol group and ethoxy group for siloxane bridge formation, which results in the formation of silica structure. Further, nucleation followed by growth are the remaining two stages in silica nanoparticles synthesis (Matsoukas and Gulari 1988).

Stober et al. has reported the synthesis of spherical and mono-dispersed silica nanoparticles for the first time in 1968 (Stöber et al. 1968). Using this method silica particles in the size range of 5–2000 nm could be synthesized by the addition of ammonia to aqueous alcohol solution of silica. Later, modified Stober methods were developed for silica nanoparticles synthesis. Rao KS obtained silica nanoparticles in the size range of 20–460 nm via the hydrolysis of tetraethyl orthosilicate in ethanol medium (Rao et al. 2005). Silica particles were also synthesized from low cost precursors such as sodium silicate solution, in the process, precipitation were carried out by the addition of acids such as hydrochloric acid with carbon dioxide (Stöber et al. 1968). Bentonite clay has also been reported to prepare silica nanoparticles via generation of sodium silicate which is then converted into sodium silicate solution with sodium hydroxide treatment further hydrolysis of this solution in the presence of nitric acid and ethanol were resulted in different size silica nanoparticle synthesis. The particle size was controlled by monitoring the concentration of silica rich clay and nitric acid (Zulfiqar et al. 2016).

Agricultural wastes such as, rice husk, rice hull, semi burned rice straw, sugarcane waste and Vietnamese rice husk are used for the green synthesis of silica nanoparticles (Zaky et al. 2008; Zhang et al. 2010; Liu et al. 2011; Rovani et al. 2018). Silica nanoparticle synthesis attempts have been made by Rovani S et al. using sugarcane waste ash, and the particles obtained were 20 nm to several micrometers range with specific surface area of 131 $m^2 g^{-1}$ and ~230 mg g^{-1} acid orange eight dye adsorption capacity (Rovani et al. 2018). Nanosized silica was synthesized using sol-gel method from Vietnamese rise husk with heat treatment, for this, ash was generated from the rise husk with thermal treatment at 600 °C for 4 h. Then sodium silicate solution were obtained from rice husk ash by treating it with sodium hydroxide solution and finally silica nanoparticles were precipitated by addition of sulphuric acid at pH 4 in the water/butanol with cationic presence. The particles with highest specific surface area (340 m^2/g) and an average size of 3 nm were also reported (Thuc and Thuc 2013).

Surface modification of silica nanoparticles are generally carried out for incorporation of a variety of structurally diverse molecules. The reagents used for surface modifications are silane coupling agents such as amino propyl methyl diethoxy silane and methacryloxy propyl triethoxy silane under non-aqueous and aqueous conditions (Kang et al. 2001; Yu et al. 2003). Kobler and Bein et al. reported formulation of highly small mesoporous silica nanoparticles by co-condensation process with phenyl triethoxysilane in the presence of triethanolamine catalyst (Kobler and Bein 2008). Silica particles of different shapes, size and pore size have been synthesized. Wang et al. has synthesized cubic shaped silica nanoparticles with the help of tartaric acid (Yu et al. 2005). Nanosized MCM-41 silica particles with several morphologies such as hexagonal shape under basic conditions have also been reported (Cai et al. 2001). Ikari et al. reported the hexagonally arrayed mesoporous silica nanoparticles with the help of binary surfactant method (Ikari et al. 2006). Mesoporous silica containing hexagonally ordered mesopores of 2 nm were synthesized using a cationic surfactant cetyl trimethyl ammonium bromide, while silica nanoparticles having bigger mesopores (20–50 nm) were designed by sulphonated aromatic polyether ether ketone (Pang et al. 2005). Sugar molecules such as glucose and fructose have been used for eco-friendly production of silica nanoparticles of mesoporous and mono-dispersed nature (Mukherjee et al. 2009), the obtained particles were of 50–1140 nm size, which were produced by reaction mixture of phenyltriethoxysilane, ammonia, water, ethanol and the non-surfactant sugar template.

Toxicity

Several reports have been published related to silica effect on human health, in a detailed study related to its toxicity, it was observed that, exposure to crystalline silica cause the fibrotic lung disease, also known as silicosis, lung cancer, emphysema and pulmonary tuberculosis (Zhang et al. 2012). Conversely amorphous form of silica is considered as safe and hence it has been a part of human diet as medical clay or as food additive (labelled E551) (Sripanyakorn et al. 2009; Jurkić et al. 2013). When used *in vivo*, amorphous form of silica is cleared more rapidly as compared to its crystalline form from the lungs which may be responsible for its lower toxicity and its use for human applications *in vivo* (Arts et al. 2007). The *in vivo* as well as *in vitro* use of mesoporous silica nanoparticles are well tolerated, when used *in vitro*, it is considered safe at lower than 100 μg/ml concentrations (Huang et al. 2008; Hudson et al. 2008; Tao et al. 2008; Lu et al. 2010), whereas lower than 200 mg/kg doses are considered safe for *in vivo* use (Lu et al. 2010). Although there is conflicting reports regarding the relationship between silica nanoparticles size and toxicity, even most of these reports indicate that physicochemical property such as, size, surface groups, dose and cell type governs the cytotoxicity of these particles. In one study, effect of different size silica nanoparticles such as 30, 48, 118 and 535 nm on viability of mouse keratinocytes was investigated and reduction in cell viability with decrease in size at 10–200 μg/ml doses was observed (Kyung et al. 2009). While in another study on human hepatocytes, the effect of 7, 20, 50 nm silica nanoparticles on cell viability at 20–640 μg/ml doses was investigated and it was found that cell viability with 20 nm particles was highest followed by 7 nm and 50 nm size particles (Lu et al. 2011). Kim and in-Yon et al. studied the biocompatibility of 20–200 nm silica nanoparticles on mouse embryonic fibroblast (NIH/3 T3), human hepatoma (HepG2) and human alveolar carcinoma (A549) cells and reported that, all the three cell type preferentially engulf silica nanoparticles of 60 nm at high doses, these particles led to disproportionate decrease in cell viability when compared with other size particles (Kim et al. 2015). Cytotoxicity and cell dysfunction is most commonly observed in dendritic cells, mast cells, lymphocytes, monocytes, macrophages and kupffer cells upon exposure to silica nanoparticles. In organ specific immunotoxicity studies, lymphocytes infiltration, granuloma formation and hydropic degeneration (condition in which cells absorb more water) in liver hepatocyte were observed, whereas in the spleen, decreased proliferation in immune cells such as B and T cells along with the altered serum IgG and IgM levels were observed, similarly in the lungs, neutrophil infiltration and in heart inflammation was observed (Chen et al. 2018). Zhao Y et al. observed that, when red blood cells were treated with different size mesoporous silica nanoparticles such as MCM-41 of ~100 nm and SBA-15 of ~600 nm, small size particles (~100 nm) readily adsorb on red blood cell surface without any cell membrane disruption whereas larger size particles (~600 nm) adsorb on cell membrane, internalized by the cells and induced cell membrane disruption which eventually leads to red blood cells hemolysis (Huang et al. 2008). It is also reported that silica nanoparticles are able to freely pass through the blood-brain barrier (BBB), and can enter into the central nervous system (Klejbor et al. 2007) hence, adverse central nervous system effects such as neurotoxicity, neuro-inflammation, neuro-degeneration were observed. Along with this, enhanced levels of amyloid beta protein were also reported which is the hallmark of Alzheimer's disease. When, silica nanoparticles were incubated with primary cultured hippocampal neurons, a rapid and persistent decalcification of endolysosome was observed which was correlated with brain amyloidogenesis and Alzheimer's disease. Therefore, use of silica nanoparticles for the treatment of Alzheimer's and other neurological disorders were not recommended (Ye et al. 2018).

6.3 Recent Advances in Inorganic Nanoparticles Mediated Natural Products Delivery

Different formulations such as liposomes, solid lipid nanoparticles, polymeric nanoparticles and/or inorganic nanoparticles are known for the delivery of natural products. Various advantages of inorganic nanoparticles are highlighted about enhancement of therapeutic potential of natural products. Figure 6.4 shows the chemical structure and Table 6.1 shows the compiled list of some natural compounds that have been delivered via inorganic nanoparticles and their efficacy has been tested *in vitro* and/or *in vivo* with positive effects.

6.3.1 Curcumin

Curcumin or diferuloylmethane, which is known as folk medicine, is a major bioactive polyphenol constituent found in the rhizome of *Curcuma longa* (Turmeric). Curcumin acts as potent anti-inflammatory and antioxidant agent and is used to specifically kill cancer cells; it can suppress tumor genesis, tumor promotion, and metastasis as reported in several preclinical studies (Shanmugam et al. 2015). The problem with the use of curcumin is that it is not soluble in water and shows low bioavailability. Attempts to improve the bioavailability of curcumin are underway employing various processes. Nanocurcumin shows improvement in bioavailability compared to conventional curcumin.

To circumvent the antibiotic resistance associated with existing antibacterials, silver nanoparticles of 25–35 nm, which contains curcumin natural product, were

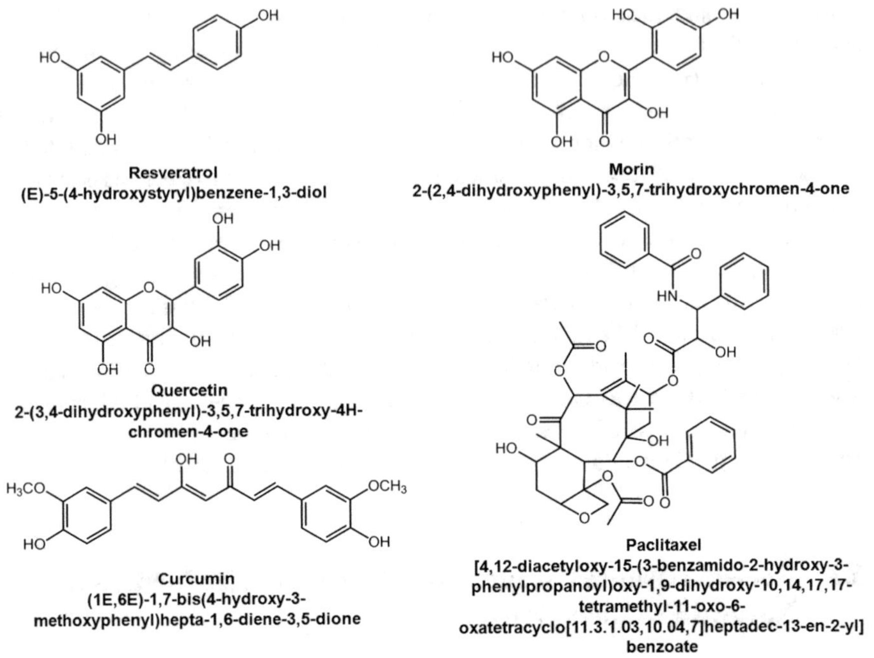

Resveratrol
(E)-5-(4-hydroxystyryl)benzene-1,3-diol

Morin
2-(2,4-dihydroxyphenyl)-3,5,7-trihydroxychromen-4-one

Quercetin
2-(3,4-dihydroxyphenyl)-3,5,7-trihydroxy-4H-chromen-4-one

Curcumin
(1E,6E)-1,7-bis(4-hydroxy-3-methoxyphenyl)hepta-1,6-diene-3,5-dione

Paclitaxel
[4,12-diacetyloxy-15-(3-benzamido-2-hydroxy-3-phenylpropanoyl)oxy-1,9-dihydroxy-10,14,17,17-tetramethyl-11-oxo-6-oxatetracyclo[11.3.1.03,10.04,7]heptadec-13-en-2-yl] benzoate

Fig. 6.4 Chemical structure of resveratrol, morin, quercetin, curcumin and paclitaxel natural products

synthesized. These curcumin silver nanoparticles were effective against wide range of microbes such as gram-positive and gram-negative bacteria and reduce the minimum effective curcumin concentration by fourfold (20 mg/L to 5 mg/L). Skin biocompatibility studies on keratinocytes (HaCaT) showed slight toxicity, whereas as an antimicrobial, curcumin silver nanoparticles showed selective toxicity towards bacteria and induces anti-inflammatory effect on human macrophages (THP-1) (Jaiswal and Mishra 2018). Abdellah et al. also studied the antibacterial effect of curcumin silver nanoparticles against *Escherichia coli* and observed improved photostability and antibacterial activity with minimal toxicity to skin cells (Abdellah et al. 2018). For the enhancement of anticancer potential of curcumin, its silver nanoformulations with average size of 17.98 nm were reported, study showed that curcumin silver nanoparticles induces cell toxicity in concentration dependent manner and had a greater cytotoxic effect on the tumor cells in comparison to plain curcumin (Garg and Garg 2018). Increased water solubility and loading efficiency was also reported with a novel in situ synthesis method in which curcumin conjugate to gold quantum cluster were developed. *In vitro* apoptotic potential in cancer cells were also supported by the *in vivo* studies in SCID cancer mouse model without showing any severe toxicity to internal organs, this showed that silver nanoparticles as an delivery vehicle for natural product could be advantageous (Khandelwal et al. 2018).

Table 6.1 Applications of metallic nanoparticles in delivering natural products

	Natural Product	Metallic nanoparticle and its modification	Purpose/Outcome	References
Inorganic metallic particles				
Silver nanoparticles				
1.	Curcumin	Silver nanoparticles	Antimicrobial and antibiofilm activity enhancement	Abdellah et al. (2018) and Jaiswal and Mishra (2018)
2.	Curcumin	Silver nanoparticles	Anticancer activity/ enhancement of anticancer activity	Garg and Garg (2018)
3.	Paclitaxel	Polyethylenimine-functionalized silver nanoparticles	HepG2 cell apoptosis	Paciotti et al. (2016)
4.	Flavonoid	Silver nanoparticles	Wound healing bio-efficacy	Sharma et al. (2019)
5.	Quercetin	Silver nanoparticles	Antibacterial and antibiofilm activities	Yu et al. (2018)
Gold nanoparticles				
6.	Curcumin	Gold- quantum clusters	Anticancer activity/ enhancement	Khandelwal et al. (2018)
7.	Paclitaxel	Gold nanoparticles	Anticancer activity	Paciotti et al. (2016)
8.	Flavonoid	Gold nanoparticles and Au-Ag bimetallic nanoparticles	Wound healing bio-efficacy	Sharma et al. (2019)
9.	Anthocyanin	PEG-gold nanoparticles	Enhanced neuroprotection	Kim et al. (2017)
10.	Rutin	Rutin conjugated gold nanoparticles	Treatment of rheumatoid arthritis	Gul et al. (2018)
11.	Vincristine sulfate	Vincristine sulfate conjugated gold nanoparticles	Enhanced antitumor efficiency	Liu et al. (2015)
12.	Betulin	Gold nanoparticles	Anti-tumor effect	Mioc et al. (2018)
13.	Quercetin	Gold nanoparticles	Anti-tumor effect, breast cancer	Balakrishnan et al. (2016) and Yilmaz et al. (2019)

(continued)

Table 6.1 (continued)

	Natural Product	Metallic nanoparticle and its modification	Purpose/Outcome	References
Inorganic non-metallic particles				
Iron oxide nanoparticles				
14.	Trans-resveratrol	Superparamagnetic Iron oxide nanoparticles	Glioma treatment/ potential cytotoxic effect	Sallem et al. (2019)
15.	Curcumin	Magnetic nanoparticles	Breast cancer therapeutics and imaging applications	Yallapu et al. (2012)
16.	Curcumin	Magnetic nanoparticles	Pancreatic cancer	Yallapu et al. (2013)
17.	Curcumin	Folate-modified and curcumin-loaded dendritic magnetite nanocarriers	Thermo-chemotherapy of cancer cells	Montazerabadi et al. (2019)
18.	Quercetin	Superparamagnetic iron oxide nanoparticles	Enhance bioavailability of quercetin, and ameliorates diabetes-related memory impairment.	Ebrahimpour et al. (2018) and Enteshari Najafabadi et al. (2018)
19.	Paclitaxel	Iron oxide nanoparticles	Breast Cancer therapy	Lugert et al. (2019)
20.	Etoposide	Magnetic polymeric microparticles	Anticancer activity	Sundar et al. (2014)
Silica nanoparticles				
21.	Curcumin	Amino-functionalized silica nanoparticles	Anticancer activity	de Oliveira et al. (2016)
22.	Carvacrol	Hybrid-silica nanoparticles	Antibacterial activity	Sokolik and Lellouche (2018)
23.	Eugenol	Mesoporous silica nanoparticles	Antimicrobial activity	Melendez-Rodriguez et al. (2019)
24.	Thymol	Biogenic silica	Biopesticide	Mattos et al. (2018)
25.	Quercetin and doxorubicin	Mesoporous silica nanoparticles	Gastric carcinoma chemotherapy	Fang et al. (2018)
26.	Morin (2′,3,4′,5,7-pentahydroxyflavone)	Mesoporous silica	Antioxidant properties	Arriagada et al. (2016)
27.	Flavonoid	Titania-functionalized silica nanoparticles	Anti-oxidant and anti-inflammatory properties	Sokolik and Lellouche (2018)
28.	Paclitaxel	Mesoporous silica nanoparticles	Anticancer activity against human lung cancer cell line A549	Wang et al. (2017)

6.3.2 *Taxol or Paclitaxel*

Taxol or Paclitaxel is first discovered by Mrs. Monroe E. Wall and Mansukh C. Wani. It is a white crystalline powder which melts at ~210 °C. Since its first isolation from the bark of Pacific yew (*Taxus brevifolia*), it still remains the drug of choice for cancer treatments. Paclitaxel was the first chemotherapeutic agent which was obtained from natural products and approved by U.S. Food and Drug Administration (FDA) for clinical use. As an potent anticancer agent, it is used for variety of cancers such as leukemia and number of solid tumors including breast, lung, ovary, gastric, brain, and prostate cancer (Jordan and Wilson 2004; Gupta et al. 2010). ***Paclitaxel*** is known for cell cycle arrest through tubulin-microtubulin equilibrium disruption and induces cancer cell death (Horinouchi et al. 2014). However paclitaxel suffers with low water solubility (~0.4 μg/ml) due to its hydrophobic nature, low permeability and serious adverse effects. Therefore, in the ***paclitaxel*** trademark formulation "Taxol" organic solvents such as polyoxyethylated castor oil (Cremophor EL) and dehydrated ethanol (50/50, v/v) have been used as solvent system for the preparation of commercial formulation, because of this it has shown its own side effects and nonlinear pharmacokinetics (Sparreboom et al. 1996). Albumin nanoparticles bound paclitaxel formulation (Abraxane) with a particle size of 130 nm is approved clinically about a decade ago for the treatment of metastatic breast cancer, and further attempts have been made for its solubility improvements along with pharmacokinetic and pharmacodynamics studies. A novel nanosystem based on polyethylenimine (PEI) decorated silver nanoparticles encapsulating ***paclitaxel*** have been synthesized and its cytotoxic effect was assessed against hepatocarcinoma (HepG2 cells), which is the third leading cause of cancer related deaths around the world. The particles were 2–3 nm in size having zeta potential values of 23 mV, which induced HepG2 cell apoptosis, triggered intracellular reactive oxygen species; thus can be used in the chemotherapy of cancer (Li et al. 2016). Gibson et al. has covalently attached ***paclitaxel*** to phenol-terminated gold nanoparticles of 2 nm size with a hexaethylene glycol linker (Gibson et al. 2007). Oh et al. developed a composite system based on gold/chitosan/pluronic nanoparticles for ***paclitaxel*** loading. For controlled release of ***paclitaxel***, gold nanoparticle synthesis with citrate reduction method were also described (Oh et al. 2008). Heo and co-workers described ***paclitaxel*** loaded gold nanoparticles synthesis with desired surface functionalization using PEG, biotin and rhodamine B linked beta-cyclodextrin, where ***paclitaxel*** formed an inclusion complex with beta-cyclodextrin. Physical characteristics of the particles showed that the gold nanoparticles were almost spherical and were in the size range of 20–40 nm. These particles had shown cytotoxicity towards HeLa, A549 and MG63 cancer cell as compared to NIH3T3 normal cells (Heo et al. 2012). Hua et al. synthesized poly(aniline-co-sodium N-(1-one-butyric acid) aniline coated Fe_3O_4 magnetic nanoparticles for paclitaxel loading in which drug was immobilized onto the surface of particles. These particles were more effective in killing PC3 and CWR22R prostate cancer cells compared to free ***paclitaxel*** and improved the aqueous solubility of ***paclitaxel*** to 312 μg/ml from

0.4 µg/ml (Hua et al. 2010). Hwu J.R. et al. have reported the conjugated ***paclitaxel*** containing gold nanoparticles and Fe_3O_4 magnetic nanoparticles tailored with maleimidyl 3-succinimidopropionate ligands for anti-cancer activity (Hwu et al. 2009). Similarly, for lung cancer delivery of ***paclitaxel***, mesoporous silica nanoparticles with a core-shell structure were reported. *In vitro* and *in vivo* studies report that ***paclitaxel*** delivery via particles not only led to an increase in dissolution rate but also improved the lung absorption of ***paclitaxel***. Nanoparticulate delivery of ***paclitaxel*** showed 2.68-fold higher concentration in the area under concentration time curve in a pulmonary inhalation study in rabbit as compared to free ***paclitaxel*** and were effective in killing human lung cancer cells (A549) (Wang et al. 2017).

6.3.3 Trans-Resveratrol

Several *in vivo* and *in vitro* reports in animal models have shown the beneficial effects of ***resveratrol***, a 3,4′,5-trihydroxy-trans-stilbene which naturally occur as polyphenolic compound in red wine, peanuts, grapes, soyabean, pomegranates and barriers (Wallerath et al. 2002; Wenzel and Somoza 2005). Studies in animals have shown that resveratrol, acts as antioxidant (Stivala et al. 2001), anti-microbial (Chalal et al. 2014), anti-inflammatory (Tili et al. 2010), estrogenomimetic and chemopreventive. Because of its antioxidant properties, it has demonstrated its therapeutic potential against skin, breast, lung, prostate, and pancreatic cancers (Aggarwal et al. 2004; Athar et al. 2007). Neuroprotective role of ***resveratrol*** has also been observed in Alzheimer's, Parkinson's disease and stroke (Ignatowicz and Baer-Dubowska 2001; Singh et al. 2019). However, poor systemic bioavailability due to its low water solubility along with very short half-life (30–45 min) limits clinical efficacy of ***resveratrol***. Various ***resveratrol*** nanoformulations have been developed to overcome these limitations, such as, gold nanoparticles and silver nanoparticles using eco-friendly synthesis methods in order to enhance their bioavailability and antibacterial efficacy. Using ***resveratrol***, which served as reducing agent, spherical-shaped nanoparticles having size range of 8.32–21.84 nm, exhibiting surface plasmon resonance at 547 nm and 412–417 nm for gold and silver respectively were reported. Antibacterial activity was observed against both gram-positive and gram-negative bacteria, the ***resveratrol*** conjugated nanocarriers exhibited enhanced antibacterial activity compared to resveratrol alone. ***Resveratrol gold nanoparticles*** demonstrated the superior antibacterial activity against *Streptococcus pneumonia* (Park et al. 2016). For the treatment of brain tumor, with the help of ***resveratrol*** derivative, Sallem et al. has reported the synthesis of surface modified superparamagnetic iron oxide nanoparticles. In the study, ***resveratrol*** derivative molecule (4′-hydroxy-4-(3-aminopropoxy) trans-stilbene was chemically synthesized and then this was grafted onto superparamagnetic iron oxide nanoparticle surface using an organosilane coupling agent and obtained particles were in the size range of 9.0 nm as reported by transmission electron microscopy. *In vitro* biological efficacy assessment showed that these particles were able to damage the plasma membrane, and limited the proliferation of cancerous cells (C6 rat glioma cells) as observed in clonogenic test (Sallem et al. 2019). Popova et al. has synthesized Trans-resveratrol encapsulated nanoporous silica particles of different structures. Nanosized BEA zeolite, nanoporous MCM-41and KIL-2 silica were used as carrier material. High loading of ***resveratrol*** were achieved in nanoparticles using solid state dry mixing and with ethanol solution, both methods were good in high ***resveratrol*** loading however, solid state dry mixing were more effective in stabilizing trans-resveratrol (Popova et al. 2014). In another study colloidal mesoporous silica nanoparticles of an average size of 283 nm encapsulating resveratrol were prepared. ***Resveratrol*** encapsulation led to solubility enhancement by ~95% and increased in vitro release as compared to pure resveratrol. The in vitro efficacy studies performed on LS147T and HT-29 colon cancer cell lines with ***resveratrol*** loaded mesoporous silica nanoparticles induces cancer cell apoptosis through PARP and cIAP1 pathways. These nanoparticle formulations also exhibited anti-inflammatory potential in lipopolysaccharide-induced NF-κB activation in RAW264.7 cells (Summerlin et al. 2016).

6.3.4 Quercetin (3, 5, 7, 3′, 4′-Pentahydroxyflavone)

Citrus fruits, green leafy vegetables as well as many seeds such as nuts, buckwheat, flowers, barks, broccoli, olive oil, apples, onions, green tea, red grapes, red wine, dark cherries and berries such as blueberries and cranberries are highly abundant with dietary bioflavonoid Quercetin. It is known for its beneficial effects such as antioxidant, antitumor and antibacterial activities, cardioprotection, anticancer, antiulcer, anti-allergic, anti-inflammatory, antiviral, and antiproliferative (D'Andrea 2015; Anand David et al. 2016). Verma et al. synthesized the poly(lactic-co-glycolic acid) coated Fe_3O_4 nanoparticles with quercetin for aerosol delivery by nebulization. Anticancer activity of quercetin was improved when formulated as magnetic core-shell nanoparticles as observed against A549 lung cancer cells. Quercetin magnetic nanoparticles enhance its water solubility and bioavailability and therefore improve the anticancer activity of the quercetin natural compound. Intranasal administration of quercetin loaded magnetic nanoparticles *in vivo* in mice reduced the viability of A549 cells significantly. In addition, development of quercetin as nanoparticles helps in enhanced solubility and dispersion along with stability against oxidation and improved biocompatibility (Verma et al. 2013). Quercetin loaded silver nanoparticles were synthesized for high antibacterial and anti-biofilm activities for the therapeutics of bovine mastitis caused by *E. coli*. The composite material developed by combining silver nanoparticles and quercetin, exhibited super antibacterial and anti-biofilm properties against multi-drug resistant *E. coli* strain, compared to free silver nanoparticles and quercetin alone (Yu et al. 2018). Gold nanoparticles conjugated with quercetin were synthesized to target breast cancer cells with specific and selective anticancer activities. These particles inhibited epithelial-mesenchymal transition and decreased angiogenesis. In Sprague-Dawely rats, quercetin conjugated gold nanoparticles inhibited breast tumor growth in 7, 12-dimethylbenz(a)anthracene (DMBA)-induced mammary carcinoma as compared to free quercetin (Balakrishnan et al. 2016). Yilmaz M et al. encapsulated quercetin in a macrocycle, p-sulfonatocalix[4]arene, which led to an increase in aqueous solubility to 62 thousand-fold with a 2.9-fold enhancement in its cytotoxicity with respect to free quercetin in SW-620 colon cancer cells. A nanohybrid system consisting of nanoparticle gold core decorated with calixarene hosts were synthesized for pH responsive quercetin release. The quercetin loaded nanohybrid significantly enhanced the colon cancer cells cytotoxicity (>50-fold). Also quercetin loaded nanoparticles showed a reduction in tumor volume in *In vivo* studies in a mouse 4T1 tumor model with no apparent particles related toxicity (Yilmaz et al. 2019). Quercetin was also conjugated to silica nanoparticles where these nanoparticles inhibit MCF-7 breast cancer cell growth through enhanced cancer cell apoptosis and cell cycle arrest (Aghapour et al. 2018). Silica/quercetin sol-gel hybrid was developed by Catauro M et al. as antioxidant dental implant material. This material was able to encapsulate high amount of quercetin with notable anti-oxidant properties (Catauro et al. 2015).

6.3.5 Flavonoids

Wound healing activates were reported using the aqueous alcoholic extract of the seeds of *Madhuca longifolia* plant. Therefore, *Madhuca longifolia* flavonoid-loaded gold, silver and gold-silver dual metallic nanoparticle formulations were reported with enhanced wound healing activities (Sharma et al. 2019). Flavonoid dihydromyricetin-conjugated silver nanoparticles for the treatment of fungal pathogenic infections caused in immunocompromised patients were also reported. The antifungal activities of these nanoparticles against *Aspergillus fumigates, Aspergillus niger, Paecilomyces formosus, Candida albicans* and *Candida parapsilosis*, as elucidated by zone of inhibition assay, were promising with reduced minimal inhibitory concentration as compared to free flavonoid which highlight the advantage of nanotechnology in natural product delivery. The results concluded that dihydromyricetin-conjugated silver nanoparticles can be used as potential alternative to commercially available antifungal agents (Ameen et al. 2018). Similarly, Morin (2′,3,4′,5,7-pentahydroxyflavone), a polyphenolic compound present in plants and vegetables (Gopal 2013), also known for several therapeutic activities such as anticancer (Sivaramakrishnan and Devaraj 2010), anti-inflammatory (Iglesias et al. 2005) and cardioprotective effects (Wu et al. 1995; Prahalathan et al. 2012; Govindasamy et al. 2014). Moreover, morin has shown skin protective effects against harmful radiations like UV-B radiation, therefore its topical formulation has been attempted for therapeutics (Lee et al. 2014; Shetty et al. 2015). However, degradation of morin in the presence of sunlight, oxygen and pH, limited its use in pharmaceutical and cosmetic formulations (Parisi et al. 2014). Therefore, amino propyl modified mesoporous silica nanoparticles loaded with morin were synthesized. The particles were ~150 nm in size and demonstrated maximum adsorption capacity for morin (20 mg g^{-1}). The amino propyl modified mesoporous silica nanoparticles also exhibited a synergistic effect i.e. antioxidant property against hydroxyl radical (Arriagada et al. 2016).

6.4 Other Natural Products Delivered Using Inorganic Particles

Camptothecin is a highly cytotoxic natural alkaloid, which acts against a broad spectrum of tumors by inhibiting the nuclear enzyme topoisomerase, but because of its poor water solubility and chemical instability its therapeutic use is limited. Therefore attempts have been made to develop camptothecin metallic nanoparticles. In such attempts, Castillo et al. has developed PEG coated iron oxide nanoparticle for camptothecin delivery to H460 lung cancer cell line. Camptothecin loaded nanoparticles exhibited remarkable pro-apoptotic activity with an increase in therapeutic efficacy (Castillo et al. 2014). Essential oils and essential oil derived compounds are usually classified as Generally Recognized As Safe (GRAS) by U.S. Food and Drug Administration and serve as biocide or antimicrobial. Carvacrol which is also a major ingredient of oregano and thyme essential oils, also bear limitations such as low aqueous solubility, easy phenol oxidation, heat/light inactivation, distinct odor, which limit its use for therapeutic applications. Therefore, hybrid silica nanoparticles with carvacrol attachment via covalent bond were prepared with great antibacterial effect against *Escherichia coli* (*E. coli*) (Sokolik and Lellouche 2018). Similarly, volatile oils such as eugenol, thymol and vanillin formulations were also reported where these were grafted onto the surface of three silica supports (fumed silica, amorphous silica and MCM-41) via aldehyde derivatization. Incorporation of essential oil into silica nanoparticles greatly enhanced their antimicrobial activity against *Listeria innocua* and *E. coli* compared to their free counterparts (Ruiz-Rico et al. 2017). Rutin, a another flavonol, abundantly found in plants, such as passion flower, buckwheat, tea, and apple. The use of rutin for various medical applications has been proposed because of its antioxidant, cytoprotective, vasoprotective, anticarcinogenic, neuroprotective and cardioprotective activities (Ganeshpurkar and Saluja 2017). Rutin conjugated gold nanoparticles were fabricated and their efficacy was tested in collagen induced arthritis in rats. It was found that these particles reduced the inflammatory response by down regulating nuclear factor-κB (NF-κB) and inducible nitric oxide synthase expression levels (Gul et al. 2018).

Another natural poly phenolic compound, Anthocyanins found in fruits, grains, and flowers, have shown antioxidant, anti-inflammatory, and anti-apoptosis properties in different preclinical studies (Ghosh and Konishi 2007; Zafra-Stone et al. 2007) hence anthocyanin-loaded PEG-gold nanoparticles were prepared and tested in $A\beta_{1-42}$-injected mouse models of Alzheimer's disease. These particles demonstrated reduction of neuroinflammation with enhanced neuroapoptotic markers suggesting its potential for the treatment of neurodegenerative diseases (Kim et al. 2017).

6.5 Conclusion

Natural product delivery via inorganic nanoparticles showed enhancement of efficacy and ameliorated problem of solubility and bioavailability. Natural products have a wide spectrum of activities and inorganic particles itself possesses antibacterial, cancer cell killing and anti-inflammatory properties. Therefore use of inorganic nanoparticles, for natural product delivery or active ingredient of natural product origin, will not only helpful in overcoming the several limitations associated with natural products but also enhance their therapeutic potential due to several associated benefits. A combination of natural product along with inorganic nanoparticles were attempted and elaborated in this chapter. These facts and reports implies the possibility of natural product-loaded inorganic nanoparticles as a therapeutic agent for the treatment of several diseases such as cancer, infections, rheumatoid arthritis and have antibacterial and antibiofilm activities. However, toxicity is a concern; therefore selection of specific delivery system or appropriate combination of natural product and formulation is very crucial desired purposes.

References

Abdellah AM, Sliem MA et al (2018) Green synthesis and biological activity of silver-curcumin nanoconjugates. Future Med Chem 10(22):2577–2588. https://doi.org/10.4155/fmc-2018-0152

Aggarwal BB, Bhardwaj A et al (2004) Role of resveratrol in prevention and therapy of cancer: preclinical and clinical studies. Anticancer Res 24(5A):2783–2840

Aghapour F, Moghadamnia AA et al (2018) Quercetin conjugated with silica nanoparticles inhibits tumor growth in MCF-7 breast cancer cell lines. Biochem Biophys Res Commun 500(4):860–865. https://doi.org/10.1016/j.bbrc.2018.04.174

Ahmed S, Saifullah et al (2016) Green synthesis of silver nanoparticles using Azadirachta indica aqueous leaf extract. J Radiat Res Appl Sci 9(1):1–7

Ali A, Zafar H et al (2016) Synthesis, characterization, applications, and challenges of iron oxide nanoparticles. Nanotechnol Sci Appl 9:49–67. https://doi.org/10.2147/NSA.S99986

Ameen F, AlYahya SA et al (2018) Flavonoid dihydromyricetin-mediated silver nanoparticles as potential nanomedicine for biomedical treatment of infections caused by opportunistic fungal pathogens. Res Chem Intermed 44(9):5063–5073

Anand David AV, Arulmoli R et al (2016) Overviews of biological importance of quercetin: a bioactive flavonoid. Pharmacogn Rev 10(20):84–89. https://doi.org/10.4103/0973-7847.194044

Ankamwar B, Chaudhary M et al (2005) Gold nanotriangles biologically synthesized using tamarind leaf extract and potential application in vapor sensing. Synth React Inorg Met Org Nano Met Chem 35(1):19–26. https://doi.org/10.1081/sim-200047527

Antony JJ, Sivalingam P et al (2015) Toxicological effects of silver nanoparticles. Environ Toxicol Pharmacol 40(3):729–732. https://doi.org/10.1016/j.etap.2015.09.003

Anuradha J, Abbasi T et al (2010) Green' synthesis of gold nanoparticles with aqueous extracts of neem (Azadirachta indica). Res J Biotechnol 5(1):75–79
Aromal SA, Philip D (2012) Benincasa hispida seed mediated green synthesis of gold nanoparticles and its optical nonlinearity. Physica E 44(7–8):1329–1334
Arriagada F, Correa O et al (2016) Morin flavonoid adsorbed on mesoporous silica, a novel antioxidant nanomaterial. PLoS One 11(11):e0164507. https://doi.org/10.1371/journal.pone.0164507
Arts JH, Muijser H et al (2007) Five-day inhalation toxicity study of three types of synthetic amorphous silicas in Wistar rats and post-exposure evaluations for up to 3 months. Food Chem Toxicol 45(10):1856–1867
Athar M, Back JH et al (2007) Resveratrol: a review of preclinical studies for human cancer prevention. Toxicol Appl Pharmacol 224(3):274–283. https://doi.org/10.1016/j.taap.2006.12.025
Attallah OA, Girgis E et al (2016) Synthesis of non-aggregated nicotinic acid coated magnetite nanorods via hydrothermal technique. J Magn Magn Mater 399:58–63
Baeza A, Colilla M et al (2015) Advances in mesoporous silica nanoparticles for targeted stimuli-responsive drug delivery. Expert Opin Drug Deliv 12(2):319–337. https://doi.org/10.1517/17425247.2014.953051
Bagherzade G, Tavakoli MM et al (2017) Green synthesis of silver nanoparticles using aqueous extract of saffron (Crocus sativus L.) wastages and its antibacterial activity against six bacteria. Asian Pac J Trop Biomed 7(3):227–233
Bagwe RP, Hilliard LR et al (2006) Surface modification of silica nanoparticles to reduce aggregation and nonspecific binding. Langmuir 22(9):4357–4362. https://doi.org/10.1021/la052797j
Bahadur A, Saeed A et al (2017) Eco-friendly synthesis of magnetite (Fe3O4) nanoparticles with tunable size: dielectric, magnetic, thermal and optical studies. Mater Chem Phys 198:229–235
Balakrishnan S, Bhat FA et al (2016) Gold nanoparticle-conjugated quercetin inhibits epithelial-mesenchymal transition, angiogenesis and invasiveness via EGFR/VEGFR-2-mediated pathway in breast cancer. Cell Prolif 49(6):678–697. https://doi.org/10.1111/cpr.12296
Banoee M, Mokhtari N et al (2010) The green synthesis of gold nanoparticles using the ethanol extract of black tea and its tannin free fraction. Iran J Mater Sci Eng 7(1):48–53
Bastus NG, Comenge J et al (2011) Kinetically controlled seeded growth synthesis of citrate-stabilized gold nanoparticles of up to 200 nm: size focusing versus Ostwald ripening. Langmuir 27(17):11098–11105. https://doi.org/10.1021/la201938u
Bhattacharya S, Srivastava A (2003) Synthesis of gold nanoparticles stabilised by metal-chelator and the controlled formation of close-packed aggregates by them. J Chem Sci 115(5–6):613–619
Bouwmeester H, Poortman J et al (2011) Characterization of translocation of silver nanoparticles and effects on whole-genome gene expression using an in vitro intestinal epithelium coculture model. ACS Nano 5(5):4091–4103. https://doi.org/10.1021/nn2007145
Brown PK, Qureshi AT et al (2013) Silver nanoscale antisense drug delivery system for photoactivated gene silencing. ACS Nano 7(4):2948–2959. https://doi.org/10.1021/nn304868y
Brunner TJ, Stark WJ, et al (2009) Nanoscale bioactive silicate glasses in biomedical applications Preface XV List of Contributors XIX
Buazar F, Baghlani-Nejazd MH et al (2016) Facile one-pot phytosynthesis of magnetic nanoparticles using potato extract and their catalytic activity. Starch-Stärke 68(7–8):796–804
Bulte JW (2009) In vivo MRI cell tracking: clinical studies. Am J Roentgenol 193(2):314–325
Cai Q, Luo Z-S et al (2001) Dilute solution routes to various controllable morphologies of MCM-41 silica with a basic medium. Chem Mater 13(2):258–263
Cao S, Zhu H (2014) Frontiers in biomaterials: the design, synthetic strategies and biocompatibility of polymer scaffolds for biomedical application. Bentham Science Publishers, Oak Park
Castillo PM, de la Mata M et al (2014) PEGylated versus non-PEGylated magnetic nanoparticles as camptothecin delivery system. Beilstein J Nanotechnol 5:1312–1319. https://doi.org/10.3762/bjnano.5.144
Catauro M, Papale F et al (2015) Silica/quercetin sol-gel hybrids as antioxidant dental implant materials. Sci Technol Adv Mater 16(3):035001. https://doi.org/10.1088/1468-6996/16/3/035001

Chalal M, Klinguer A et al (2014) Antimicrobial activity of resveratrol analogues. Molecules 19(6):7679–7688. https://doi.org/10.3390/molecules19067679
Chen YS, Hung YC et al (2009) Assessment of the in vivo toxicity of gold nanoparticles. Nanoscale Res Lett 4(8):858–864. https://doi.org/10.1007/s11671-009-9334-6
Chen L, Liu J et al (2018) The toxicity of silica nanoparticles to the immune system. Nanomedicine (Lond) 13(15):1939–1962
Cho WS, Cho M et al (2009a) Acute toxicity and pharmacokinetics of 13 nm-sized PEG-coated gold nanoparticles. Toxicol Appl Pharmacol 236(1):16–24. https://doi.org/10.1016/j.taap.2008.12.023
Cho WS, Kim S et al (2009b) Comparison of gene expression profiles in mice liver following intravenous injection of 4 and 100 nm-sized PEG-coated gold nanoparticles. Toxicol Lett 191(1):96–102. https://doi.org/10.1016/j.toxlet.2009.08.010
Connor EE, Mwamuka J et al (2005) Gold nanoparticles are taken up by human cells but do not cause acute cytotoxicity. Small 1(3):325–327. https://doi.org/10.1002/smll.200400093
Daisy P, Saipriya K (2012) Biochemical analysis of Cassia fistula aqueous extract and phytochemically synthesized gold nanoparticles as hypoglycemic treatment for diabetes mellitus. Int J Nanomedicine 7:1189–1202. https://doi.org/10.2147/IJN.S26650
D'Andrea G (2015) Quercetin: a flavonol with multifaceted therapeutic applications? Fitoterapia 106:256–271. https://doi.org/10.1016/j.fitote.2015.09.018
Das VL, Thomas R et al (2014) Extracellular synthesis of silver nanoparticles by the Bacillus strain CS 11 isolated from industrialized area. 3 Biotech 4(2):121–126
De Jong WH, Hagens WI et al (2008) Particle size-dependent organ distribution of gold nanoparticles after intravenous administration. Biomaterials 29(12):1912–1919. https://doi.org/10.1016/j.biomaterials.2007.12.037
de Oliveira LF, Bouchmella K et al (2016) Functionalized silica nanoparticles as an alternative platform for targeted drug-delivery of water insoluble drugs. Langmuir 32(13):3217–3225. https://doi.org/10.1021/acs.langmuir.6b00214
Demir A, Topkaya R et al (2013) Green synthesis of superparamagnetic Fe3O4 nanoparticles with maltose: its magnetic investigation. Polyhedron 65:282–287
Ebrahimpour S, Esmaeili A et al (2018) Effect of quercetin-conjugated superparamagnetic iron oxide nanoparticles on diabetes-induced learning and memory impairment in rats. Int J Nanomedicine 13:6311–6324. https://doi.org/10.2147/IJN.S177871
Edison TJI, Sethuraman M (2012) Instant green synthesis of silver nanoparticles using Terminalia chebula fruit extract and evaluation of their catalytic activity on reduction of methylene blue. Process Biochem 47(9):1351–1357
El Mahdy MM, Eldin TA et al (2015) Evaluation of hepatotoxic and genotoxic potential of silver nanoparticles in albino rats. Exp Toxicol Pathol 67(1):21–29. https://doi.org/10.1016/j.etp.2014.09.005
El-Kassas HY, Aly-Eldeen MA et al (2016) Green synthesis of iron oxide (Fe 3 O 4) nanoparticles using two selected brown seaweeds: characterization and application for lead bioremediation. Acta Oceanol Sin 35(8):89–98
Enteshari Najafabadi R, Kazemipour N et al (2018) Using superparamagnetic iron oxide nanoparticles to enhance bioavailability of quercetin in the intact rat brain. BMC Pharmacol Toxicol 19(1):59. https://doi.org/10.1186/s40360-018-0249-7
Fang J, Zhang S et al (2018) Quercetin and doxorubicin co-delivery using mesoporous silica nanoparticles enhance the efficacy of gastric carcinoma chemotherapy. Int J Nanomedicine 13:5113–5126. https://doi.org/10.2147/IJN.S170862
Feng Q, Liu Y et al (2018) Uptake, distribution, clearance, and toxicity of iron oxide nanoparticles with different sizes and coatings. Sci Rep 8(1):2082. https://doi.org/10.1038/s41598-018-19628-z
Fernandez-Fernandez A, Manchanda R et al (2011) Theranostic applications of nanomaterials in cancer: drug delivery, image-guided therapy, and multifunctional platforms. Appl Biochem Biotechnol 165(7–8):1628–1651. https://doi.org/10.1007/s12010-011-9383-z

Franci G, Falanga A et al (2015) Silver nanoparticles as potential antibacterial agents. Molecules 20(5):8856–8874. https://doi.org/10.3390/molecules20058856

Ganeshpurkar A, Saluja AK (2017) The pharmacological potential of Rutin. Saudi Pharm J 25(2):149–164. https://doi.org/10.1016/j.jsps.2016.04.025

Gao S, Shi Y et al (2008) Biopolymer-assisted green synthesis of iron oxide nanoparticles and their magnetic properties. J Phys Chem C 112(28):10398–10401

Garg S, Garg A (2018) Encapsulation of curcumin in silver nanoparticle for enhancement of anticancer drug delivery. Int J Pharm Sci Res 9(3):1160–1166

Ge F, Li M-M et al (2012) Effective removal of heavy metal ions Cd2+, Zn2+, Pb2+, Cu2+ from aqueous solution by polymer-modified magnetic nanoparticles. J Hazard Mater 211:366–372

Ghosh D, Konishi T (2007) Anthocyanins and anthocyanin-rich extracts: role in diabetes and eye function. Asia Pac J Clin Nutr 16(2):200–208

Gibson JD, Khanal BP et al (2007) Paclitaxel-functionalized gold nanoparticles. J Am Chem Soc 129(37):11653–11661. https://doi.org/10.1021/ja075181k

Glavee GN, Klabunde KJ et al (1995) Chemistry of borohydride reduction of iron (II) and iron (III) ions in aqueous and nonaqueous media. Formation of nanoscale Fe, FeB, and Fe2B powders. Inorg Chem 34(1):28–35

Go M-R, Bae S-H et al (2017) Interactions between food additive silica nanoparticles and food matrices. Front Microbiol 8:1013

Gokduman K, Bestepe F et al (2018) Dose-, treatment- and time-dependent toxicity of superparamagnetic iron oxide nanoparticles on primary rat hepatocytes. Nanomedicine (Lond) 13(11):1267–1284. https://doi.org/10.2217/nnm-2017-0387

Gonçalves M (2018) Sol-gel silica nanoparticles in medicine: a natural choice. Design, synthesis and products. Molecules 23(8):2021

Gopal JV (2013) Morin hydrate: botanical origin, pharmacological activity and its applications: a mini-review. Pharm J 5(3):123–126

Govindasamy C, Alnumair KS et al (2014) GW25-e5392 Morin, a flavonoid, on lipid peroxidation and antioxidant status in experimental myocardial ischemic rats. J Am Coll Cardiol 64(16 Supplement):C56

Gul A, Kunwar B et al (2018) Rutin and rutin-conjugated gold nanoparticles ameliorate collagen-induced arthritis in rats through inhibition of NF-kappaB and iNOS activation. Int Immunopharmacol 59:310–317. https://doi.org/10.1016/j.intimp.2018.04.017

Guo L, Huang Q et al (2001) Iron nanoparticles: synthesis and applications in surface enhanced Raman scattering and electrocatalysis. Phys Chem Chem Phys 3(9):1661–1665. https://doi.org/10.1039/b009951l

Gupta SC, Kim JH et al (2010) Regulation of survival, proliferation, invasion, angiogenesis, and metastasis of tumor cells through modulation of inflammatory pathways by nutraceuticals. Cancer Metastasis Rev 29(3):405–434. https://doi.org/10.1007/s10555-010-9235-2

Hadrup N, Lam HR (2014) Oral toxicity of silver ions, silver nanoparticles and colloidal silver–a review. Regul Toxicol Pharmacol 68(1):1–7. https://doi.org/10.1016/j.yrtph.2013.11.002

Hench LL, West JK (1990) The sol-gel process. Chem Rev 90(1):33–72

Heo DN, Yang DH et al (2012) Gold nanoparticles surface-functionalized with paclitaxel drug and biotin receptor as theranostic agents for cancer therapy. Biomaterials 33(3):856–866. https://doi.org/10.1016/j.biomaterials.2011.09.064

Horinouchi H, Yamamoto N et al (2014) A phase 1 study of linifanib in combination with carboplatin/paclitaxel as first-line treatment of Japanese patients with advanced or metastatic non-small cell lung cancer (NSCLC). Cancer Chemother Pharmacol 74(1):37–43. https://doi.org/10.1007/s00280-014-2478-9

Horst MF, Coral DF et al (2017) Hybrid nanomaterials based on gum Arabic and magnetite for hyperthermia treatments. Mater Sci Eng C Mater Biol Appl 74:443–450. https://doi.org/10.1016/j.msec.2016.12.035

Hua MY, Yang HW et al (2010) Magnetic-nanoparticle-modified paclitaxel for targeted therapy for prostate cancer. Biomaterials 31(28):7355–7363. https://doi.org/10.1016/j.biomaterials.2010.05.061

Huang DM, Chung TH et al (2008) Internalization of mesoporous silica nanoparticles induces transient but not sufficient osteogenic signals in human mesenchymal stem cells. Toxicol Appl Pharmacol 231(2):208–215. https://doi.org/10.1016/j.taap.2008.04.009

Hudson SP, Padera RF et al (2008) The biocompatibility of mesoporous silicates. Biomaterials 29(30):4045–4055. https://doi.org/10.1016/j.biomaterials.2008.07.007

Hwu JR, Lin YS et al (2009) Targeted paclitaxel by conjugation to iron oxide and gold nanoparticles. J Am Chem Soc 131(1):66–68. https://doi.org/10.1021/ja804947u

Iglesias CV, Aparicio R et al (2005) Effects of morin on snake venom phospholipase A2 (PLA2). Toxicon 46(7):751–758. https://doi.org/10.1016/j.toxicon.2005.07.017

Ignatowicz E, Baer-Dubowska W (2001) Resveratrol, a natural chemopreventive agent against degenerative diseases. Pol J Pharmacol 53(6):557–569

Ikari K, Suzuki K et al (2006) Structural control of mesoporous silica nanoparticles in a binary surfactant system. Langmuir 22(2):802–806

Iravani S, Korbekandi H et al (2014) Synthesis of silver nanoparticles: chemical, physical and biological methods. Res Pharm Sci 9(6):385

Jaiswal S, Mishra P (2018) Antimicrobial and antibiofilm activity of curcumin-silver nanoparticles with improved stability and selective toxicity to bacteria over mammalian cells. Med Microbiol Immunol 207(1):39–53. https://doi.org/10.1007/s00430-017-0525-y

Jordan MA, Wilson L (2004) Microtubules as a target for anticancer drugs. Nat Rev Cancer 4(4):253–265. https://doi.org/10.1038/nrc1317

Jurkić LM, Cepanec I et al (2013) Biological and therapeutic effects of ortho-silicic acid and some ortho-silicic acid-releasing compounds: new perspectives for therapy. Nutr Metab 10(1):2

Kajani AA, Zarkesh-Esfahani SH et al (2016) Anticancer effects of silver nanoparticles encapsulated by Taxus baccata extracts. J Mol Liq 223:549–556

Kang S, Hong SI et al (2001) Preparation and characterization of epoxy composites filled with functionalized nanosilica particles obtained via sol–gel process. Polymer 42(3):879–887

Kasaai MR (2015) Nanosized particles of silica and its derivatives for applications in various branches of food and nutrition sectors. J Nanotechnol 2015:1–6. https://doi.org/10.1155/2015/852394

Kasthuri J, Kathiravan K et al (2009a) Phyllanthin-assisted biosynthesis of silver and gold nanoparticles: a novel biological approach. J Nanopart Res 11(5):1075–1085

Kasthuri J, Veerapandian S et al (2009b) Biological synthesis of silver and gold nanoparticles using apiin as reducing agent. Colloids Surf B: Biointerfaces 68(1):55–60. https://doi.org/10.1016/j.colsurfb.2008.09.021

Kerekes L, Hakl J et al (2002) Study of magnetic relaxation in partially oxidized nanocrystalline iron. Czechoslov J Phys 52(1):A89–A92

Khan M, Khan M et al (2013) Green synthesis of silver nanoparticles mediated by Pulicaria glutinosa extract. Int J Nanomedicine 8:1507–1516. https://doi.org/10.2147/IJN.S43309

Khan M, Khan ST et al (2014) Antibacterial properties of silver nanoparticles synthesized using Pulicaria glutinosa plant extract as a green bioreductant. Int J Nanomedicine 9:3551–3565. https://doi.org/10.2147/IJN.S61983

Khandelwal P, Alam A et al (2018) Retention of anticancer activity of curcumin after conjugation with fluorescent gold quantum clusters: an in vitro and in vivo xenograft study. ACS Omega 3(5):4776–4785. https://doi.org/10.1021/acsomega.8b00113

Khataee A, Kayan B et al (2017) Ultrasound-assisted removal of Acid Red 17 using nanosized Fe3O4-loaded coffee waste hydrochar. Ultrason Sonochem 35(Pt A):72–80. https://doi.org/10.1016/j.ultsonch.2016.09.004

Kim JH, Kim JH et al (2009) Intravenously administered gold nanoparticles pass through the blood-retinal barrier depending on the particle size, and induce no retinal toxicity. Nanotechnology 20(50):505101. https://doi.org/10.1088/0957-4484/20/50/505101

Kim I-Y, Joachim E et al (2015) Toxicity of silica nanoparticles depends on size, dose, and cell type. Nanomedicine 11(6):1407–1416
Kim MJ, Rehman SU et al (2017) Enhanced neuroprotection of anthocyanin-loaded PEG-gold nanoparticles against Abeta1-42-induced neuroinflammation and neurodegeneration via the NF-KB /JNK/GSK3beta signaling pathway. Nanomedicine (Lond) 13(8):2533–2544. https://doi.org/10.1016/j.nano.2017.06.022
Kittler S, Greulich C et al (2010) Toxicity of silver nanoparticles increases during storage because of slow dissolution under release of silver ions. Chem Mater 22(16):4548–4554. https://doi.org/10.1021/cm100023p
Klabunde KJ, Stark J et al (1996) Nanocrystals as stoichiometric reagents with unique surface chemistry. J Phys Chem 100(30):12142–12153
Klejbor I, Stachowiak EK et al (2007) ORMOSIL nanoparticles as a non-viral gene delivery vector for modeling polyglutamine induced brain pathology. J Neurosci Methods 165(2):230–243. https://doi.org/10.1016/j.jneumeth.2007.06.011
Kobler J, Bein T (2008) Porous thin films of functionalized mesoporous silica nanoparticles. ACS Nano 2(11):2324–2330
Kouvaris P, Delimitis A et al (2012) Green synthesis and characterization of silver nanoparticles produced using Arbutus unedo leaf extract. Mater Lett 76:18–20
Koźlecki T, Polowczyk I et al (2016) Improved synthesis of nanosized silica in water-in-oil microemulsions. J Nanopart 2016:1–9
Kumar A, Pandey AK et al (2011) Cellular uptake and mutagenic potential of metal oxide nanoparticles in bacterial cells. Chemosphere 83(8):1124–1132. https://doi.org/10.1016/j.chemosphere.2011.01.025
Kumar A, Ma H et al (2012) Gold nanoparticles functionalized with therapeutic and targeted peptides for cancer treatment. Biomaterials 33(4):1180–1189. https://doi.org/10.1016/j.biomaterials.2011.10.058
Kumar B, Smita K et al (2016) Phytosynthesis and photocatalytic activity of magnetite (Fe3O4) nanoparticles using the Andean blackberry leaf. Mater Chem Phys 179:310–315
Kyung OY, Grabinski CM et al (2009) Toxicity of amorphous silica nanoparticles in mouse keratinocytes. J Nanopart Res 11(1):15–24
Le Ouay B, Stellacci F (2015) Antibacterial activity of silver nanoparticles: a surface science insight. Nano Today 10(3):339–354
Lee J, Shin YK et al (2014) Protective mechanism of morin against ultraviolet B-induced cellular senescence in human keratinocyte stem cells. Int J Radiat Biol 90(1):20–28. https://doi.org/10.3109/09553002.2013.835502
Li F, Vipulanandan C et al (2003) Microemulsion and solution approaches to nanoparticle iron production for degradation of trichloroethylene. Colloids Surf A Physicochem Eng Asp 223(1–3):103–112
Li S, Shen Y et al (2007) Green synthesis of silver nanoparticles using Capsicum annuum L. extract. Green Chem 9(8):852–858
Li Y, Guo M et al (2016) Polyethylenimine-functionalized silver nanoparticle-based co-delivery of paclitaxel to induce HepG2 cell apoptosis. Int J Nanomedicine 11:6693–6702. https://doi.org/10.2147/IJN.S122666
Li P, Li S et al (2017) Green synthesis of β-CD-functionalized monodispersed silver nanoparticles with ehanced catalytic activity. Colloids Surf A Physicochem Eng Asp 520:26–31
Lide DR (2004) CRC handbook of chemistry and physics. CRC press, Boca Raton
Lipka J, Semmler-Behnke M et al (2010) Biodistribution of PEG-modified gold nanoparticles following intratracheal instillation and intravenous injection. Biomaterials 31(25):6574–6581. https://doi.org/10.1016/j.biomaterials.2010.05.009
Liu S, Han MY (2010) Silica-coated metal nanoparticles. Chem Asian J 5(1):36–45
Liu Y, Guo Y et al (2011) A sustainable route for the preparation of activated carbon and silica from rice husk ash. J Hazard Mater 186(2–3):1314–1319

Liu Y, He M et al (2015) Delivery of vincristine sulfate-conjugated gold nanoparticles using liposomes: a light-responsive nanocarrier with enhanced antitumor efficiency. Int J Nanomedicine 10:3081–3095. https://doi.org/10.2147/IJN.S79550

Loeschner K, Hadrup N et al (2011) Distribution of silver in rats following 28 days of repeated oral exposure to silver nanoparticles or silver acetate. Part Fibre Toxicol 8:18. https://doi.org/10.1186/1743-8977-8-18

Lu J, Liong M et al (2010) Biocompatibility, biodistribution, and drug-delivery efficiency of mesoporous silica nanoparticles for cancer therapy in animals. Small 6(16):1794–1805. https://doi.org/10.1002/smll.201000538

Lu X, Qian J et al (2011) In vitro cytotoxicity and induction of apoptosis by silica nanoparticles in human HepG2 hepatoma cells. Int J Nanomedicine 6:1889

Luborsky F, Paine T (1960) Angular variation of the magnetic properties of elongated single-domain iron particles. J Appl Phys 31(5):S66–S68

Lugert S, Unterweger H et al (2019) Cellular effects of paclitaxel-loaded iron oxide nanoparticles on breast cancer using different 2D and 3D cell culture models. Int J Nanomedicine 14:161–180. https://doi.org/10.2147/IJN.S187886

Lührs A-K, Geurtsen W (2009) The application of silicon and silicates in dentistry: a review. In: Biosilica in evolution, morphogenesis, and nanobiotechnology. Springer, Heidelberg, pp 359–380

Lunge S, Singh S et al (2014) Magnetic iron oxide (Fe3O4) nanoparticles from tea waste for arsenic removal. J Magn Magn Mater 356:21–31

Martino A, Stoker M et al (1997) The synthesis and characterization of iron colloid catalysts in inverse micelle solutions. Appl Catal A Gen 161(1–2):235–248

Matsoukas T, Gulari E (1988) Dynamics of growth of silica particles from ammonia-catalyzed hydrolysis of tetra-ethyl-orthosilicate. J Colloid Interface Sci 124(1):252–261

Mattos BD, Tardy BL et al (2018) Controlled biocide release from hierarchically-structured biogenic silica: surface chemistry to tune release rate and responsiveness. Sci Rep 8(1):5555. https://doi.org/10.1038/s41598-018-23921-2

Melendez-Rodriguez B, Figueroa-Lopez KJ et al (2019) Electrospun antimicrobial films of poly(3-hydroxybutyrate-co-3-hydroxyvalerate) containing eugenol essential oil encapsulated in mesoporous silica nanoparticles. Nanomaterials (Basel) 9(2). https://doi.org/10.3390/nano9020227

Mioc M, Pavel IZ et al (2018) The cytotoxic effects of betulin-conjugated gold nanoparticles as stable formulations in Normal and melanoma cells. Front Pharmacol 9:429. https://doi.org/10.3389/fphar.2018.00429

Mittal AK, Bhaumik J et al (2014) Biosynthesis of silver nanoparticles: elucidation of prospective mechanism and therapeutic potential. J Colloid Interface Sci 415:39–47

Montazerabadi A, Beik J et al (2019) Folate-modified and curcumin-loaded dendritic magnetite nanocarriers for the targeted thermo-chemotherapy of cancer cells. Artif Cells Nanomed Biotechnol 47(1):330–340. https://doi.org/10.1080/21691401.2018.1557670

Mukherjee I, Mylonakis A et al (2009) Effect of nonsurfactant template content on the particle size and surface area of monodisperse mesoporous silica nanospheres. Microporous Mesoporous Mater 122(1–3):168–174

Nagajyothi P, Lee S-E et al (2012) Green synthesis of silver and gold nanoparticles using Lonicera japonica flower extract. Bull Kor Chem Soc 33(8):2609–2612

Nagaraj B, Malakar B et al (2012) Environmental benign synthesis of gold nanoparticles from the flower extracts of Plumeria alba Linn, (Frangipani) and evaluation of their biological activities. Int J Drug Dev Res 4(1):144–150

Naka Y, Komori Y et al (2010) One-pot synthesis of organo-functionalized monodisperse silica particles in W/O microemulsion and the effect of functional groups on addition into polystyrene. Colloids Surf A Physicochem Eng Asp 361(1–3):162–168

Nakkala JR, Mata R et al (2016) The antioxidant and catalytic activities of green synthesized gold nanoparticles from Piper longum fruit extract. Process Saf Environ Prot 100:288–294

Namanga J, Foba J et al (2013) Synthesis and magnetic properties of a superparamagnetic nanocomposite pectin-magnetite nanocomposite. J Nanomater 2013:87
Nanda A, Saravanan M (2009) Biosynthesis of silver nanoparticles from Staphylococcus aureus and its antimicrobial activity against MRSA and MRSE. Nanomedicine 5(4):452–456
Narayanan KB, Sakthivel N (2008) Coriander leaf mediated biosynthesis of gold nanoparticles. Mater Lett 62(30):4588–4590
Nel AE, Madler L et al (2009) Understanding biophysicochemical interactions at the nano-bio interface. Nat Mater 8(7):543–557. https://doi.org/10.1038/nmat2442
Nghiem THL, Nguyen TT et al (2012) Capping and in vivo toxicity studies of gold nanoparticles. Adv Nat Sci Nanosci Nanotechnol 3(1):015002
Niidome T, Yamagata M et al (2006) PEG-modified gold nanorods with a stealth character for in vivo applications. J Control Release 114(3):343–347. https://doi.org/10.1016/j.jconrel.2006.06.017
Niraimathee V, Subha V et al (2016) Green synthesis of iron oxide nanoparticles from Mimosa pudica root extract. Int J Environ Sustain Dev 15(3):227–240
Oh KS, Kim RS et al (2008) Gold/chitosan/pluronic composite nanoparticles for drug delivery. J Appl Polym Sci 108(5):3239–3244
Paciotti GF, Zhao J et al (2016) Synthesis and evaluation of paclitaxel-loaded gold nanoparticles for tumor-targeted drug delivery. Bioconjug Chem 27(11):2646–2657. https://doi.org/10.1021/acs.bioconjchem.6b00405
Pan Y, Neuss S et al (2007) Size-dependent cytotoxicity of gold nanoparticles. Small 3(11):1941–1949. https://doi.org/10.1002/smll.200700378
Pandey S, Oza G et al (2012) Green synthesis of highly stable gold nanoparticles using Momordica charantia as nano fabricator. Arch Appl Sci Res 4(2):1135–1141
Pang J, Na H et al (2005) Effect of ionic polymer on cetyltrimethyl ammonium bromide templated synthesis of mesoporous silica. Microporous Mesoporous Mater 86(1–3):89–95
Parida UK, Bindhani BK et al (2011) Green synthesis and characterization of gold nanoparticles using onion (Allium cepa) extract. World J Nano Sci Eng 1(04):93
Parisi OI, Puoci F et al (2014) Polyphenols and their formulations: different strategies to overcome the drawbacks associated with their poor stability and bioavailability. In: Polyphenols in human health and disease. Elsevier, Burlington, pp 29–45
Park S, Cha SH et al (2016) Antibacterial nanocarriers of resveratrol with gold and silver nanoparticles. Mater Sci Eng C Mater Biol Appl 58:1160–1169. https://doi.org/10.1016/j.msec.2015.09.068
Patil MP, Singh RD, Koli PB, Patil KT, Jagdale BS, Tipare AR, Kim GD (2018) Antibacterial potential of silver nanoparticles synthesized using Madhuca longifolia flower extract as a green resource. Microb Pathog 121:184–189. https://doi.org/10.1016/j.micpath.2018.05.040
Phull A-R, Abbas Q et al (2016) Antioxidant, cytotoxic and antimicrobial activities of green synthesized silver nanoparticles from crude extract of Bergenia ciliata. Future J Pharm Sci 2(1):31–36
Phumying S, Labuayai S et al (2013) Aloe vera plant-extracted solution hydrothermal synthesis and magnetic properties of magnetite (Fe 3 O 4) nanoparticles. Appl Phys A 111(4):1187–1193
Pinzaru I, Coricovac D et al (2018) Stable PEG-coated silver nanoparticles – a comprehensive toxicological profile. Food Chem Toxicol 111:546–556. https://doi.org/10.1016/j.fct.2017.11.051
Popova M, Szegedi A et al (2014) Preparation of resveratrol-loaded nanoporous silica materials with different structures. J Solid State Chem 219:37–42
Prahalathan P, Kumar S et al (2012) Morin attenuates blood pressure and oxidative stress in deoxycorticosterone acetate-salt hypertensive rats: a biochemical and histopathological evaluation. Metabolism 61(8):1087–1099. https://doi.org/10.1016/j.metabol.2011.12.012
Raja S, Ramesh V et al (2017) Green biosynthesis of silver nanoparticles using Calliandra haematocephala leaf extract, their antibacterial activity and hydrogen peroxide sensing capability. Arab J Chem 10(2):253–261

Rajendran SP, Sengodan K (2017) Synthesis and characterization of zinc oxide and iron oxide nanoparticles using Sesbania grandiflora leaf extract as reducing agent. J Nanosci 2017:1–7. https://doi.org/10.1155/2017/8348507

Rao KS, El-Hami K et al (2005) A novel method for synthesis of silica nanoparticles. J Colloid Interface Sci 289(1):125–131

Rivera-Rangel RD, González-Muñoz MP et al (2018) Green synthesis of silver nanoparticles in oil-in-water microemulsion and nano-emulsion using geranium leaf aqueous extract as a reducing agent. Colloids Surf A Physicochem Eng Asp 536:60–67

Roohi F, Lohrke J et al (2012) Studying the effect of particle size and coating type on the blood kinetics of superparamagnetic iron oxide nanoparticles. Int J Nanomedicine 7:4447–4458. https://doi.org/10.2147/IJN.S33120

Rosenholm JM, Sahlgren C et al (2011) Multifunctional mesoporous silica nanoparticles for combined therapeutic, diagnostic and targeted action in cancer treatment. Curr Drug Targets 12(8):1166–1186

Rovani S, Santos JJ et al (2018) Highly pure silica nanoparticles with high adsorption capacity obtained from sugarcane waste ash. ACS Omega 3(3):2618–2627

Ruiz-Rico M, Pérez-Esteve É et al (2017) Enhanced antimicrobial activity of essential oil components immobilized on silica particles. Food Chem 233:228–236

Sallem F, Haji R et al (2019) Elaboration of trans-resveratrol derivative-loaded superparamagnetic Iron oxide nanoparticles for glioma treatment. Nanomaterials (Basel) 9(2). https://doi.org/10.3390/nano9020287

Seip CT, O'Connor CJ (1999) The fabrication and organization of self-assembled metallic nanoparticles formed in reverse micelles. Nanostruct Mater 12(1–4):183–186

Semmler-Behnke M, Kreyling WG et al (2008) Biodistribution of 1.4- and 18-nm gold particles in rats. Small 4(12):2108–2111. https://doi.org/10.1002/smll.200800922

Shaik M, Khan M et al (2018) Plant-extract-assisted green synthesis of silver nanoparticles using Origanum vulgare L. extract and their microbicidal activities. Sustainability 10(4):913

Shanmugam MK, Rane G et al (2015) The multifaceted role of curcumin in cancer prevention and treatment. Molecules 20(2):2728–2769. https://doi.org/10.3390/molecules20022728

Sharma M, Yadav S et al (2019) Biofabrication and characterization of flavonoid-loaded Ag, Au, Au-Ag bimetallic nanoparticles using seed extract of the plant Madhuca longifolia for the enhancement in wound healing bio-efficacy. Prog Biomater 8(1):51–63. https://doi.org/10.1007/s40204-019-0110-0

Shetty PK, Venuvanka V et al (2015) Development and evaluation of sunscreen creams containing morin-encapsulated nanoparticles for enhanced UV radiation protection and antioxidant activity. Int J Nanomedicine 10:6477–6491. https://doi.org/10.2147/IJN.S90964

Shi H, Tan L et al (2014) Green synthesis of Fe3O4 nanoparticles with controlled morphologies using urease and their application in dye adsorption. Dalton Trans 43(33):12474–12479. https://doi.org/10.1039/c4dt01161a

Shivaji S, Madhu S et al (2011) Extracellular synthesis of antibacterial silver nanoparticles using psychrophilic bacteria. Process Biochem 46(9):1800–1807

Shrifian-Esfahni A, Salehi MT et al (2015) Chitosan-modified superparamgnetic iron oxide nanoparticles: design, fabrication, characterization and antibacterial activity. Chemik 69(1):19–32

Shukla R, Bansal V et al (2005) Biocompatibility of gold nanoparticles and their endocytotic fate inside the cellular compartment: a microscopic overview. Langmuir 21(23):10644–10654. https://doi.org/10.1021/la0513712

Silva GA (2004) Introduction to nanotechnology and its applications to medicine. Surg Neurol 61(3):216–220

Singh AP, Singh R et al (2019) Health benefits of resveratrol: evidence from clinical studies. Med Res Rev. https://doi.org/10.1002/med.21565

Sivaramakrishnan V, Devaraj SN (2010) Morin fosters apoptosis in experimental hepatocellular carcinogenesis model. Chem Biol Interact 183(2):284–292. https://doi.org/10.1016/j.cbi.2009.11.011

Sokolik CG, Lellouche J-P (2018) Hybrid-silica nanoparticles as a delivery system of the natural biocide carvacrol. RSC Adv 8(64):36712–36721

Sonavane G, Tomoda K et al (2008) Biodistribution of colloidal gold nanoparticles after intravenous administration: effect of particle size. Colloids Surf B Biointerfaces 66(2):274–280. https://doi.org/10.1016/j.colsurfb.2008.07.004

Sparreboom A, van Tellingen O et al (1996) Nonlinear pharmacokinetics of paclitaxel in mice results from the pharmaceutical vehicle Cremophor EL. Cancer Res 56(9):2112–2115

Sre PR, Reka M et al (2015) Antibacterial and cytotoxic effect of biologically synthesized silver nanoparticles using aqueous root extract of Erythrina indica lam. Spectrochim Acta A Mol Biomol Spectrosc 135:1137–1144

Sripanyakorn S, Jugdaohsingh R et al (2009) The comparative absorption of silicon from different foods and food supplements. Br J Nutr 102(6):825–834

Stensberg MC, Wei Q et al (2011) Toxicological studies on silver nanoparticles: challenges and opportunities in assessment, monitoring and imaging. Nanomedicine (Lond) 6(5):879–898. https://doi.org/10.2217/nnm.11.78

Stivala LA, Savio M et al (2001) Specific structural determinants are responsible for the antioxidant activity and the cell cycle effects of resveratrol. J Biol Chem 276(25):22586–22594. https://doi.org/10.1074/jbc.M101846200

Stöber W, Fink A et al (1968) Controlled growth of monodisperse silica spheres in the micron size range. J Colloid Interface Sci 26(1):62–69

Summerlin N, Qu Z et al (2016) Colloidal mesoporous silica nanoparticles enhance the biological activity of resveratrol. Colloids Surf B Biointerfaces 144:1–7. https://doi.org/10.1016/j.colsurfb.2016.03.076

Sundar VD, Dhanaraju MD et al (2014) Fabrication and characterization of etoposide loaded magnetic polymeric microparticles. Int J Drug Deliv 6(1):24

Tan T, Liu S et al (2011) 5.14-microemulsion preparative methods (overview) A2-Andrews, David L. In: Comprehensive nanoscience and technology. Academic Press, Amsterdam, pp 399–441

Tao Z, Morrow MP et al (2008) Mesoporous silica nanoparticles inhibit cellular respiration. Nano Lett 8(5):1517–1526. https://doi.org/10.1021/nl080250u

Teimuri-Mofrad R, Hadi R et al (2017) Green synthesis of gold nanoparticles using plant extract: mini-review. Nanochem Res 2(1):8–19

Thuc CNH, Thuc HH (2013) Synthesis of silica nanoparticles from Vietnamese rice husk by sol–gel method. Nanoscale Res Lett 8(1):58

Tili E, Michaille JJ et al (2010) Resveratrol decreases the levels of miR-155 by upregulating miR-663, a microRNA targeting JunB and JunD. Carcinogenesis 31(9):1561–1566. https://doi.org/10.1093/carcin/bgq143

Tong L, Wei Q et al (2009) Gold nanorods as contrast agents for biological imaging: optical properties, surface conjugation and photothermal effects. Photochem Photobiol 85(1):21–32. https://doi.org/10.1111/j.1751-1097.2008.00507.x

Tran QH, Le A-T (2013) Silver nanoparticles: synthesis, properties, toxicology, applications and perspectives. Adv Nat Sci Nanosci Nanotechnol 4(3):033001

Turkevich J, Stevenson PC et al (1951) A study of the nucleation and growth processes in the synthesis of colloidal gold. Discuss Faraday Soc 11(0):55–75. https://doi.org/10.1039/df9511100055

Vallet-Regi M, Ramila A et al (2001) A new property of MCM-41: drug delivery system. Chem Mater 13(2):308–311

Vallet-Regi M, Balas F et al (2007) Mesoporous materials for drug delivery. Angew Chem Int Ed Engl 46(40):7548–7558. https://doi.org/10.1002/anie.200604488

Venkateswarlu S, Rao YS et al (2013) Biogenic synthesis of Fe3O4 magnetic nanoparticles using plantain peel extract. Mater Lett 100:241–244

Venkateswarlu S, Kumar BN et al (2014) Bio-inspired green synthesis of Fe3O4 spherical magnetic nanoparticles using Syzygium cumini seed extract. Phys B Condens Matter 449:67–71

Venkateswarlu S, Kumar BN et al (2019) A novel green synthesis of Fe3O4 magnetic nanorods using Punica Granatum rind extract and its application for removal of Pb (II) from aqueous environment. Arab J Chem 12(4):588–596. https://doi.org/10.1016/j.arabjc.2014.09.006

Verma NK, Crosbie-Staunton K et al (2013) Magnetic core-shell nanoparticles for drug delivery by nebulization. J Nanobiotechnol 11(1). https://doi.org/10.1186/1477-3155-11-1

Vidal-Vidal J, Rivas J et al (2006) Synthesis of monodisperse maghemite nanoparticles by the microemulsion method. Colloids Surf A Physicochem Eng Asp 288(1–3):44–51

Wallerath T, Deckert G et al (2002) Resveratrol, a polyphenolic phytoalexin present in red wine, enhances expression and activity of endothelial nitric oxide synthase. Circulation 106(13):1652–1658

Wang Y-XJ (2015) Current status of superparamagnetic iron oxide contrast agents for liver magnetic resonance imaging. World J Gastroenterol 21(47):13400

Wang J, Sun J et al (2003) One-step hydrothermal process to prepare highly crystalline Fe3O4 nanoparticles with improved magnetic properties. Mater Res Bull 38(7):1113–1118

Wang Z, Zhu H et al (2009) One-pot green synthesis of biocompatible arginine-stabilized magnetic nanoparticles. Nanotechnology 20(46):465606. https://doi.org/10.1088/0957-4484/20/46/465606

Wang J, Shah ZH et al (2014) Silica-based nanocomposites via reverse microemulsions: classifications, preparations, and applications. Nanoscale 6(9):4418–4437. https://doi.org/10.1039/c3nr06025j

Wang T, Liu Y et al (2017) Effect of paclitaxel-mesoporous silica nanoparticles with a core-shell structure on the human lung cancer cell line A549. Nanoscale Res Lett 12(1):66. https://doi.org/10.1186/s11671-017-1826-1

Wen H, Dan M et al (2017) Acute toxicity and genotoxicity of silver nanoparticle in rats. PLoS One 12(9):e0185554. https://doi.org/10.1371/journal.pone.0185554

Wenzel E, Somoza V (2005) Metabolism and bioavailability of trans-resveratrol. Mol Nutr Food Res 49(5):472–481. https://doi.org/10.1002/mnfr.200500010

Wetzel S, Bon RS et al (2011) Biology-oriented synthesis. Angew Chem Int Ed 50(46):10800–10826

Wu TW, Fung KP et al (1995) Molecular properties and myocardial salvage effects of morin hydrate. Biochem Pharmacol 49(4):537–543

Wu W, He Q et al (2008) Magnetic iron oxide nanoparticles: synthesis and surface functionalization strategies. Nanoscale Res Lett 3(11):397

Wu F, Harper BJ et al (2017) Differential dissolution and toxicity of surface functionalized silver nanoparticles in small-scale microcosms: impacts of community complexity. Environ Sci Nano 4(2):359–372. https://doi.org/10.1039/c6en00324a

Yallapu MM, Othman SF et al (2012) Curcumin-loaded magnetic nanoparticles for breast cancer therapeutics and imaging applications. Int J Nanomedicine 7:1761–1779. https://doi.org/10.2147/IJN.S29290

Yallapu MM, Ebeling MC et al (2013) Novel curcumin-loaded magnetic nanoparticles for pancreatic cancer treatment. Mol Cancer Ther 12(8):1471–1480. https://doi.org/10.1158/1535-7163.MCT-12-1227

Yan F, Jiang J et al (2014) Synthesis and characterization of silica nanoparticles preparing by low-temperature vapor-phase hydrolysis of SiCl4. Ind Eng Chem Res 53(30):11884–11890

Ye Y, Hui L et al (2018) Effects of silica nanoparticles on endolysosome function in primary cultured neurons. Can J Physiol Pharmacol 999:1–9

Yilmaz M, Karanastasis AA et al (2019) Inclusion of quercetin in gold nanoparticles decorated with supramolecular hosts amplifies its tumor targeting properties. ACS Appl Bio Mater. https://doi.org/10.1021/acsabm.8b00748

Yu Y-Y, Chen C-Y et al (2003) Synthesis and characterization of organic–inorganic hybrid thin films from poly (acrylic) and monodispersed colloidal silica. Polymer 44(3):593–601

Yu K, Guo Y et al (2005) Synthesis of silica nanocubes by sol–gel method. Mater Lett 59(29–30):4013–4015

Yu J, Xu D et al (2016) Facile one-step green synthesis of gold nanoparticles using Citrus maxima aqueous extracts and its catalytic activity. Mater Lett 166:110–112

Yu L, Shang F et al (2018) The anti-biofilm effect of silver-nanoparticle-decorated quercetin nanoparticles on a multi-drug resistant Escherichia coli strain isolated from a dairy cow with mastitis. PeerJ 6:e5711. https://doi.org/10.7717/peerj.5711

Zafra-Stone S, Yasmin T et al (2007) Berry anthocyanins as novel antioxidants in human health and disease prevention. Mol Nutr Food Res 51(6):675–683

Zaky R, Hessien M et al (2008) Preparation of silica nanoparticles from semi-burned rice straw ash. Powder Technol 185(1):31–35

Zhang G, Yang Z et al (2009) Influence of anchoring ligands and particle size on the colloidal stability and in vivo biodistribution of polyethylene glycol-coated gold nanoparticles in tumor-xenografted mice. Biomaterials 30(10):1928–1936. https://doi.org/10.1016/j.biomaterials.2008.12.038

Zhang H, Zhao X et al (2010) A study on the consecutive preparation of d-xylose and pure superfine silica from rice husk. Bioresour Technol 101(4):1263–1267

Zhang H, Dunphy DR et al (2012) Processing pathway dependence of amorphous silica nanoparticle toxicity: colloidal vs pyrolytic. J Am Chem Soc 134(38):15790–15804. https://doi.org/10.1021/ja304907c

Zheng Y-h, Cheng Y et al (2006) Synthesis and magnetic properties of Fe3O4 nanoparticles. Mater Res Bull 41(3):525–529

Zhou W, He W et al (2009) Biosynthesis and magnetic properties of mesoporous Fe3O4 composites. J Magn Magn Mater 321(8):1025–1028

Zhou X, Zhang N et al (2017) Silicates in orthopedics and bone tissue engineering materials. J Biomed Mater Res A 105(7):2090–2102

Zhu X, Tian S et al (2012) Toxicity assessment of iron oxide nanoparticles in zebrafish (Danio rerio) early life stages. PLoS One 7(9):e46286. https://doi.org/10.1371/journal.pone.0046286

Zhu M, Zhu Y et al (2014) Mesoporous silica nanoparticles/hydroxyapatite composite coated implants to locally inhibit osteoclastic activity. ACS Appl Mater Interfaces 6(8):5456–5466

Zulfiqar U, Subhani T et al (2016) Synthesis and characterization of silica nanoparticles from clay. J Asian Ceramic Soc 4(1):91–96

Chapter 7
Nanotechnology Based Targeting Strategies for the Delivery of Camptothecin

Santwana Padhi and Anindita Behera

Abstract Camptothecin is a potent anti-cancer drug which has shown appreciable anti-tumor activity against a broad spectrum of cancers such as breast, ovarian, colon, lung and stomach. The water insolubility property, rapid conversion of its bioactive lactone form to inactive carboxylate form under physiological condition, incidence of drug resistance and the associated off-target side effects due to extended use of campothecin restricts its widespread clinical usage. The development of novel treatment modalities is therefore the demand of era. The last two decades have already evidenced the explosive growth of nanotechnology channeling into plethora of innovations in therapeutic domains, the main dominance being the ability to target the tumor tissues either passively or actively.

This chapter deals in brief with the origin and mode of action of campothecin, the drawbacks associated with the use of the conventional drug (campothecin), the structural modifications carried out to overcome the stability and solubility issues as well as the novel targeted drug delivery platforms (passive and active targeting) that has been explored for the treatment of solid tumors. The structure activity relationship of different campothecin derivatives are elaborately discussed. The mechanism of action of passive and active targeting illustrates the wide applicability of campothecin as an anticancer lead against different solid tumors and cell lines.

Keywords Camptothecin · Nanotherapeutics · Topoisomerase – I · EPR · Stability · Solubility · Passive targeting · Active targeting · Solid tumor

---

S. Padhi
KIIT Technology Business Incubator, KIIT Deemed to be University, Bhubaneswar, Odisha, India

A. Behera (✉)
School of Pharmaceutical Sciences, Siksha 'O' Anusandhan Deemed to be University, Bhubaneswar, Odisha, India
e-mail: anindita02@gmail.com

A. Saneja et al. (eds.), *Sustainable Agriculture Reviews 44*, Sustainable Agriculture Reviews 44, https://doi.org/10.1007/978-3-030-41842-7_7

7.1 Introduction

Cancer is known to be a highly complex disease caused by defect or instability of the genes causing alteration in cellular pathology such as abnormal growth and replication of the cells due to genetic mutation, chromosome translocation, gene malfunction and disabled apoptosis (Mansoori et al. 2007). It is a major public health problem worldwide being the leading cause of death in developed countries and second leading cause of death in developing countries. An estimated 12.66 million people were diagnosed with cancer across the world in 2008, and 7.56 million people died from the disease (Ferlay et al. 2010). By 2030, the global burden is expected to grow to 21.4 million new cancer cases and 13.2 million cancer deaths (Bray et al. 2012). In spite of the fast progression of the disease and the increased global burden, the conventional treatment modality for cancer is restricted to surgical resection, radiation, chemotherapy which has limited efficacy because of lack of specificity, inadequate drug concentration at tumor sites, poor drug delivery, drug resistance and significant damage to noncancerous tissues (Padhi et al. 2018). These limitations results in poor diagnosis and ineffective treatment. To address the above stated limitations, structural modifications in the molecule needs to be undertaken and various novel and innovative drug delivery vehicles with accurate diagnosis, higher therapeutic indices and better safety profile needs to be developed.

Over the past few decades, there have been significant advances in the field of nanotechnology based therapeutics especially its integration with oncology. This field of nanotechnology or more precisely nanomedicines has gained wider acceptance and has emerged as a solution to the drawbacks of conventional therapy bringing a revolution in the both cancer diagnostics and therapeutics. These nanotherapeutics offer various advantages over the conventional deliveries such as stability, high specificity, controlled release, high payload, improved solubility and bioavailability (Heath and Davis 2008). Its capability to by-pass multidrug resistance and ability to accommodate and transport molecules of varied physical and chemical nature makes it a promising drug delivery system for an anticancer loaded nanoparticle based therapeutics such as Nab-paclitaxel or Abraxane®. It is a solvent-free formulation of paclitaxel loaded albumin nanoparticles (Abraxis Biosciences Ltd) that has been approved by USFDA for the treatment of metastatic breast cancer while many other molecules are in various phases of clinical trials (Glück 2014). Apart from therapeutic use, nanotherapeutics can also serve as valuable diagnostic tool for early detection of cancer or biomarkers so that the treatment can be initiated at early stage of disease progression.

The pathophysiological condition and the anatomical changes that occur in the tumor tissue and its microenvironment allow several advantages for the delivery of nanotherapeutics which can be engineered to achieve site specific delivery via passive or active targeting. The physiology of the normal and tumor tissue differ in various aspects. Passive targeting takes the advantage of these changes to deliver the drug at the tumor sites. Normal vasculature is highly ordered, 8–10 μm in diameter and uniformly structured. In contrast, tumor blood vessels are 20–100 μm in diameter, leaky in nature, with abnormal, distended capillaries with sluggish and irregular flow patterns (Giaccia 1998). They also possess wide fenestrations of 200 nm–1.2 μm allowing the nanoparticles to permeate easily into the extravascular spaces resulting in their high accumulation inside the tumors. This phenomenon of action nanoparticle is more commonly referred to as the enhanced permeability effect or EPR effect (Padhi et al. 2015). Another form of passive targeting involves the conjugation of drug to tumor specific molecule which when comes in contact with the tumor microenvironment converts into an active substance (Sinha 2006). The tumor is also known to overexpress certain receptors or antigens, and this fact can be utilized to functionalize the nanotherapeutics to recognize the receptors on the tumor surface (active targeting) leading to its preferential accumulation within the tumor via receptor mediated endocytosis (Dauty et al. 2002).

Camptothecin is a potent anti-cancer drug which inhibits the enzyme, DNA topoisomerase I, during the S-phase of cell cycle (Zunino et al. 2002). DNA topoisomerase I is an enzyme which is essential for replication and transcription of DNA. The stabilizing action of the resultant DNA topoisomerase I complex results in apoptosis of cancerous cells. It is known to be active against a broad spectrum of cancer, namely ovarian, stomach, colon, cervical, breast and lung cancers (Zhang 2017). As evidenced from the *in vitro* studies, campothecin is known to be a potent chemotherapeutic drug but its efficacy is not proven in clinical studies due to its associated dose limiting toxicity and reported instability. It exists in two inter-convertible forms, lactone and carboxylate. The bioactive lactone form is rapidly hydrolyzed to inactive carboxylate form at physiological pH. The carboxylate form is known to be tenfolds less active and is rapidly cleared from the systemic circulation upon binding with plasma proteins (Mross 2004). The sodium carboxylate salt of camptothecin was used in the early stages of clinical trials as campothecin has a property of poor water solubility. But the usage of this form was discontinued due to the associated side effects such as hemorrhagic cystitis, neutropenia and thrombocytopenia. When the fact that camptothecin acts by inhibiting topoisomerase I was established, campothecin and its derivatives were widely used for the treatment of solid tumors. Campothecin needs to be administered in low dose, multiple times to portray its therapeutic efficacy. Hence various nanoparticulate delivery systems needs to be employed to maintain the lactone ring stability and improve the solubility profile of the drug molecule.

The present chapter focuses on the origin as well as mode of action of campothecin, the structural modifications that can be done to improve the stability and solubility issues associated with it and the passive and active targeting strategies that can be employed to target the tumor tissues.

7.2 Origin of Camptothecin

Camptothecin is one of the most used anticancer agent of the present era. It was first discovered and isolated by two scientists i.e. Wall and Wani in 1966 from the bark of *Camptotheca acuminata Decne*, a chinese Happy Tree, during the search for some new steroids from the plant source (Wall et al. 1966). Campothecin is a monoterpene pentacyclic quinolone alkaloid containing a planar pentacyclic ring system that includes a pyrrolidine nucleus (B ring), a quinolinic nucleus (A and B ring), a lactam nucleus (D ring), a lactam system containing a stereocenter with (S) configuration and an α – hydroxy group in the E ring (Fig. 7.1). The research on campothecin grew rapidly but the associated drawbacks of poor solubility, stability and unpredictable adverse drug – drug interaction stagnated its further development. In the meanwhile, FDA approved the molecule for Phase II clinical trial for colon cancer, but the adverse drug reactions resulted in its withdrawal from the phase trial in 1972 (Martino et al. 2017). In 1980, DNA Topoisomerase I (Top I) was identified as the cellular target for anticancer agents, which opened up avenues for further studies on campothecin. To meet the requirement of high quantity of campothecin for several pre-clinical and clinical usage, several biological sources were identified other than *C. acuminata* and also several extraction processes were introduced. The list of natural sources of campothecin is listed in Table 7.1.

7.3 Mode of Action of Camptothecin

Campothecin is an anti-cancer molecule which acts by targeting the human DNA topoisomerase I (Topo-I). It inhibits the enzyme by blocking the re-joining step of the cleavage reaction of Topo-I, which causes accumulation of a covalent reaction intermediate. The intermediate is the cleavable complex and the primary

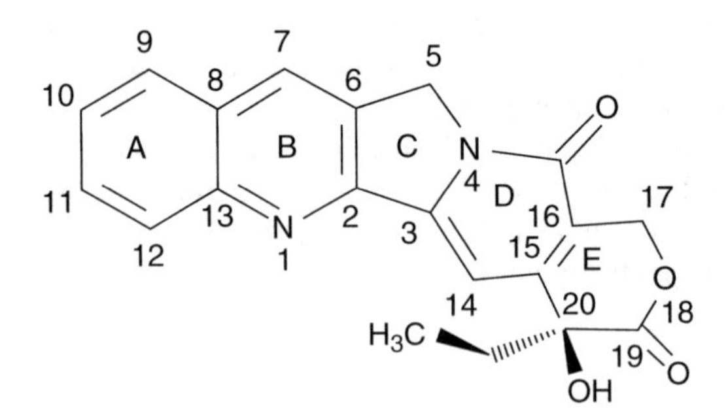

Fig. 7.1 Chemical structure of Camptothecin [A monoterpene pentacyclic quinolone alkaloid containing a planar pentacyclic ring system containing a pyrrolidine nucleus (B ring), a quinolinic nucleus (A and B ring), a lactam nucleus (D ring), a lactam system containing a stereocenter with (S) configuration and an α – hydroxy group in the E ring]

Table 7.1 Biological sources of Campothecin

Biological source	Parts of the source containing Campothecin	Family	References
Camptotheca acuminata Decne	Bark	Nyssaceae	Li and Liu (2004)
Nothapodytes nimmoniana	Shoots, tender stems, leaves, bark and Endophytes	Icacinaceae	Govindachari and Viswanathan (1972)
Ervatamia heyneana	Wood and stem bark	Apocynaceae	Gunasekera et al. (1979)
Merriliodendron megacarpam	Leaves and stem	Icacinaceae	Arisawa et al. (1981)
Mostuca brunonis Didr.	Whole plant extract and roots	Loganiaceae	Dai et al. (1999)
Camptotheca lowreyana	Glandular trichomes of young and old leaves	Nyssaceae	Zhou et al. (2000)
Pyrenacantha klaineana	Stems	Icacinaceae	Zhou et al. (2000)
Pyrenacantha volubilis	Fruits	Icacinaceae	Ramesha et al. (2013)
Chonemorpha grandiflora	Stem with bark and callus culture	Apocynaceae	Malpathak et al. (2010)
Sarcostigma kleinii	Leaf, stem bark and fruit	Icacinaceae	Ramesha et al. (2013)
Ophiorrhiza mungos Linn.	Entire plant	Rubiaceae	Kumar et al. (2018)
Ophiorrhiza pumila champ	Leaves, young roots, hairy roots, entire plant	Rubiaceae	Saito et al. (2001)
Ophiorrhiza liukiuensis	Hairy roots	Rubiaceae	Asano et al. (2004)
Ophiorrhiza kuroiwai	Hairy roots	Rubiaceae	Asano et al. (2004)
Ophiorrhiza japonica	Entire plant	Rubiaceae	Ya-ut et al. (2011)
Ophiorrhiza shendurunii	Entire plant	Rubiaceae	Rajan et al. (2013)
Ophiorrhiza grandiflora	Stem	Rubiaceae	Rajan et al. (2013)
Ophiorrhiza eriantha	Stem	Rubiaceae	Rajan et al. (2013)
Ophiorrhiza trichocarpon Blume	Flowers	Rubiaceae	Rajan et al. (2013)
Ophiorrhiza pectinata	Entire plant	Rubiaceae	Rajan et al. (2013)
Ophiorrhiza filistipula	Leaves	Rubiaceae	Arbain et al. (1993)
Gomphandra comosa	Fruit	Stemonuraceae	Ramesha et al. (2013)

(continued)

Table 7.1 (continued)

Biological source	Parts of the source containing Campothecin	Family	References
Gomphandra coriacea	Leaf, twig, seed coat, endosperm and root	Stemonuraceae	Ramesha et al. (2013)
Gomphandra polymorpha	Fruits	Stemonuraceae	Ramesha et al. (2013)
Gomphandra tetrandra	Leaf, stem bark	Stemonuraceae	Ramesha et al. (2013)
Iodes cirrhosa	Fruits	Icacinaceae	Ramesha et al. (2013)
Iodes hookeriana	Leaf, bark, stem, fruits	Icacinaceae	Ramesha et al. (2013)
Miquelia dentate	Twig, leaf, seed coat, fruit coat, Cotyledon, root, bud, fruit, mature and immature fruit	Icacinaceae	Ramesha et al. (2013)
Miquelia kleinii	Fruit	Icacinaceae	Ramesha et al. (2013)
Natsiatum herpecticum	Leaf, twig and fruit	Icacinaceae	Ramesha et al. (2013)
Apodytes dimidiata	Leaf and stem bark	Icacinaceae	Ramesha et al. (2013)
Codiocarpus andamanicus	Leaf and fruit	Icacinaceae	Ramesha et al. (2013)

mechanism of action of campothecin involves the cell death at the S-phase of cell cycle. In the S-phase of cell cycle (which involves the DNA replication), campothecin causes potential collisions between advancing replication forks and the intermediate Topo-I cleavable complex (Ryan et al. 1994). The collision of campothecin with the transcription machinery triggers the formation of long lived covalent Topo-I – DNA complexes, which is responsible for cytotoxicity of campothecin. When cell death occurs due to cytotoxicity of campothecin, there are two ways of repair of Topo-I mediated DNA damage by modification of covalent intermediate of Topo-I – DNA complex. The two possible pathways of damage repair are

- Ubiquitin/26S proteasome pathway which causes degradation of Topo-I (Campothecin induced Topo-I down-regulation)
- SUMO (Small Ubiquitin like MOdifier) conjugation to Topo-I.
- Formation of ternary complex of Topo-I – campothecin – DNA complex occurs by covalent binding. Campothecin binds at the interface between Topo-I and DNA and inhibits at the relegation step in the S-phase. The binding of campothecin is an uncompetitive type inhibition because campothecin doesn't bind with either the enzyme or the DNA substrate, but interacts with the enzyme – DNA complex to form a reversible non-productive complex (Yamauchi 2011; Hertzberg et al. 1989a, b).

- Campothecin kills the cells by S-phase specific cytotoxicity. The reversible ternary complex is non-lethal in nature, but their collision with the advancing replication forks causes the cell death. The collision leads to three events following the collision (Avemann et al. 1988).

  - The double-strand of DNA break,
  - The replication fork is arrested irreversibly, and
  - The breakdown of Topo-I-linked DNA break at the site of collision.

  The collision causes the cell death, only when the ternary cleavable complex is formed on the strand complementary to the leading strand of DNA synthesis. If the concentration of campothecin is high, the cytotoxicity leads to death of non-S-phase cells (Liu et al. 2006).
- Campothecin also causes interference in the synthesis of RNA at the level of transcription elongation. This arrest leads to accumulation of RNA polymerase elongation complexes. The ternary complex of Topo–I–campothecin–DNA arrests the transcription leading to premature termination. The inhibition of transcription by collision of replication fork depends on the orientation of the cleavable ternary complex relative to transcribing RNA polymerase. It is only possible if the cleavable ternary complex is formed on the template strand of transcription. The collision between RNA polymerase and ternary complexes on the template strand of transcription arrests the transcription of RNA and subsequently the reversible ternary cleavable complexes are converted into irreversible Topo-I linked single stand breaks (Wu and Liu 1997). The collision which results in the formation of irreversible Topo-I linked single strand facilitates the ubiquitin/26S proteasome pathway which destroys Topo-I. After the degradation of Topo-I, single strand break gets repaired before transcription coupled repair (TCR). The enzyme tyrosyl – DNA phosphodiesterase (TDP 1) helps in removal of residual peptides that are attached covalently to DNA. The DNA can also be repaired by TCR independent of TDP 1.
- Campothecin causes rapid attachment of SUMO – 1 (Small Ubiquitin-like Modifier) to Topo – I. Human SUMO–1is a ubiquitin like protein and have about 18% similarity to ubiquitin. They show similar type of activation and conjugation but involves different set of enzymes. The UCB9 is the only distinct enzyme for SUMO–1. Activation and attachment of SUMO–1 is also known as SUMOylation. Ubiquitination and SUMOylation of Topo-I show some similarities and dissimilarities in their reaction. The similarities are that both reactions are dependent on formation of ternary complex and are not affected by replication inhibitors. But the dissimilarities are as following (Saitoh et al. 1998):

  - Ubiquitination of Topo – I is transcription dependent whereas SUMOylation of Topo-I is independent of transcription.
  - Ubiquitination reaction is specific for the phosphorylated Topo-I enzymes, whereas SUMOylation is specific for dephosphorylated Topo-I (Phosphorylated Topo-I has been shown to be associated with transcription).

- Ubiquitination of Topo-I enzyme is found to be defective in most of the human tumor cells, but SUMOylation of Topo-I is normal in both tumor and normal cells

If both ubiquitination and SUMOylation of Topo-I is considered, then the former occurs within the transcribed regions, while the later occurs outside the transcribed regions. The difference is with the phosphorylated condition of Topo –I. SUMOylation of Topo-I leads to relocation of the enzyme in other cellular compartment, so it no longer participates and forms the cleavable ternary complex. So both down-regulation and relocation of Topo-I are the effective ways to repair the Topo-I-mediated DNA damage (D'Arpa and Liu 1995).

7.4 Structure Activity Relationship (SAR) of Camptothecin

Since the invention of campothecin, much research work has been carried out to overcome the problems associated with its solubility and stability. The stability of campothecin is expected to be lowered due to the hydrolysis of the lactam group into inactive carboxylate form in biological conditions for which the bioavailability of campothecin is lowered (Burke et al. 1995). The hydrolysis of lactam ring is due to the presence of α – hydroxy group which accelerates the hydrolysis by intramolecular hydrogen bonding (Bom et al. 1999). To overcome the problems associated with solubility, stability and efficacy, several structural modifications have been carried out, which are enlisted in the Table 7.2.

7.5 Targeted Delivery of Camptothecin

Conventional chemotherapeutics face the challenges in bioavailability due to the pharmacokinetic factors like absorption, distribution, metabolism and excretion. The anticancer drugs administered by conventional methods are absorbed through biological membrane, undergo first pass metabolism, strongly bind to plasma proteins and are easily excreted by kidney which leads to lowering of therapeutic efficacy. To overcome these associated problems and to increase the efficacy and bioavailability of the anticancer moieties, targeted drug delivery is employed. This is also known as smart drug delivery (Muller and Keck 2004).

Targeted drug delivery system is utilized to enhance the stay of the drug in the tumor tissues with more target specific and localized action. There are some advantages of targeted drug delivery systems over the conventional drug delivery system which are enlisted below:

Table 7.2 SAR of camptothecin

Modification of ring	Position of carbon	Substituted functional group	Change in the activity	References
A	C10	Hydroxy & Methoxy	More active and potent than campothecin	Wall et al. (1986)
	C11	Hydroxy		
	C9 or C10	Amino	Increased activity and potency	Wani et al. (1980a, b)
	C12	Amino	Inactive	Wani et al. (1986)
	C10 &11	Dimethoxy		
	C11	$N(CH_3)_2$, CF_3, COOH, CH_2NH_2, CH_2OH, CHO, CH_3	Inactive against Leukemia	Wani et al. (1987a, b)
	C9	Amino	More potent than campothecin	
	C10	Amino & nitro		
	C10 & 11	Methylene dioxy		
	C9	Dimethyl Amino	Increase in water solubility of the analogue (Topotecan) than campothecin	Kingsbury et al. (1991)
	C10	Hydroxy		
	C7	Hydroxy	7-ethyl-10-[4-(1-piperidino)-1-piperidino] carbonyl oxy camptothecin (Irinotecan) had increased water solubility and was active against leukemia, cancer of lungs, ovary and cervix	Sawada et al. (1991)
	C10	[4-(1-piperidino)-1-piperidino]		
	C10	Hydroxy Campothecin derivatives	Active as prodrug and specific activity towards carboxyl esterase	Takayama et al. (1998)
	C9	Alkylidene and alkyl groups	Unsubstituted alkylidene showed better cytotoxicity as compared to campothecin	Candiani et al. (1997)
			Polar amide group caused decrease in cytotoxic activity	
	C10		Decrease in cytotoxic activity	

(continued)

Table 7.2 (continued)

Modification of ring	Position of carbon	Substituted functional group	Change in the activity	References
	C9	Amino	Proved to be more potent than campothecin in pre-clinical studies. The compound had less solubility than topotecan and irinotecan and the phase II trial showed less activity and myelosuppression in colorectal and lung cancer	Giovanella et al. 1989; Pitot et al. (2000)
	C9	Nitro (Rubitecan)	Less water soluble analogue and formulated for oral administration against ovary and primary peritoneal cancer	Verschraegen et al. (1999)
	C11	Aza	Potent inhibitor of Topo-I with improved water solubility as compared to campothecin	Uehling et al. (1995)
B	C7	7 – substituted water soluble analogues	Evaluated for cytotoxicity on five different tumour cell lines. Isopropylamino analogue was found to be more potent than campothecin	Jew et al. (1996)
	C7	t-butyldimethylsilyl	Silatecan displayed more stability in human blood serum and was found to be 25 times more lipophilic than campothecin. It showed same cytotoxic potency on ovarian, colon, prostate and breast cancers as compared to topotecan	Bom et al. (2000)
	C10	Hydroxy		
	C7	Alkyl or alkenyl chain or cyano or carbethoxy groups	Showed potent cytotoxic activity *in vitro* against human non-small-cell lung carcinoma H460 cell-line	Dallavalle et al. (2000)
		Cyano	Showed high *in vitro* cytotoxicity against H460, a topotecan-resistant cell subline and cisplatin-resistant ovarian carcinoma subline	

			In vivo evaluation of the antitumor activity, 7- cyano substituted campothecin showed significantly more effectiveness than topotecan in the H460 tumor model and its activity was comparable in small-cell-lung carcinoma and colon carcinoma model as compared to topotecan	
	C7	Imino methyl derivatives	Increased cytotoxicity against H460 non-small-lung carcinoma cell lines as compared to topotecan	Dallavalle et al. (2001a)
	C7	Oxy imino methyl derivatives	Increased *in vitro* efficacy	Dallavalle et al. (2001b)
	C7	Five conjugates composed of a paclitaxel and a campothecin derivative joined by an imine linkage	Evaluated for cytotoxic effect and DNA Topo – I Inhibition. These analogues showed high potency of tumor cell replication. Few analogues were more potent than paclitaxel and campothecin against HCT-8 cell lines of colon adenocarcinoma.	Ohtsu et al. (2003)
		Complete modifications of ring B to form rigid analogues which is required for desired activity of campothecin.	Showed complete DNA fragmentation	Lackey et al. (1995)
		8-ethyl-2-(5-fluoro-2-oxo-1,2-dihydroindol-3-ylidene)-8-hydroxy-2,3,5,8-tetrahydro-6-oxa-3a-azacyclopenta (b) naphthalene-1,4,7-trione		

(continued)

Table 7.2 (continued)

Modification of ring	Position of carbon	Substituted functional group		Change in the activity	References
C	C5	Amino		*In vitro* cytotoxicity effect was reduced	Subrahmanyam et al. (1999a, b)
		Methylamine or Hydroxylamine		Good activity and better water solubility	
	C5	Carbo methoxy methyl analogue	10 – hydroxy derivative in A ring	Better activity than the analogues containing unsubstituted and 9 -nitro in ring A	Subrahmanyam et al. (1999a)
		Acetone derivatives	9 – nitro derivative in A ring	More potent than the analogues containing unsubstituted and 10 -hydroxy in ring A	
		Substituted Alkyl ester derivatives		More potent than simple alkyl ester derivatives	
		Cyclic amide analogues		More activity than unsubstituted amide derivatives	
	C5	Alkoxy-campothecin analogues having substituted A ring and B ring		Tested for *in vitro* cytotoxicity in human tumour cell-lines. 12 – nitro substituted in A ring is inactive and 9 – hydroxy-5-ethoxy analogue is the most potent analogue.	Subrahmanyam et al. (2000)
D	Modification of D ring with benzene ring			The analogues were found to be less active (decrease in 40–60 fold) or inactive as compared to campothecin. So the pyridine ring was the required nucleus for the binding of campothecin to enzyme – DNA cleavable complex	Nicholas et al. (1990)
Modification of B, D & E ring	Ester moiety in place of the E ring lactone or a methyl ester at C14 or a saturated deaza B ring, or contain a·combination of these permutations			The drug was less active (35–50-fold) or inactive against Topo-I relaxation reaction, which confirmed the requirement for a lactone group in the E ring	Crow and Crothers (1992)

			Steric requirements at position 14 is considered to be crucial for activity. The planarity of the A and B rings of campothecin is also required for the drug to inhibit Topo-I	
E	C20	Deoxy, Chloro & Hemilactol	Comparatively inactive *in vivo*	Wani et al. (1987a, b)
		R & S configuration	20 (R) analogue is 10–100 times less active as compared to 20 (S) analogue	
	Hydrolysis of lactone ring		The sodium camptothecin displayed antitumor activity but was 10 times less potent than campothecin	Wani et al. (1980a, b)
	Modification of ethyl and hydroxyl group of E ring & open chain analogue of E ring		Four analogues showed higher toxicity and potency as compared to campothecin.	Sugasawa et al. (1976)
			Analogues lacking hydroxyl group at C – 20 were found to be inactive	
	Synthesis of open chain analogues		All the analogues were evaluated for antitumor activity and found to be less active than campothecin.	Adamovics and Hutchinson (1979)
			The pyrrolidinamide analogue was found to be inactive *in vivo* in 388 and L1210 assay but found active *in vitro* in L1210 cell cytotoxicity assay	
	C20 hydroxyl group is retained but lacked the oxygen of lactone nucleus of E ring		The lactam derivative neither inhibited Topo-I nor was cytotoxic. The analogue differed from campothecin in two aspects:	Hertzberg et al. (1989a, b)
			(i) the lactam ring was more stable to hydrolysis than the lactone ring and	

(continued)

Table 7.2 (continued)

Modification of ring	Position of carbon	Substituted functional group	Change in the activity	References
			(ii) the lactam NH was a hydrogen bond donor while the lactone oxygen was a hydrogen bond acceptor.	
			Thio lactone derivative whose E ring more closely resembled the lactone ring in campothecin was expected to show more activity than lactone, but sulfur was a poor hydrogen bond acceptor than oxygen and also increased steric bulk and ring size. The inactivity of this analogue justified strongly the requirement for a β-hydroxy lactone ring for bioactivity.	
	Homologues of β-hydroxy lactone		The analogue was more stable than campothecin and inhibited DNA relaxation by Topo-I with high potency	Lavergne et al. (1997)
			The analogue inhibited the growth of L1210 cells with IC_{50} of 16.2 nM, which was a considerable increased potency as compared to campothecin (126 nm) and topotecan (601 nM).	
Hexacyclic CAMPOTHECIN analogues	Novel hexacyclic CAMPOTHECIN analogues	By forming bridges at 7 and 9 position of CAMPOTHECIN, a hexacyclic ring named as F ring	The analogue was two times more bioactive than 10 – hydroxy – 7- ethyl campothecin	Sugimori et al. (1994)
		Introduction of a heteroatom at position 2 & 3 of F ring.	These analogues showed same order of activity as 10 – hydroxy – 7- ethyl campothecin.	
		3 – oxa, 3 – aza & 3 – thio analogues	2 – oxa derivative was two times active than 10 – hydroxy – 7- ethyl campothecin and 2 – aza showed similar activity.	
		2 – oxa & 2 – aza derivatives		

	To study the effect of size of ring F, the hexacyclic ring is replaced with 5 & 7 membered F ring		5 membered F ring with S configuration was two times active than 10 – hydroxy – 7 – ethyl campothecin. 7 membered F ring having no substitution in ring A was equivalent to twice the activity as 10 – hydroxy – 7 – ethyl campothecin.	Searle and Williams (1992)
	Synthesis of new derivatives of CAMPOTHECIN with higher water solubility	C7 – 7-[(4-methylpiperazino) Methyl]	Found to be potent inhibitor of topo-I	Luzzio et al. (1995)
		C10 – C11 – 10,11-(methylene dioxy/ethylenedioxy)-20 (S) campothecin trifluoroacetate	In human tumor cell cytotoxicity studies, these compounds showed potent antitumor activity against ovarian, melanoma, breast and colon cancer	
	C7	Quaternary ammonium salt derivatives of 10,11-(methylenedioxy) camptothecin	Water solubility was higher than topotecan and it was two times more potent than topotecan	Lackey et al. (1996)
			The analogue was found to be 2 times more potent in colon adenocarcinoma, 4 times more potent in ovarian adenocarcinoma and 7 times more potent against SKVLB cell-lines as compared to topotecan	
	A ring and F ring modified hexacyclic analogues of campothecin	Substitution with electron-withdrawing groups at position 5 of hexacyclic ring (5-hydroxy, 5-methoxy, 5-chloro, and 5-fluoro)	Found to be more potent than 10 – hydroxy – 7 – ethyl campothecin.	Sugimori et al. (1998)

(continued)

Table 7.2 (continued)

Modification of ring	Position of carbon	Substituted functional group	Change in the activity	References
	A series of C7 – N alkylamino ethyl – 10, 11 – methylenedioxy- and Ethylenedioxy – campothecin		Methylenedioxy derivative displayed more cytotoxicity than ethylenedioxy campothecin.	Jew et al. (1999)
			Few derivatives were 100 times more potent against CAOV-3 cell lines (ovarian cancer) than campothecin	
	A ring modified hexacyclic campothecin analogues containing 1,4-oxazine ring		Cytotoxicity of the analogues were compared with respect to campothecin, topotecan and 10 – hydroxy – 7 – ethyl campothecin in six different human cancer cell lines.	Kim et al. (2001)
			Four analogues were found to be twofold more potent than topotecan and as potent as campothecin towards cancer cell-lines.	
			In the Topo-I cleavable complex enzyme assay, the analogue was found to be eightfold more potent than campothecin, fourfold more potent than topotecan and as potent as 10 – hydroxy – 7 – ethyl campothecin	

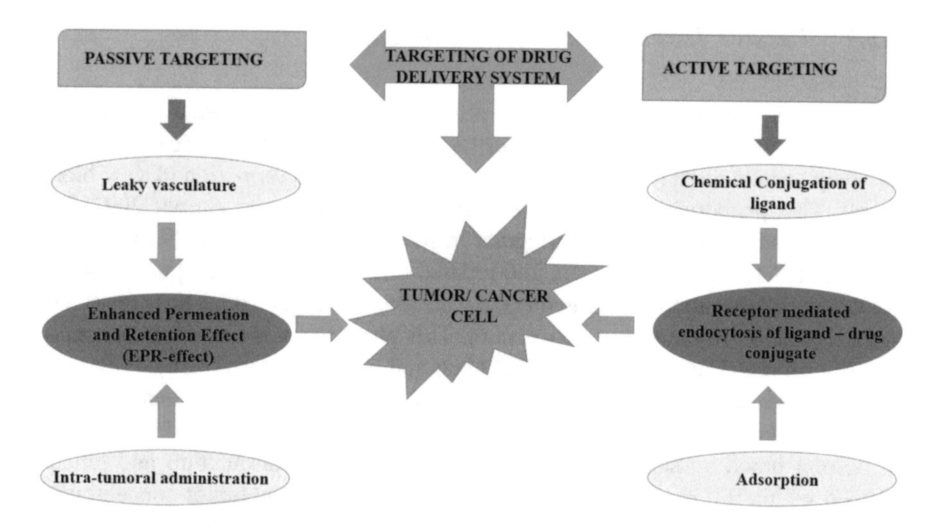

Fig. 7.2 Mechanism of targeting of anticancer agent [In targeting of a drug for solid tumors, two approaches can be adopted (i) Passive targeting where the drug reaches the target cancerous or tumor cells by leaky vasculature between the tissues or by direct Intra-tumoral administration, (ii) Active targeting involves the chemical conjugation of the anticancer drug with receptor specific ligands or conjugation by adsorption. Passive targeting acts by Enhanced Permeation and retention or EPR effect which causes the effective increase in drug concentration at the site of tumor whereas active targeting involves the conjugation of the drug with a suitable ligand which is specific to overexpressed receptors for the specific type of tumor. The drug-ligand conjugate enters into the tumor cell by endocytosis and delivering the drug inside the cell, so in active targeting the healthy tissues are not affected by the anticancer drugs]

– The frequency of drug administration is reduced so more uniform effect of the drug, and reduction of drug side effects are observed.
– Fluctuation in circulating drug levels is reduced up-to a great extent.

Targeted drug delivery is based on a method which delivers a fixed amount of drug for a prolonged period of time at the targeted diseased tissues in the body. This maintains the plasma and tissue drug levels and also don't affect the noncancerous or healthy tissues. So targeted delivery system improves efficacy and reduces side-effects.

For a targeted delivery system, the design criteria such as drug properties, side-effects of the drugs, route of administration, targeted site and the disease condition should be considered.

There are two kinds of targeted drug delivery as illustrated Fig. 7.2

- Passive targeting
- Active targeting

7.5.1 Passive Targeting of Camptothecin

After a solid tumor reaches a certain size, the surrounding normal vasculature becomes insufficient to fuel it for its further proliferation. When these cells start to die, they secrete certain growth factors which support the formation of new blood vessel. This phenomena is termed as angiogenesis (Bates and Harper 2002).The newly formed blood vessels are irregular in structure which eventually forms wide fenestrations ranging in size from 200 to 2000 nm. When blood components get in contact with these abnormal structures, the discontinuous gaps offer little resistance to extravasation to the tumor interstitium. This phenomena is termed as enhanced permeation retention or the EPR effect (Jain and Stylianopoulos 2010). Polymeric nano-vehicles, due to their large size as compared to chemotherapeutic soluble drugs cannot penetrate the tight endothelial junctions which is the usual architecture of normal blood vessels, but can extravagate the tumor vasculature and remain trapped within it (Greish 2007). With due course of time, the tumor concentration reaches several folds higher as compared to the plasma due to the lack of efficient lymphatic drainage which forms the basis for the application for EPR – based selective anticancer drug delivery. For the concept of EPR to be effective, the plasma concentration of the drug should be sufficiently high as measured by the area under the time-concentration curve and the polymeric nanotherapeutics should have a molecular weight greater than the renal threshold (i.e., 40 kDa) so that they do not get cleared up through renal clearance and accumulate in tumor tissues following intravenous (I.V) injection to give a desired therapeutic response (Maeda et al. 2001). Higher drug concentration in tumor is achieved as long as the drug remains in the circulation (Maeda 1991).

Many a times the enhanced permeability and retention (EPR) effect is overrated resulting in differences between rapidly growing rodent tumors, slowly growing rodent tumors and human tumors. It is a highly heterogeneous phenomena which varies from a tumor model to human model. The tumors do not have a structured architecture. There are regions in which endothelial cells are not leaky, the interstitial fluid pressure is high or tumor cell density is high which in turn results in limiting the penetration of even small sized molecules in the interstitium where as in some other parts particles as large as 200 nm are able to extravagate (Lammers et al. 2012).

Among various delivery approaches to improve the solubility and lactone ring stability of campothecin, nanoparticulate delivery systems that incorporate campothecin have been extensively investigated for their drug-loading capacity, controlled release and tumor-targeting ability.

The phagocytic clearance of campothecin-loaded solid lipid nanoparticles (SLNs) by the reticulo endothelial system (RES) was effectively reduced by a transient RES blocking with pre-injection of blank SLNs. The campothecin – SLNs with size of 156 ± 16 nm and a narrow size distribution were formulated which demonstrated a cumulative % campothecin release of 35% up to 12 h and about 85% for 120 h. It was observed that campothecin – SLNs intravenously injected into tumor-bearing BALB/c mice were significantly taken up by RES organs than by the tumor tissues even if they were surface modified with poloxamer and PEG. Extensive accumulation of nanoparticles in the RES organs would be favorable for treating tumors with the RES as a target site, but would be challenging if non-RES tissues were the target. But significant reduction in RES uptake and increased accumulation in tumor tissue was observed following intravenous administration with pre-injection of blank SLNs into tumor-bearing mice. The observed result may be attributed to the transient RES blocking by pre-injected blank SLNs, because it has been reported elsewhere that the RES in animals can be temporarily blocked by injection of blank colloidal carriers or RES blocking agents (Jang et al. 2016).

The ability of systemically administered poly (lactic-co-glycolic acid) nanoparticles (PLGA NPs) to deliver hydrophobic payloads (campothecin) to intracranial glioma was investigated. It was reported that PLGA NPs accumulated 10 times higher in tumor as compared to native campothecin. Upon encapsulation in PLGA NPs, it was observed that a dose of 20 mg/kg of campothecin was well tolerated. Campothecin-loaded NPs were efficient in slowing the growth of intracranial GL261 tumors in immune competent C57 albino mice, providing a significant survival benefit as compared to mice receiving saline, free campothecin or low dose of campothecin NPs (median survival of 36.5 days compared to 28, 32, 33.5 days, respectively). These data established the feasibility of treating intracranial glioma with systemically administered nanoparticles loaded campothecin (Householder et al. 2015).

Cırpanlı et al. reported about the encapsulation of campothecin in nanoparticulate delivery systems either using amphiphilic cyclodextrins, poly (lactide-co-glycolide) (PLGA) or poly-caprolactone (PCL) with an aim to maintain the bioactive lactone form and prevent the drug from hydrolysis to the inactive carboxylate form. The nanoparticles were formulated utilizing with nanoprecipitation technique. The formulated particle sizes were in the range of 130–280 nm and surface charges were negative Cyclodextrin based nanoformulations portrayed an initial burst release of 158 ng/mL, and the remaining campothecin was released with a controlled release profile over 288 h. While for PLGA nanoparticles, 12 ng/mL of the campothecin was released within 1 h followed by release of campothecin lactone form within a period of more than 48 h. For PCL nanoparticles, 13 ng/mL of the campothecin was released within 3 h and the campothecin lactone form release was completed in 24 h. 6-O-Capro-β-cyclodextrin (1.44 μg/60 μL campothecin) and concentrated 6-O-Capro-β-cyclodextrin (2.88 μg/60 μL campothecin) nanoparticles significantly modified the growth or lethality of the 9 L glioma, as the median survival time was observed to be 26 days for the untreated group and between 27 and 33 days for cyclodextrin nanoparticle groups. These observed results indicated the fact that campothecin-loaded amphiphilic cyclodextrin nanoparticles may be utilized as a promising carrier system for the effective delivery of campothecin when compared to polymeric analogues (Çırpanlı et al. 2011).

In another study, chitosan/glycerol-2-phosphate (β-GP) was employed to encapsulate campothecin and the resultant biodegradable implant was evaluated in mouse fibrosarcoma (RIF-l) which was implanted subcutaneously in C3H mice. It was observed that the implant containing 4.5% camptothecin by weight was more effective in delaying the tumor growth as compared to systemically delivered camptothecin and those injected with blank chitosan demonstrated no tumor growth inhibition. The observed effect may be corroborated with the controlled release of campothecin from the implant to the tumor tissues and the exposure of tumor cells to drug concentrations for a prolonged period of time which results in more cytotoxicity (Berrada et al. 2005).

Micelle as a drug delivery vehicle was also employed for the successful entrapment of campothecin. In one study, the core of the micelle constituted of hydrophobic poly (p-dioxanone) in which campothecin was entrapped with a hydrophilic shell of chitosan. The release of entrapped campothecin in cell lines such as L929 fibroblast and HeLa (human cervical cell line) was higher at pH 5 than at pH 7.4. The observed effect was due to the efficient internalization of the micelles containing campothecin into the cancer cells which might be explained by the electrostatic interactions between positively charged micelles and negatively charged cell membranes (Tang et al. 2013).

A conjugate of 20(S)-camptothecin and a cyclodextrin based polymer known as CRLX101 was formulated for the controlled release of campothecin in tumors for an extended period of time. The expression of proteins such as topoisomerase-1 (correlated with poor patient survival), Ki-67 (marker for proliferating cells), CaIX (associated with dysplasia), CD31 (responsible for leukocyte trafficking across the endothelial layer) and VEGF (related to angiogenesis) were observed to be decreased after the treatment with CRLX101. The severity of adverse events such as anemia, fatigue, cystitis and thrombocytopenia observed in phase 1/2a in CRLX101 treated patients were far less in severity as compared to systemic administration of campothecin. These results reinforce the fact regarding the efficiency of CRLX101 as an anti-tumor agent in patients with solid tumor malignancies (Gaur et al. 2014).

7.5.2 *Active Targeting of Camptothecin*

Active targeting enhances the efficacy of the drug loaded in nano-systems and make it more specific to target site. The active targeting is achieved either by conjugating a specific ligand for the over-expressed receptors on the tumor surface or by simple adsorption process. The ligand brings about the receptor mediated endocytosis in the cell membrane leading to an increase in uptake of the drug molecule in the tumor cells as illustrated in Fig. 7.3 (Galvin et al. 2011). A number of studies have reported the effective functionalization of ligands onto the surface of nano-systems which have demonstrated enhanced therapeutic efficacy when compared to native drug or passively targeted systems.

Biotin (vitamin H or vitamin B_7) is a growth promoter at the cellular level so it has been used as a suitable targeting ligand. Biotin receptors are overexpressed on the surface of several cancer cells, such as leukemia (L1210FR), mastocytoma (P815), ovarian (OV 2008, ID8), colon (Colo-26), lung (M109), renal (RENCA,

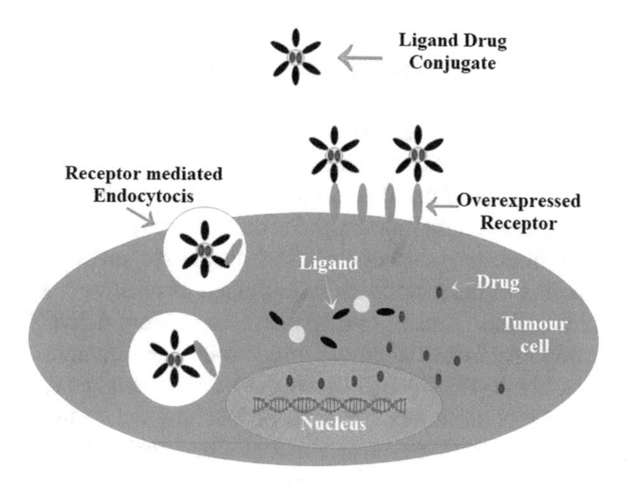

Fig. 7.3 Uptake of anticancer drug by the tumor cell by receptor mediated endocytosis [Anticancer drug forms a conjugate with receptor specific ligands, which attaches to overexpressed receptors in case of cancer or tumor cells and enters the cell by endocytosis. After entering into the cell the drug is delivered to the specific cell organelle]

RD0995), and breast (4 T1, JC, MMT06056) cancer cell lines. Zi et al. synthesized a series of biotinylated conjugates of campothecin derivatives by esterification and click chemistry which were tested on five human tumor cell lines including HL-60 (leukemia), SMMC-7721 (hepatoma), A-549 (lung cancer), MCF-7 (breast cancer) and SW480 (colon cancer). It has been reported elsewhere that biotin-conjugated drug molecules were able to increase the uptake of anticancer drugs in tumor cells (Shi et al. 2014; Yellepeddi et al. 2013). In the synthesis of derivatives of campothecin, the lactone ring (E ring) was kept intact to be assessed for anticancer activity and simultaneously biotin was conjugated and the conjugates of biotin-campothecin derivatives were tested for five different human cancer cell lines using 3-(4,5-dimethylthiazol-2-yl)-2,5-diphenyltetrazolium bromide (MTT) assay. 12 novel derivatives were designed which showed appreciable cytotoxicity in all the five cell lines. SAR study of the derivatives suggested that the length of linker between the biotin and campothecin, the 1, 2, 3-triazole ring, the 20 – hydroxy (OH^-) group in the E ring, and the 7-ethyl group of the campothecin analogues were responsible for the potency of the anticancer activity of the synthesized conjugates (Zi et al. 2019).

In another study, Zhou et al. proposed a carrier free drug delivery of campothecin (prodrug) in a pH sensitive microenvironment. The prodrug is the chemically modified derivative of the active drug which on exposure to suitable and specific environment inside the body gets converted into an active form having improved solubility, increased stability and permeability with reduced toxicity. Prodrugs with a constant hydrophilic-hydrophobic ratio get self-assembled by π-π and hydrogen-bond interactions to form nanodrugs with favorable properties. Small molecular nanodrugs (SMNs) are having some advantageous properties like facile fabrication, well-defined structure, high drug loading content and minimum carrier material. In this study, a prodrug of campothecin, camptothecin- 20 (S) – glycinate (Campothecin-NH_2) was self-assembled into nanofibers of 100 nanometers (nm) in width and length of several micrometers (μm). The cellular uptake and tumor penetration behavior of the nanofibers were observed by time-lapse video microscopy. It was observed that these formulated nanofibers entered rapidly into the cancer cells by penetrating the cell membrane and efficiently released the active drug. The surface of the nanofibers was further coated with a pH-responsive PEG layer which improved the stability of the nanofibers and shielded its positive charge to minimize non-specific interactions. These pH-responsive nanofibers actively delivered the encapsulated campothecin in the acidic tumor microenvironment. The pegylated campothecin nanofibers (PEG-campothecin-NF) were stabilized to avoid aggregation to reduce non-specific interactions. After reaching acidic tumor interstitium, PEG layer got detached from the surface of nanofibers which exposed the underlying positive charges, thereby facilitating uptake by tumor cells. These findings confirmed the fact that nanofibers can reach deep into tumors, efficiently enter cancer cells and are a promising carrier-free drug delivery system. (Zhou et al. 2019).

Dharap et al. designed three conjugates of campothecin, where campothecin was used as an anticancer agent – apoptosis inductor. The three conjugates were campothecin with PEG (Campothecin-PEG) and with two synthetic peptides similar to luteinizing hormone-releasing hormone (LHRH) and BCL-2 homology 3 (BH3) peptide. The peptides were the targeting ligands which are known to suppress the anti-apoptotic defense mechanism. The three conjugates, Campothecin-PEG, Campothecin- PEG – BH3 and Campothecin – PEG – LHRH were evaluated in human ovarian cancer cell lines, A2780 for cytotoxicity, expression of genes encoding BCL-2, BCL-XL, SMAC, APAF-1 proteins and caspases 3 and 9, the activity of caspases 3 and 9 and apoptosis induction. PEG was selected as a carrier for the nanosystem whereas the synthetic peptides were selected for the increasing the antitumor activity and for suppression of anti-apoptotic defense of campothecin in conjugate form. BH3 bound to anti-apoptotic BCL-2 family proteins helps in decreasing resistance in ovarian cancer cells to different chemotherapeutic agents. LHRH receptors are over-expressed in ovarian and some other cancer cells, and are not found in healthy tissues. So LHRH was a potential target to deliver the campothecin specifically to ovarian cancer cells and the cellular uptake was also increased. In this study, PEG used as a carrier increased the solubility and bioavailability and cytotoxicity of campothecin. BH3 prevented the activation of cellular anti-apoptotic defense and along with campothecin the pro-apoptotic action of BH3 peptide was enhanced whereas LHRH conjugation helped in the specific binding of the campothecin conjugate to ovarian cancer cells and resulted in increase in cellular uptake of campothecin by cancer cells by receptor mediated endocytosis. It was observed that Campothecin-PEG-LHRH conjugate had the highest toxicity as compared to Campothecin, Campothecin-PEG and Campothecin-PEG-BH3 (Dharap 2003).

Chen et al. developed a folate receptor (FR) targeted theranostics loaded with campothecin in acoustic nanodroplet (FA-Campothecin-ND) formulation and evaluated it in FR positive KB and FR-negative HT-1080 cell lines and mouse xenograft tumor models. The ND formulation was based on lipid-stabilized low-boiling perfluorocarbon (PFC) that can undergo acoustic droplet vaporization (ADV) under ultrasound (US) exposure. Acoustic droplet vaporization is a technique in which the superheated perfluorocarbon (PFC) can be converted into microbubbles. When a target ligand (e.g., antibodies, aptamers, and peptides) is conjugated to acoustic droplets, ADV occurs selectively near the target cell. This phenomenon causes membrane damage or increase in permeation by a process of mechanical stretching. High concentration of folic acid is essential for DNA replication in malignant cells hence folate receptors are over-expressed in breast, ovary, uterus, colon, lung and kidney cancers. In normal healthy cells these receptors are present in very low density. So a folate receptor targeted delivery formulation was developed to deliver campothecin. The ligand based FA-Campothecin-ND acted by internalization of the drug through receptor mediated endocytosis. Folate stimulated the endocytosis and ruptured the endosomes to deliver the loaded drug. The extreme hydrophobicity and lipophobicity of PFCs makes the inner side of the surfactant monolayer of acoustic droplets a preferable site for loading and delivery of campothecin in a stable form. The results of this study from the *in vitro, in vivo* and mouse xenograft model suggested that the ND formulation loaded with campothecin, not only functioned as targeted ultrasound contrast agent but also provided selective antitumor activity at FR positive tumors (Chen et al. 2015).

Knežević & Lin reported the synthesis and characterization of a series of magnetic analogues of mesoporous silica nanoparticles (MSNs). These are core–shell type of materials with magnetic iron oxide nanoparticles as the core and a mesoporous silica framework as the shell of the nanospheres. Magnetic measurements confirmed the MSNs were super-paramagnetic in nature due to the presence of super-paramagnetic maghemite nanoparticles in the core of MSNs. In this research study, the loading and release of anticancer drugs, 9-aminoacridine and campothecin was demonstrated. Additionally, a cadmium sulphide (CdS) nanoparticle-capped campothecin loaded magnetic mesoporous silica based drug delivery system was also developed. Cadmium-containing nanoparticles inhibited DNA repair and caused free radical-induced DNA damage, mitochondrial damage, induction of apoptosis, and disruption of intracellular calcium signaling. UV irradiation caused the release of campothecin is measured which was measured by fluorescence spectroscopy (Knežević and Lin 2013).

Omar et al. designed a novel drug delivery system containing a short, star shaped hydrophilic polyethylene glycol (PEG) polymer as backbone and hydrophobic drug campothecin (PEG4-campothecin) in the core. The bio-conjugated formulation had a great potential as drug delivery system due to its high number of active end groups per polymer unit and the conjugate self-assembled into stable spherical nanoparticles of 200 nm in size. A sustained release pattern was observed for campothecin drug without burst effect. The biological evaluation of PEG4-campothecin against HeLa cells showed improved cellular uptake and enhanced cytotoxicity compared to free campothecin. The uptake of campothecin by the tumor cells was achieved with minimum toxicity. The covalent bonding between the drug and the polymer was labile under physiological condition of body causing the drug release by destruction of the amphiphilic structure (Omar et al. 2017).

Bahadur et al. developed a drug delivery system (HCN) based on a polymer – drug conjugate with an intracellularly cleavable linker. The drug delivery system was targeted to human epidermal growth factor receptor 2 (HER2). The polymer used was poly [2-(pyridin-2-yldisulfanyl)]-graft-poly (ethylene glycol) (PDSG). The drug delivery system was designed based on the interaction of redox potential sensitive disulfide bonds with herceptin (HER2 antibody). The hydrophobic campothecin was first thiolated and then grafted onto the PDSG polymer to form C-PDSG through thiol–disulfide exchange reaction. The drug-polymer conjugate released the drug due to difference in the redox potential between intracellular and extracellular environment. The redox potential difference triggered the release of the conjugate in such a way that about 62% of drug was released within 20 min. For the study, both HER2 positive cancer cells and HER2 negative cells were selected. HCN was found to be stable in physiological environment and super-sensitive to the stimulus of elevated intracellular redox potential. Further confocal microscopy showed that HCN could specifically kill HER2 – positive cancer cells whereas no effect was observed for HER2 – negative cancer cells. (Remant Bahadur et al. 2014).

Muniesa et al. entrapped campothecin in a mercapto-functionalized silica hybrid which consisted of a non-porous core and a mesoporous shell. The resultant nanoparticles were in a size range of 50–60 nm with a negative surface charge which confirmed the stability of the colloids in aqueous medium. In vitro biological study was carried out in of this delivery HeLa cervix cancer cell line which displayed similar cytotoxic behavior for both the pure drug and the nanosystem. Further, the inclusion of the fluorophore Rhodamine-B (RhB) in the core of the nanovehicle facilitated the imaging at the subcellular level. Confocal microscopy confirmed the fact that the nanosystems entered the cells by a process of endocytosis but were able to escape from the endo-lysosomes and enter the cytosolic compartment to release the entrapped campothecin (Muniesa et al. 2013).

7.6 Conclusion

Tumors are heterogeneous and complex structures which show prominent variations in the tumor microenvironment. It has been observed that both active and passive targeting approaches have their own short-comings. There are significant drawbacks in passive targeting approach that result in very low drug payload in the tumor tissues leading to a decrease in therapeutic efficacy and it also fails to discriminate between healthy and diseased tissues which lead to incidences of off-target side effects. In case of active targeting approach, the effective release of the entrapped chemotherapeutics in the tumor microenvironment is a question to be addressed. Extensive research needs to be followed upon to develop these targeted nanotherapeutics to favorably modify their bio-distribution profile and increase the efficacy at the targeted site. The creation of stronger and more predictive pre-clinical animal models, an in-depth understanding of the tumor microenvironment and the implementation of GLP and standardization policies in academia are needed to bridge the gap between benches to bed side. Major breakthrough in the field of nanotechnology might enhance the efficacy of treatment modalities for the treatment of many diseases, including cancer, and contribute to a growing arsenal of passive and active targeted nanomedicines to target the dreadful disease condition.

References

Adamovics J, Hutchinson C (1979) Prodrug analogs of the antitumor alkaloid camptothecin. J Med Chem 22(3):310–314. https://doi.org/10.1021/jm00189a018

Arbain D, Putra D, Sargent M (1993) The alkaloids of *Ophiorrhiza filistipula*. Aust J Chem 46(7):977. https://doi.org/10.1071/ch9930977

Arisawa M, Gunasekera S, Cordell G, Farnsworth N (1981) Plant anticancer agents XXI. Constituents of *Merrilliodendron megacarpum∗*. Planta Med 43(12):404–407. https://doi.org/10.1055/s-2007-971533

Asano T, Watase I, Sudo H, Kitajima M, Takayama H, Aimi N et al (2004) Camptothecin production by in vitro cultures of Ophiorrhiza liukiuensis and O. kuroiwai. Plant Biotechnol 21(4):275–281. https://doi.org/10.5511/plantbiotechnology.21.275

Avemann K, Knippers R, Koller T, Sogo J (1988) Camptothecin, a specific inhibitor of type I DNA topoisomerase, induces DNA breakage at replication forks. Mol Cell Biol 8(8):3026–3034. https://doi.org/10.1128/mcb.8.8.3026

Bahadur KCR, Chandrashekaran V, Cheng B, Chen H, Peña M, Zhang J et al (2014) Redox potential ultrasensitive nanoparticle for the targeted delivery of camptothecin to HER2-positive cancer cells. Mol Pharm 11(6):1897–1905. https://doi.org/10.1021/mp5000482

Bates D, Harper S (2002) Regulation of vascular permeability by vascular endothelial growth factors. Vasc Pharmacol 39(4–5):225–237. https://doi.org/10.1016/s1537-1891(03)00011-9

Berrada M, Serreqi A, Dabbarh F, Owusu A, Gupta A, Lehnert S (2005) A novel non-toxic camptothecin formulation for cancer chemotherapy. Biomaterials 26(14):2115–2120. https://doi.org/10.1016/j.biomaterials.2004.06.013

Bom D, Curran D, Chavan A, Kruszewski S, Zimmer S, Fraley K, Burke T (1999) Novel A, B, E-ring-modified camptothecins displaying high lipophilicity and markedly improved human blood stabilities. J Med Chem 42(16):3018–3022. https://doi.org/10.1021/jm9902279

Bom D, Curran D, Kruszewski S, Zimmer S, Thompson Strode J, Kohlhagen G et al (2000) The novel Silatecan 7-tert-Butyldimethylsilyl-10-hydroxycamptothecin displays high lipophilicity, improved human blood stability, and potent anticancer activity. J Med Chem 43(21):3970–3980. https://doi.org/10.1021/jm000144o

Bray F, Jemal A, Grey N, Ferlay J, Forman D (2012) Global cancer transitions according to the human development index (2008–2030): a population-based study. Lancet Oncol 13(8):790–801. https://doi.org/10.1016/s1470-2045(12)70211-5

Burke T, Munshi C, Mi Z, Jiang Y (1995) The important role of albumin in determining the relative human blood stabilities of the camptothecin anticancer drugs. J Pharm Sci 84(4):518–519. https://doi.org/10.1002/jps.2600840426

Candiani I, Bedeschi A, Cabri W, Zarini F, Visentin G, Capolongo L et al (1997) Synthesis and cytotoxic activity of alkylidene- and alkyl-substituted camptothecins. Bioorg Med Chem Lett 7(7):847–850. https://doi.org/10.1016/s0960-894x(97)00105-4

Chen W, Kang S, Lin J, Wang C, Chen R, Yeh C (2015) Targeted tumor theranostics using folate-conjugated and camptothecin-loaded acoustic nanodroplets in a mouse xenograft model. Biomaterials 53:699–708. https://doi.org/10.1016/j.biomaterials.2015.02.122

Çırpanlı Y, Allard E, Passirani C, Bilensoy E, Lemaire L, Çalış S, Benoit J (2011) Antitumoral activity of camptothecin-loaded nanoparticles in 9L rat glioma model. Int J Pharm 403(1–2):201–206. https://doi.org/10.1016/j.ijpharm.2010.10.015
Crow R, Crothers D (1992) Structural modifications of camptothecin and effects on topoisomerase I inhibition. J Med Chem 35(22):4160–4164. https://doi.org/10.1021/jm00100a022
D'Arpa P, Liu L (1995) Cell cycle-specific and transcription-related phosphorylation of mammalian topoisomerase I. Exp Cell Res 217(1):125–131. https://doi.org/10.1006/excr.1995.1071
Dai J, Hallock Y, Cardellina J, Boyd M (1999) 20-O-β-Glucopyranosyl camptothecin from mostueabrunonis: a potential camptothecin pro-drug with improved solubility. J Nat Prod 62(10):1427–1429. https://doi.org/10.1021/np990100m
Dallavalle S, Delsoldato T, Ferrari A, Merlini L, Penco S, Carenini N et al (2000) Novel 7-substituted camptothecins with potent antitumor activity. J Med Chem 43(21):3963–3969. https://doi.org/10.1021/jm000944z
Dallavalle S, Ferrari A, Biasotti B, Merlini L, Penco S, Gallo G et al (2001a) Novel 7-oxyiminomethyl derivatives of camptothecin with potent in vitro and in vivo antitumor activity. J Med Chem 44(20):3264–3274. https://doi.org/10.1021/jm0108092
Dallavalle S, Ferrari A, Merlini L, Penco S, Carenini N, De Cesare M et al (2001b) Novel cytotoxic 7-iminomethyl and 7-aminomethyl derivatives of camptothecin. Bioorg Med Chem Lett 11(3):291–294. https://doi.org/10.1016/s0960-894x(00)00649-1
Dauty E, Remy J, Zuber G, Behr J (2002) Intracellular delivery of nanometric DNA particles via the folate receptor. Bioconjug Chem 13(4):831–839. https://doi.org/10.1021/bc0255182
Dharap S (2003) Molecular targeting of drug delivery systems to ovarian cancer by BH3 and LHRH peptides. J Control Release 91(1–2):61–73. https://doi.org/10.1016/s0168-3659(03)00209-8
Ferlay J, Shin H, Bray F, Forman D, Mathers C, Parkin D (2010) Estimates of worldwide burden of cancer in 2008: GLOBOCAN 2008. Int J Cancer 127(12):2893–2917. https://doi.org/10.1002/ijc.25516
Galvin P, Thompson D, Ryan K, McCarthy A, Moore A, Burke C et al (2011) Nanoparticle-based drug delivery: case studies for cancer and cardiovascular applications. Cell Mol Life Sci 69(3):389–404. https://doi.org/10.1007/s00018-011-0856-6
Gaur S, Wang Y, Kretzner L, Chen L, Yen T, Wu X et al (2014) Pharmacodynamic and pharmacogenomic study of the nanoparticle conjugate of camptothecin CRLX101 for the treatment of cancer. Nanomedicine 10(7):1477–1486. https://doi.org/10.1016/j.nano.2014.04.003
Giaccia A (1998) Cancer therapy and tumor physiology. Science 279(5347):10e–15e. https://doi.org/10.1126/science.279.5347.10e
Giovanella B, Stehlin J, Wall M, Wani M, Nicholas A, Liu L et al (1989) DNA topoisomerase I–targeted chemotherapy of human colon cancer in xenografts. Science 246(4933):1046–1048. https://doi.org/10.1126/science.2555920
Glück S (2014) Nab -paclitaxel for the treatment of aggressive metastatic breast cancer. Clin Breast Cancer 14(4):221–227. https://doi.org/10.1016/j.clbc.2014.02.001
Govindachari T, Viswanathan N (1972) Alkaloids of *Mappia foetida*. Phytochemistry 11(12):3529–3531. https://doi.org/10.1016/s0031-9422(00)89852-0
Greish K (2007) Enhanced permeability and retention of macromolecular drugs in solid tumors: a royal gate for targeted anticancer nanomedicines. J Drug Target 15(7–8):457–464. https://doi.org/10.1080/10611860701539584
Gunasekera S, Badawi M, Cordell G, Farnsworth N, Chitnis M (1979) Plant anticancer agents X. isolation of camptothecin and 9-methoxycamptothecin from Ervatamia heyneaya. J Nat Prod 42(5):475–477. https://doi.org/10.1021/np50005a006
Heath J, Davis M (2008) Nanotechnology and cancer. Annu Rev Med 59(1):251–265. https://doi.org/10.1146/annurev.med.59.061506.185523
Hertzberg R, Caranfa M, Hecht S (1989a) On the mechanism of topoisomerase I inhibition by camptothecin: evidence for binding to an enzyme-DNA complex. Biochemistry 28(11):4629–4638. https://doi.org/10.1021/bi00437a018

Hertzberg R, Caranfa M, Holden K, Jakas D, Gallagher G, Mattern M et al (1989b) Modification of the hydroxylactone ring of camptothecin: inhibition of mammalian topoisomerase I and biological activity. J Med Chem 32(3):715–720. https://doi.org/10.1021/jm00123a038

Householder K, DiPerna D, Chung E, Wohlleb G, Dhruv H, Berens M, Sirianni R (2015) Intravenous delivery of camptothecin-loaded PLGA nanoparticles for the treatment of intracranial glioma. Int J Pharm 479(2):374–380. https://doi.org/10.1016/j.ijpharm.2015.01.002

Jain R, Stylianopoulos T (2010) Delivering nanomedicine to solid tumors. Nat Rev Clin Oncol 7(11):653–664. https://doi.org/10.1038/nrclinonc.2010.139

Jang D, Moon C, Oh E (2016) Improved tumor targeting and antitumor activity of camptothecin loaded solid lipid nanoparticles by pre-injection of blank solid lipid nanoparticles. Biomed Pharmacother 80:162–172. https://doi.org/10.1016/j.biopha.2016.03.018

Jew S, Kim H, Kim M, Roh E, Cho Y, Kim J et al (1996) Synthesis and antitumor activity of 7-substituted 20(RS)-camptothecin analogues. Bioorg Med Chem Lett 6(7):845–848. https://doi.org/10.1016/0960-894x(96)00131-x

Jew S, Kim H, Kim M, Roh E, Hong C, Kim J et al (1999) Synthesis and in vitro cytotoxicity of hexacyclic camptothecin analogues. Bioorg Med Chem Lett 9(22):3203–3206. https://doi.org/10.1016/s0960-894x(99)00555-7

Kim D, Ryu D, Lee J, Lee N, Kim Y, Kim J et al (2001) Synthesis and biological evaluation of novel A-ring modified hexacyclic camptothecin analogues. J Med Chem 44(10):1594–1602. https://doi.org/10.1021/jm000475

Kingsbury W, Boehm J, Jakas D, Holden K, Hecht S, Gallagher G et al (1991) Synthesis of water-soluble (aminoalkyl) camptothecin analogs: inhibition of topoisomerase I and antitumor activity. J Med Chem 34(1):98–107. https://doi.org/10.1021/jm00105a017

Knežević N, Lin V (2013) A magnetic mesoporous silica nanoparticle-based drug delivery system for photosensitive cooperative treatment of cancer with a mesopore-capping agent and mesopore-loaded drug. Nanoscale 5(4):1544. https://doi.org/10.1039/c2nr33417h

Kumar G, Fayad A, Nair A (2018) Ophiorrhiza mungos var. angustifolia – estimation of camptothecin and pharmacological screening. Plant Sci Today 5(3):113–120. https://doi.org/10.14719/pst.2018.5.3.395

Lackey K, Besterman J, Fletcher W, Leitner P, Morton B, Sternbach D (1995) Rigid analogs of camptothecin as DNA topoisomerase I inhibitors. J Med Chem 38(6):906–911. https://doi.org/10.1021/jm00006a008

Lackey K, Sternbach D, Croom D, Emerson D, Evans M, Leitner P et al (1996) Water soluble inhibitors of topoisomerase I: quaternary salt derivatives of camptothecin. J Med Chem 39(3):713–719. https://doi.org/10.1021/jm950507y

Lammers T, Kiessling F, Hennink W, Storm G (2012) Drug targeting to tumors: principles, pitfalls and (pre-) clinical progress. J Control Release 161(2):175–187. https://doi.org/10.1016/j.jconrel.2011.09.063

Lavergne O, Lesueur-Ginot L, Rodas F, Bigg D (1997) BN 80245: an E-ring modified camptothecin with potent anti-proliferative and topoisomerase I inhibitory activities. Bioorg Med Chem Lett 7(17):2235–2238. https://doi.org/10.1016/s0960-894x(97)00398-3

Li Z, Liu Z (2004) Camptothecin accumulation in *Camptotheca acuminata* seedlings in response to acetylsalicylic acid treatment. Can J Plant Sci 84(3):885–889. https://doi.org/10.4141/p03-138

Liu L, Desai S, Li T, Mao Y, Sun M, Sim S (2006) Mechanism of action of camptothecin. Ann N Y Acad Sci 922(1):1–10. https://doi.org/10.1111/j.1749-6632.2000.tb07020.x

Luzzio M, Besterman J, Emerson D, Evans M, Lackey K, Leitner P et al (1995) Synthesis and antitumor activity of novel water soluble derivatives of camptothecin as specific inhibitors of topoisomerase I. J Med Chem 38(3):395–401. https://doi.org/10.1021/jm00003a001

Maeda H (1991) SMANCS and polymer-conjugated macromolecular drugs: advantages in cancer chemotherapy. Adv Drug Deliv Rev 6(2):181–202. https://doi.org/10.1016/0169-409x(91)90040-j

Maeda H, Sawa T, Konno T (2001) Mechanism of tumor-targeted delivery of macromolecular drugs, including the EPR effect in solid tumor and clinical overview of the prototype polymeric drug SMANCS. J Control Release 74(1–3):47–61. https://doi.org/10.1016/s0168-3659(01)00309-1

Malpathak N, Kulkarni A, Patwardhan A, Lele U (2010) Production of camptothecin in cultures of *Chonemorpha grandiflora*. Pharm Res 2(5):296. https://doi.org/10.4103/0974-8490.72327

Mansoori G, Mohazzabi P, McCormack P, Jabbari S et al (2007) Nanotechnology in cancer prevention, detection and treatment: bright future lies ahead. World Rev Sci Technol Sustain Dev 4(2/3):226. https://doi.org/10.1504/wrstsd.2007.013584

Martino E, Della Volpe S, Terribile E, Benetti E, Sakaj M, Centamore A et al (2017) The long story of camptothecin: from traditional medicine to drugs. Bioorg Med Chem Lett 27(4):701–707. https://doi.org/10.1016/j.bmcl.2016.12.085

Mross K (2004) A phase I clinical and pharmacokinetic study of the camptothecin glycoconjugate, BAY 38-3441, as a daily infusion in patients with advanced solid tumors. Ann Oncol 15(8):1284–1294. https://doi.org/10.1093/annonc/mdh313

Muller R, Keck C (2004) Challenges and solutions for the delivery of biotech drugs – a review of drug nanocrystal technology and lipid nanoparticles. J Biotechnol 113(1–3):151–170. https://doi.org/10.1016/j.jbiotec.2004.06.007

Muniesa C, Vicente V, Quesada M, Sáez-Atiénzar S, Blesa J, Abasolo I et al (2013) Glutathione-sensitive nanoplatform for monitored intracellular delivery and controlled release of Camptothecin. RSC Adv 3(35):15121. https://doi.org/10.1039/c3ra41404c

Nicholas A, Wani M, Manikumar G, Wall M, Kohn K, Pommier Y (1990) Plant antitumor agents. 29. Synthesis and biological activity of ring D and ring E modified analogs of camptothecin. J Med Chem 33(3):972–978. https://doi.org/10.1021/jm00165a014

Ohtsu H, Nakanishi Y, Bastow K, Lee F, Lee K (2003) Antitumor agents 216. Synthesis and evaluation of paclitaxel–camptothecin conjugates as novel cytotoxic agents1. Bioorg Med Chem 11(8):1851–1857. https://doi.org/10.1016/s0968-0896(03)00040-3

Omar R, Bardoogo Y, Corem-Salkmon E, Mizrahi B (2017) Amphiphilic star PEG-Camptothecin conjugates for intracellular targeting. J Control Release 257:76–83. https://doi.org/10.1016/j.jconrel.2016.09.025

Padhi S, Mirza M, Verma D, Khuroo T, Panda A, Talegaonkar S et al (2015) Revisiting the nanoformulation design approach for effective delivery of topotecan in its stable form: an appraisal of its in vitro behavior and tumor amelioration potential. Drug Deliv 23(8):2827–2837. https://doi.org/10.3109/10717544.2015.1105323

Padhi S, Kapoor R, Verma D, Panda A, Iqbal Z (2018) Formulation and optimization of topotecan nanoparticles: in vitro characterization, cytotoxicity, cellular uptake and pharmacokinetic outcomes. J Photochem Photobiol B Biol 183:222–232. https://doi.org/10.1016/j.jphotobiol.2018.04.022

Pitot H, Knost J, Mahoney M, Kugler J, Krook J, Hatfield A et al (2000) A north central cancer treatment group phase II trial of 9-aminocamptothecin in previously untreated patients with measurable metastatic colorectal carcinoma. Cancer 89(8):1699–1705. https://doi.org/10.1002/1097-0142(20001015)89:8<1699::aid-cncr8>3.0.co;2-t

Rajan R, Varghese S, Kurup R, Gopalakrishnan R, Venkataraman R, Satheeshkumar K, Baby S (2013) Search for Camptothecin-yielding Ophiorrhiza species from southern Western Ghats in India: a HPTLC-densitometry study. Ind Crop Prod 43:472–476. https://doi.org/10.1016/j.indcrop.2012.07.054

Ramesha B, Suma H, Senthilkumar U, Priti V, Ravikanth G, Vasudeva R et al (2013) New plant sources of the anti-cancer alkaloid, camptothecine from the Icacinaceae taxa, India. Phytomedicine 20(6):521–527. https://doi.org/10.1016/j.phymed.2012.12.003

Ryan A, Squires S, Strutt H, Evans A, Johnson R (1994) Different fates of camptothecin-induced replication fork-associated double-strand DNA breaks in mammalian cells. Carcinogenesis 15(5):823–828. https://doi.org/10.1093/carcin/15.5.823

Saito K, Sudo H, Yamazaki M, Koseki-Nakamura M, Kitajima M, Takayama H, Aimi N (2001) Feasible production of camptothecin by hairy root culture of *Ophiorrhiza pumila*. Plant Cell Rep 20(3):267–271. https://doi.org/10.1007/s002990100320

Saitoh H, Sparrow D, Shiomi T, Pu R, Nishimoto T, Mohun T, Dasso M (1998) UBC9p and the conjugation of SUMO-1 to RanGAP1 and RanBP2. Curr Biol 8(2):121–124. https://doi.org/10.1016/s0960-9822(98)70044-2

Sawada S, Okajima S, Aiyama R, Nokata K, Furuta T, Yokokura T et al (1991) Synthesis and antitumor activity of 20(S)-Camptothecin derivatives: carbamate-linked, water-soluble derivatives of 7-Ethyl-10-hydroxycamptothecin. Chem Pharm Bull 39(6):1446–1454. https://doi.org/10.1248/cpb.39.1446

Searle M, Williams D (1992) The cost of conformational order: entropy changes in molecular associations. J Am Chem Soc 114(27):10690–10697. https://doi.org/10.1021/ja00053a002

Shi J, Wu P, Jiang Z, Wei X (2014) Synthesis and tumor cell growth inhibitory activity of biotinylated annonaceous acetogenins. Eur J Med Chem 71:219–228. https://doi.org/10.1016/j.ejmech.2013.11.012

Sinha R (2006) Nanotechnology in cancer therapeutics: bioconjugated nanoparticles for drug delivery. Mol Cancer Ther 5(8):1909–1917. https://doi.org/10.1158/1535-7163.mct-06-0141

Subrahmanyam D, Sarma V, Venkateswarlu A, Sastry T, Kulakarni A, Srinivasa Rao D, Krishna Reddy K (1999a) In vitro cytotoxicity of 5-aminosubstituted 20(S)-camptothecins. Part 1. Bioorg Med Chem 7(9):2013–2020. https://doi.org/10.1016/s0968-0896(99)00130-3

Subrahmanyam D, Venkateswarlu A, Rao K, Sastry T, Vandana G, Kumar S (1999b) Novel C-ring analogues of 20(S)-camptothecin-part-2: synthesis and cytotoxicity of 5-C-substituted 20(S)-camptothecin analogues. Bioorg Med Chem Lett 9(12):1633–1638. https://doi.org/10.1016/s0960-894x(99)00268-1

Subrahmanyam D, Sarma V, Venkateswarlu A, Sastry T, Srinivas A, Krishna C et al (2000) Novel C-ring analogues of 20(S)-camptothecin. Part 3: synthesis and their in vitro cytotoxicity of A-, B- and C-ring analogues. Bioorg Med Chem Lett 10(4):369–371. https://doi.org/10.1016/s0960-894x(00)00005-6

Sugasawa T, Toyoda T, Uchida N, Yamaguchi K (1976) Experiments on the synthesis of dl-camptothecin. 4. Synthesis and antileukemic activity of dl-camptothecin analogues. J Med Chem 19(5):675–679. https://doi.org/10.1021/jm00227a019

Sugimori M, Ejima A, Ohsuki S, Uoto K, Mitsui I, Matsumoto K et al (1994) Antitumor agents. VII. Synthesis and antitumor activity of novel Hexacyclic Camptothecin analogs. J Med Chem 37(19):3033–3039. https://doi.org/10.1021/jm00045a007

Sugimori M, Ejima A, Ohsuki S, Uoto K, Mitsui I, Kawato Y et al (1998) Synthesis and antitumor activity of ring A- and F-modified hexacyclic camptothecin analogues. J Med Chem 41(13):2308–2318. https://doi.org/10.1021/jm970765q

Takayama H, Watanabe A, Hosokawa M, Chiba K, Satoh T, Aimi N (1998) Synthesis of a new class of camptothecin derivatives, the long-chain fatty acid esters of 10-hydroxycamptothecin, as a potent prodrug candidate, and their in vitro metabolic conversion by carboxylesterases. Bioorg Med Chem Lett 8(5):415–418. https://doi.org/10.1016/s0960-894x(98)00039-0

Tang D, Song F, Chen C, Wang X, Wang Y (2013) A pH-responsive chitosan-b-poly (p-dioxanone) nanocarrier: formation and efficient antitumor drug delivery. Nanotechnology 24(14):145101. https://doi.org/10.1088/0957-4484/24/14/145101

Uehling D, Nanthakumar S, Croom D, Emerson D, Leitner P, Luzzio M et al (1995) Synthesis, topoisomerase I inhibitory activity, and in vivo evaluation of 11-azacamptothecin analogs. J Med Chem 38(7):1106–1118. https://doi.org/10.1021/jm00007a008

Verschraegen C, Gupta E, Loyer E, Kavanagh J, Kudelka A, Freedman R et al (1999) A phase II clinical and pharmacological study of oral 9-nitrocamptothecin in patients with refractory epithelial ovarian, tubal or peritoneal cancer. Anti-Cancer Drugs 10(4):375–384. https://doi.org/10.1097/00001813-199904000-00005

Wall M, Wani M, Cook C, Palmer K, McPhail A, Sim G (1966) Plant antitumor agents. I. the isolation and structure of Camptothecin, a novel alkaloidal leukemia and tumor inhibitor from *Camptotheca acuminata*. J Am Chem Soc 88(16):3888–3890. https://doi.org/10.1021/ja00968a057

Wall M, Wani M, Natschke S, Nicholas A (1986) Plant antitumor agents. Isolation of 11-hydroxycamptothecin from *Camptotheca acuminata* Decne: total synthesis and biological activity. J Med Chem 29(8):1553–1555. https://doi.org/10.1021/jm00158a044

Wani M, Ronman P, Lindley J, Wall M (1980a) ChemInform abstract: plant antitumor agents. 18. Synthesis and biological activity of camptothecin and analogs. Chemischer Informationsdienst 11(38). https://doi.org/10.1002/chin.198038307

Wani M, Ronman P, Lindley J, Wall M (1980b) Plant antitumor agents. Synthesis and biological activity of camptothecin analogs. J Med Chem 23(5):554–560. https://doi.org/10.1021/jm00179a016

Wani M, Nicholas A, Wall M (1986) Plant antitumor agents. Synthesis and antileukemic activity of camptothecin analogs. J Med Chem 29(11):2358–2363. https://doi.org/10.1021/jm00161a035

Wani M, Nicholas A, Wall M (1987a) Plant antitumor agents. Resolution of a key tricyclic synthon, 5′(RS)-1,5-dioxo-5′-hydroxy-2′H,5′H,6′H-6′-oxopyrano[3′,4′-f].DELTA.6,8-tetrahydroindolizine: total synthesis and antitumor activity of 20(S)- and 20(R)-camptothecin. J Med Chem 30(12):2317–2319. https://doi.org/10.1021/jm00395a024

Wani M, Nicholas A, Manikumar G, Wall M (1987b) Plant antitumor agents. Total synthesis and antileukemic activity of ring substituted camptothecin analogs. Structure-activity correlations. J Med Chem 30(10):1774–1779. https://doi.org/10.1021/jm00393a016

Wu J, Liu L (1997) Processing of topoisomerase I cleavable complexes into DNA damage by transcription. Nucleic Acids Res 25(21):4181–4186. https://doi.org/10.1093/nar/25.21.4181

Yamauchi T (2011) Camptothecin induces DNA strand breaks and is cytotoxic in stimulated normal lymphocytes. Oncol Rep 25(2):347. https://doi.org/10.3892/or.2010.1100

Ya-ut P, Chareonsap P, Sukrong S (2011) Micropropagation and hairy root culture of *Ophiorrhiza alata* craib for camptothecin production. Biotechnol Lett 33(12):2519–2526. https://doi.org/10.1007/s10529-011-0717-2

Yellepeddi V, Vangara K, Palakurthi S (2013) Poly (amido) amine (PAMAM) dendrimer–cisplatin complexes for chemotherapy of cisplatin-resistant ovarian cancer cells. J Nanopart Res 15(9):1897. https://doi.org/10.1007/s11051-013-1897-6

Zhang S (2017) Cancer therapy with co-delivery of Camptothecin. J Drug Deliv Ther 7(3). https://doi.org/10.22270/jddt.v7i3.1450

Zhou B, Hoch J, Johnson R, Mattern M, Eng W, Ma J et al (2000) Use of COMPARE analysis to discover new natural product drugs: isolation of Camptothecin and 9-methoxycamptothecin from a new source. J Nat Prod 63(9):1273–1276. https://doi.org/10.1021/np000058r

Zhou Z, Piao Y, Hao L, Wang G, Zhou Z, Shen Y (2019) Acidity-responsive shell-sheddable camptothecin-based nanofibers for carrier-free cancer drug delivery. Nanoscale 11(34):15907–15916. https://doi.org/10.1039/c9nr03872h

Zi C, Yang L, Xu F, Dong F, Ma R, Li Y et al (2019) Synthesis and antitumor activity of biotinylated camptothecin derivatives as potent cytotoxic agents. Bioorg Med Chem Lett 29(2):234–237. https://doi.org/10.1016/j.bmcl.2018.11.049

Zunino F, Dallavalle S, Laccabue D, Beretta G, Merlini L, Pratesi G (2002) Current status and perspectives in the development of Camptothecins. Curr Pharm Des 8(27):2505–2520. https://doi.org/10.2174/1381612023392801

Printed in the United States
by Baker & Taylor Publisher Services